PHYSIOLOGY AND ANATOMY

PHYSIOLOGY AND ANATOMY

Editor

Bernard Gorbold

scitus
academics

Physiology and Anatomy
Edited by **Bernard Gorbold**

ISBN: 9781681179308
Library of Congress Control Number: 2017948960

SCITUS Academics LLC,
616, Corporate Way, Suite 2, 4766,
Valley Cottage, NY 10989

www.scitusacademics.com

Printed in United States of America.

Preface

The term anatomy refers to the science that deals with the form and structure of animals. Physiology deals with the study of functions of the body or any of its parts. A thorough knowledge of the structure of an animal imparts a lot of information about the various functions it is capable of performing. All animals are made up of cells, some of which are specialized to carry out different functions. Simple animals, such as sponges, are made up of only a few types of cell. In more complex animals, cells are organized into tissues, such as muscles and nerves that are necessary for movement. Tissues can form organs, such as the heart, which is used to pump blood around the circulatory system. An understanding of physiology underpins advances in biomedical research, so research in animal physiology impacts broadly on medicine, agriculture, industry and environmental science. In line with the broad scope of this topic, the research uses a huge range of techniques and model organisms and approaches span the study of single cells in culture and the ways they respond to their substrate and environment, through to the study of cellular physiology in living animals and the way whole organ systems develop and respond to environmental stress.

This volume 'Physiology and Anatomy' describes the structure of the animal body and the way in which it works. It describes the principles of physiology that underlie all animals, rather than focusing exclusively on humans. It provides a comprehensive overview of physiological systems in an efficient and concise manner by moving between description of general principles of physiology and specific examples of these principles from a variety of animal species. Veterinary nurses need to have a firm grasp of the normal structure of an animal's body and how it functions before they can understand the effect diseases and injuries have and the best ways to treat them. It provides the learner the relevant knowledge of the structure and the function of the animal body to enable the learner to provide better animal care.

Table of Contents

CHAPTER 1

Physiological Effects of Ergot Alkaloid and Indole-Diterpene Consumption on Sheep under Hot and Thermoneutral Ambient Temperature Conditions

Michelle L. E. Henry *, Stuart Kemp, Frank R. Dunshea and Brian J. Leury

Faculty of Veterinary and Agricultural Sciences, University of Melbourne, Parkville 3010, Australia;

SIMPLE SUMMARY

Perennial ryegrass endophyte toxicosis in sheep, caused by ingestion of alkaloids, is characterized by staggers and heat stress resulting in poor production or, in some cases, death. This study investigated effects of alkaloid consumption in sheep housed under thermoneutral versus hot ambient temperature on some production and physiological responses. Alkaloid ingestion and extreme heat produced a similar heat stress response and, when combined, this response was exacerbated. Productivity and health and welfare may be compromised when sheep are consuming alkaloids, and this effect is most severe when alkaloid consumption occurs under hot environmental conditions.

ABSTRACT

A controlled feeding study was undertaken to determine the physiological and production effects of consuming perennial ryegrass alkaloids (fed via seed) under extreme heat in sheep. Twenty-four Merino ewe weaners (6 months; initial BW 30.8 ± 1.0 kg) were selected and the treatment period lasted 21 days following a 14 day acclimatisation period. Two levels of two factors were used. The first factor was alkaloid, fed at a nil (NilAlk) or moderate level (Alk; 80 μg/kg LW ergovaline and 20.5 μg/kg·LW lolitrem B). The second factor was ambient temperature applied at two levels; thermoneutral (TN; constant 21–22 °C) or heat (Heat; 9:00 AM–5:00 PM at 38 °C; 5:00 PM–9:00 AM at 21–22 °C), resulting in four treatments, NilAlk TN, NilAlk Heat, Alk TN and Alk Heat. Alkaloid consumption reduced dry matter intake (p = 0.008), and tended to reduce liveweight (p = 0.07). Rectal temperature and respiration rate were increased by both alkaloid and heat ($p < 0.05$ for all). Respiration rate increased to severe levels when alkaloid and heat were combined, indicating the short term effects which may be occurring in perennial ryegrass toxicosis (PRGT) areas during severe weather conditions, a novel finding. When alkaloid ingestion and heat were administered separately, similar physiological responses occurred, indicating alkaloid ingestion causes a similar heat stress response to 38 °C heat.

Keywords: alkaloid; perennial ryegrass toxicosis; heat stress; physiology; sheep

1. INTRODUCTION

Perennial ryegrass (PRG) (*Lolium perenne*) is the primary forage used in the cool temperate, winter-spring rainfall region of Australia because of its persistence and productivity [1]. This persistence and productivity is largely due to the presence of an endophyte (*Epichloe festucaevar Lolii.*), an internal living fungi which exists in a symbiotic relationship with the plant. The endophyte produces alkaloids during stressful times of the year (mainly summer and autumn) for the plant to reduce insect attack and overgrazing and increase plant persistence and productivity [2]. One important alkaloid, ergovaline, has been found to produce heat stress [3,4,5], reduced prolactin secretion [6,7] and reduces animal production [4,5,8]. The compound lolitrem B causes tremors and staggers which has been well documented in research undertaken in New Zealand

[5,9,10,11]. When animals consume infected forage, the condition perennial ryegrass toxicosis (PRGT) can occur.

Perennial ryegrass toxicosis (PRGT), is a significant animal disease in Australia. During a severe outbreak in 2002, there were 29,000 confirmed sheep deaths, however, it is thought this number only represented a third of farms affected [1]. Body temperatures were observed to rise above 41 °C and respiration rates increased, causing heat stress in sheep [1]. Previous work undertaken under grazing conditions in Australia have shown heat stress (increased respiration rate) when grazing Coopworth ewes on Victorian perennial ryegrass and subterranean clover mix [12]. The cost of PRGT in Australia has been estimated to be $100 million per annum [13].

A previous study undertaken by the authors [14] investigated the interaction between alkaloid consumption and moderate heat stress (maximum ambient temperature 33 °C) and demonstrated a strong interaction between alkaloid consumption and moderate heat conditions. However, southern Australia experiences summer conditions which can reach and exceed 40 °C, far in excess of that studied previously. Furthermore, the concentration of alkaloids in PRG peaks during summer raising the question of the effects of alkaloids in sheep under hot ambient temperature conditions. Research undertaken in New Zealand which has focused on the heat stress effects of PRGT in sheep has focused on grazing studies in natural New Zealand environmental conditions [15]. While this research has found evidence of hyperthermia, these symptoms have only been observed occasionally in the field [15]. Further, summer and autumn ambient temperature conditions in Australia are found to be harsher, as compared to ambient temperature conditions in New Zealand. The aim of this study was to investigate the short-term effects of PRG alkaloids ergovaline and lolitrem B in sheep housed under thermoneutral *vs.* hot ambient temperature, as experienced in summer and autumn in Australia, on some production and physiological responses.

2. MATERIALS AND METHODS

The experimental protocol was reviewed and approved by the University of Melbourne, Science, Optometry and Vision Sciences and Land and Environment Animal ethics Committee (704727.1).

2.1. Animal Management

Twenty four Merino ewe weaners aged 6 months (30.8 ± 1.0 kg) were sourced from The University of Melbourne (Victoria, Australia), Dookie campus. Sheep were housed in group pens for seven days (initial acclimatisation period), followed by a seven day acclimatisation period to the metabolism crates (steel crates with a wire mesh floor measuring 1.5 m long and 0.6 m wide and raised 0.5 m from the ground) and controlled temperature room before a 21 day treatment period. Sheep were fed a high fibre pellet (9.9 MJ/kg ME and 11.6% CP) with the addition of whole barley (12.8 MJ/kg ME and 12.0% CP), throughout all stages of the experiment, twice daily, ad libitum. Ad libitum feed intake was maintained by increasing feed offered by 10% to sheep which consumed 90% of feed the prior day.

2.2. Experimental Protocol

Sheep were PRG naïve at the beginning of the experiment. Alkaloid was administered via whole wild-type endophyte infected PRG seed and this was mixed with whole barley to improve palatability of the seed. Most sheep consumed the mix of seed and barley presented, however, there was some individual sheep variability, with some sheep sorting through the feed, avoiding the seed. The use of this simple mixing method proved to be the most effective way of monitoring and ensuring sheep consumed the seed presented. A previous study using goats found only a small percentage (1.6%) of whole perennial ryegrass seed passes through to the faeces [15] so seed was fed whole. Additionally, previous studies by the authors have found feeding seed whole produces consistent PRGT effects [14,16]. Lolitrem B levels are higher than ergovaline in plant tissue [17], in the current study seed was fed which contained lower levels of lolitrem B compared to ergovaline. This method of alkaloid administration was chosen due to the study focusing on the effects of ergovaline, additionally, fresh pasture or hay was not available for use. Sheep were stratified by liveweight, into groups of four sheep and then within each group randomly assigned to a treatment group. Alkaloid concentration was fed at a Nil level (NilAlk) where sheep received no PRG seed (since, at the time of the experiment, no nil endophyte infected seed was available), and a moderate level where sheep received alkaloid (Alk) via whole infected PRG seed at a concentration of 80 μg/kg LW ergovaline and 20.5 μg/kg LW lolitrem B. The seed fed contained 11.8 ppm lolitrem B and 25.9 ppm ergovaline (not including ergovalinine). This level is very high as compared to levels in pasture which may fluctuate between 1.0–1.6 ppm ergovaline and 2.4–3.9 ppm lolitrem B at peak times in summer and autumn in Australian pastures [18]. Two temperature controlled rooms were used, one was kept at thermoneutral conditions (TN; 16–18 °C) and the other was heated (Heat; 38 °C between 9:00 AM and 4:00 PM and 21–22 °C between 8:00 PM and 9:00 AM).This resulted in four treatment groups, NilAlk TN, NilAlk Heat, Alk TN and Alk Heat. Two groups of six sheep moved through each room.

Alkaloid dosing levels were based on ergovaline and lolitrem B concentrations used in previous studies [14,16], when moderately high concentrations of alkaloids were fed, resulting in compromised animal production and a heat stress response. However, in these studies sheep did not consume all alkaloid offered, possibly because of feed aversion. Therefore, this experiment aimed at feeding a low to moderate dose of lolitrem B and ergovaline to ensure the consumption of alkaloid.

Sheep were fed seed/barley at 8:00 AM while pellets were fed twice daily (8:00 AM and 4:00 PM) with the seed/barley mix provided first to ensure consumption. Feed and water intake were measured daily. Alkaloid intake was estimated by measuring the amount of seed in the feed refusal each day. Liveweight was measured on day 0 and 21, prior to feeding. Urine and faecal separators were placed on the metabolism crates on days 5 and 11–17, and 24 h 10% volume/weight samples of faeces were collected and volume of urine was recorded. Faecal samples were weighed and dried at 100 °C for 24 h or until samples reached a stable weight and the dry matter (DM) content was determined.

Rectal temperature, respiration rate and skin temperature (measured on the back and leg) were measured at three hourly intervals commencing at 8:00 AM and finishing at 5:00 PM on days 2, 5, 9, and 12 of the treatment period. Rectal temperature was measured using a hand held digital thermometer (Vega Technologies Inc., Taiwan, China). Respiration rate was measured by counting the number of breaths taken, via flank movements, in 30 s intervals. Skin temperature was measured on the back and hind leg by parting the wool and holding a hand held digital thermometer (Vega Technologies Inc., Taiwan, China) to the skin for 30 s.

Ambient temperature and relative humidity were measured using a temperature and humidity probe with a data logger attached (DigiTech, New South Wales, Australia).

2.3. Statistical Analysis

Statistical analysis was undertaken using the Genstat statistical package (16th Edition, VSN International, Hemel Hempstead, UK). For all parameters significance was analysed using general analysis of variance. The model included fixed effects of alkaloid (NilAlk *vs.* Alk), temperature (TN *vs.* Heat) and time (hour, day or week depending on the parameter), and the random effect was sheep ID. For DMI, faecal water, urine output, and physiological parameters the data was analysed for weekly responses. To analyse changes over a 9 h period on day 12 for physiological parameters the model included fixed effects of alkaloid, temperature and time (hour), and the random effect was sheep ID. Covariates were also included in the analyses using baseline data or day 0 data where appropriate.

3. RESULTS

3.1. Environmental Conditions

Mean climatic conditions for the study are presented in Figure 1. Ambient temperature remained at TN conditions on day 0 for all treatments.

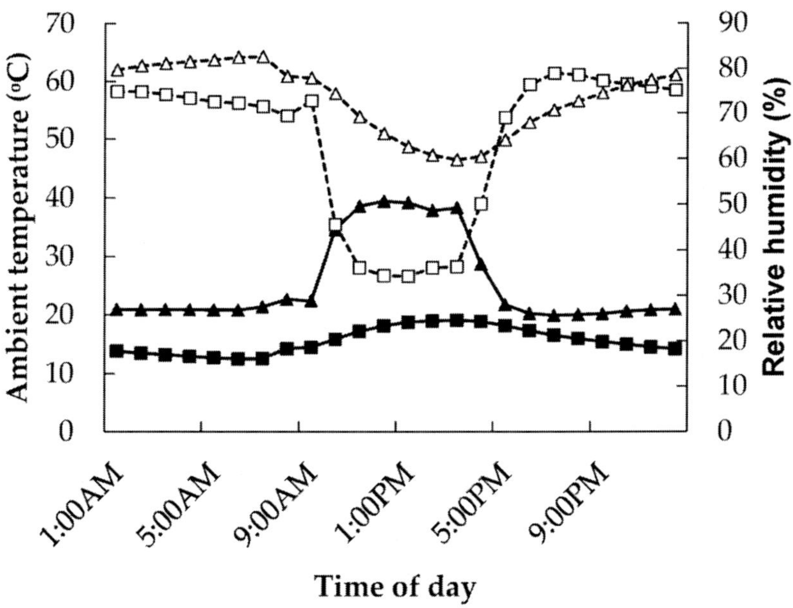

Figure 1. Ambient temperature (▲, ■) and relative humidity (Δ, □) measured under constant thermoneutral temperatures (16–18 °C) (■, □, respectively) and under variable heated conditions (38 °C 9:00 AM–5:00 PM and 21–22 °C 5:00 PM–9:00 AM) (▲, Δ, respectively).

During days 1–21 between 9:00 AM and 4:00 PM, in the Heat treatment, the means ± SE for ambient temperature and relative humidity were 38.0 ± 0.11 °C and 29.0% ± 0.22%, respectively. Between 4:00 PM and 9:00 AM, mean ambient temperature and relative humidity for the Heat treatments were 21.4 ± 0.04 °C and 56.7% ± 0.17%, respectively.

3.2. Production Response

Sheep consumed 88% of the infected seed offered in the Alk TN treatment and 68% of infected seed in the Alk Heat treatment ($p < 0.001$).

DMI was lower in the Alk treatment compared to the NilAlk treatment (1179 and 1010 g/d; for NilAlk and Alk, respectively; sed 57.1; $p = 0.008$; Figure 2a) but was not altered by temperature ($p = 0.53$). There was a significant alkaloid × week interaction ($p < 0.001$; Figure 2a) such that DMI in the NilAlk treatment remained relatively stable while DMI decreased over time in the Alk treatment. There was an interaction between alkaloid × temperature × week, such that the Alk Heat treatment decreased DMI to the lowest level compared with the other treatments ($p = 0.02$; Figure 2a).

Average daily gain over the experimental period tended to be less for the Alk treatment ($p = 0.07$; Table 1) but was not affected by temperature ($p = 0.44$; Table 1). There was a trend for final liveweight to be lower in the Alk treatment (32.2 and 30.8 kg for NilkAlk and Alk, respectively; sed 1.0; $p = 0.07$), while temperature had no effect (31.8 and 31.2 kg, for TN and Heat, respectively; sed 0.74; $p = 0.44$). There were no significant interactions for liveweight or ADG.

There was a tendency for water intake to increase in the Alk treatment ($p = 0.09$; Figure 2b), while heat increased water intake over the entire treatment period ($p = 0.02$). There was an interaction between alkaloid and week, such that water intake increased over the treatment period in the Alk treatment ($p = 0.008$). Urine output was variable in the Alk treatment over the experimental period, increasing largely over the first week ($p = 0.04$; Figure 2c). Faecal water and DMD were not altered by alkaloid or temperature (Table 1).

3.3. Physiological Response

3.3.1. Overall Experimental Period

Data presented for all parameters has been averaged over all sampling days and times.

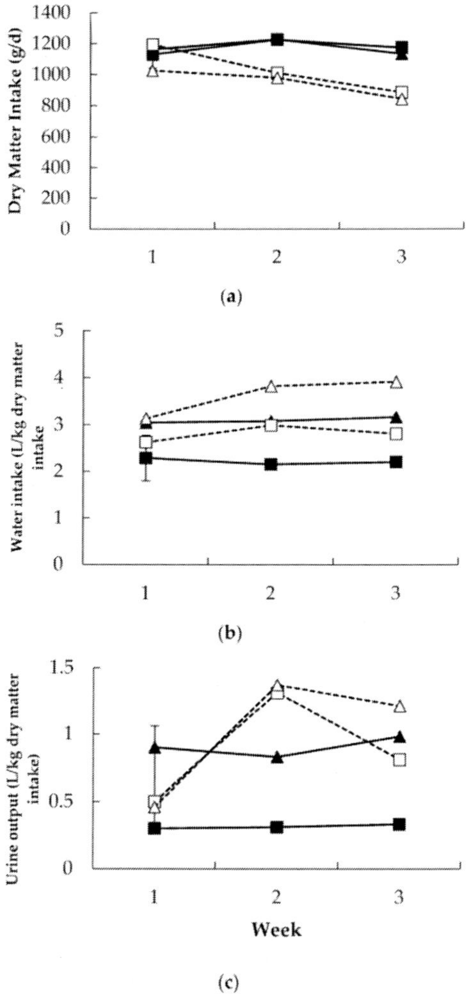

Figure 2. Relationships between (a) dry matter intake, (b) water intake, and (c) urine output and week of study in Merino sheep fed diets containing no (■, ▲) or added perennial ryegrass alkaloids (□, Δ) under either thermoneutral (squares) or heat (triangles) conditions. The *p*-values for the effect of alkaloid, temperature, week and the interaction between week and temperature, week and alkaloid and week, alkaloid and temperature were 0.008, 0.53, <0.001, 0.35, <0.001 and 0.02 for dry matter intake, 0.09, 0.02, 0.02, 0.17, 0.008, 0.68 for water intake and 0.25, 0.39, 0.09, 0.65, 0.04 and 0.87 for urine output, respectively. The standard error of the difference for the three-way interaction is displayed on the data from the sheep receiving no alkaloids under thermoneutral conditions. There were no other significant interactions.

Table 1. Overall average daily gain (ADG), faecal water (%) and dry matter digestibility (DMD) for each treatment.

Treatment/Temperature	NilAlk		Alk		Sed [1]	*p* Value		
	TN	Heat	TN	Heat		Alk [2]	Temp [3]	AlkxTemp
ADG (grams/day)	83	83	34	−34	61	0.07	0.44	0.44
Faecal water (%)	59	58	62	57	2.9	0.64	0.15	0.31
DMD (%)	57	55	58	58	2.2	0.17	0.58	0.46

[1] Standard error of the difference; [2] NilAlk *vs.* Alk; [3] Temperature (TN *vs.* Heat).

Data for rectal temperature and respiration rate are presented in Table 2. Overall, rectal temperature was higher in the Alk treatment (39.73 and 40.13 °C, for NilAlk and Alk, respectively; sed 0.07; $p < 0.001$) and in the Heat treatment (39.80 and 40.07 °C, for TN and Heat, respectively; sed 0.07; $p = 0.002$). There was an alkaloid x week interaction ($p = 0.001$; Table 2) such that rectal temperature decreased over time in the Alk treatment compared with the NilAlk treatment. Overall, respiration rate was higher in the Alk treatment (105 and 153 breaths/min, for NilAlk and Alk, respectively; sed 7.3; $p < 0.001$) and Heat treatment (102 and 156 breaths/min, for TN and Heat, respectively; sed 7.3; $p < 0.001$). There was a tendency for a temperature × week ($p = 0.07$) interaction such that respiration rate increased over time in the Heat treatment but not in the TN treatment.

Table 2. Rectal temperature, respiration rate, back skin temperature and leg skin temperature averaged over week 1 and 2 of the experimental period for each treatment.

Treatment/Temperature	NilAlk		Alk		Sed [1]	*p* Value		
	TN	Heat	TN	Heat		Alk [2]	Temp [3]	Week
Rectal temperature (°C)								
W1	39.55	39.95	40.12	40.33	0.11	<0.001	0.002	<0.001
W2	39.56	39.88	39.95	40.11				
Respiration rate (breaths/min)								
W1	77	132	135	174	11.5	<0.001	<0.001	0.77
W2	77	134	119	184				
Back skin temperature (°C)								
W1	38.40	39.05	38.82	39.31	0.13	0.004	<0.001	0.002
W2	38.32	39.01	38.63	39.17				
Leg skin temperature (°C)								
W1	38.19	38.98	38.66	39.18	0.12	0.008	<0.001	<0.001
W2	38.09	38.87	38.27	38.95				

[1] Standard error of the difference; [2] NilAlk *vs.* Alk; [3] Temperaure (TN *vs.* Heat).

Data for skin temperatures are presented in Table 2. Overall back skin temperature was higher in the Alk treatment (38.70 °C and 38.98 °C, for NilAlk and Alk, respectively; sed 0.08; p = 0.004) and Heat treatment (38.53 and 39.13 °C, for TN and Heat, respectively; sed 0.08; $p < 0.001$). Similarly, leg skin temperature was higher in the Alk (38.54 and 38.76 °C, for NilAlk and Alk, respectively; sed 0.07; p = 0.008) and Heat treatments (38.30 °C and 39.00 °C, for TN and Heat, respectively; sed 0.07; $p < 0.001$). There was a significant alkaloid × week interaction for leg skin temperature (p = 0.01; Table 2) such that temperature decreased over time in the Alk treatment but not in the NilAlk treatment.

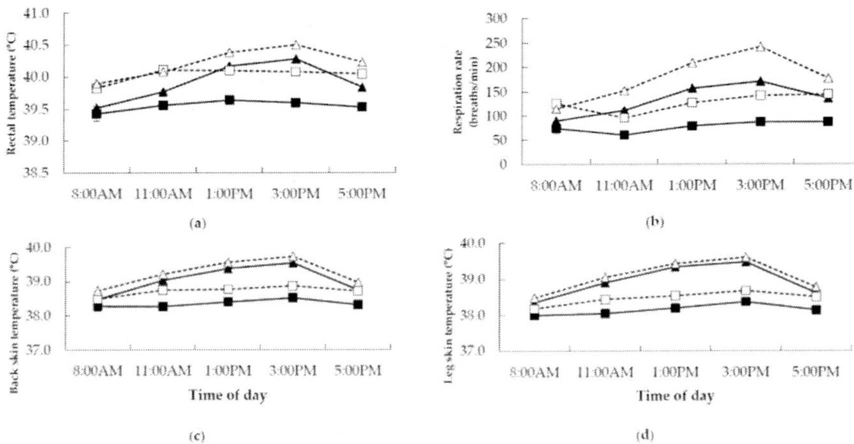

Figure 3. Relationships between (a) rectal temperature, (b) respiration rate, (c) back skin temperature, and (d) leg skin temperature and time of day in Merino sheep fed diets containing no (■,▲) or added perennial ryegrass alkaloids (□, Δ) under either thermoneutral (squares) or heat (triangles) conditions on day 12 of treatment. The p-values for the effect of alkaloid, temperature and time and the interaction between alkaloid and time, temperature and time and alkaloid and temperature and time were <0.001, 0.002, <0.001, 0.35, <0.001 and 0.33 for rectal temperature, <0.001, <0.001, <0.001, 0.009, <0.001 and 0.044 for respiration rate, 0.004, <0.001, <0.001, 0.82, <0.001 and 0.39 for back skin temperature and 0.008, <0.001, <0.001, 0.75, <0.001 and 0.86 for leg skin temperature, respectively. The standard error of the difference for the three-way interaction is displayed on the data from the sheep receiving no alkaloids under thermoneutral conditions. There were no other significant interactions.

3.3.2. Temporal Patterns in Physiological Response

Over the nine hour sampling period on day 12, rectal temperature increased in the Alk and Heat treatments ($p = 0.008$ and $p < 0.001$, respectively; Figure 3a). There was a temperature × time ($p < 0.001$) interaction such that the Heat treatment was variable over time compared to the TN treatment.

Respiration rate increased in the Alk and Heat treatments ($p < 0.001$; Figure 3b). There was an alkaloid × time interaction ($p = 0.009$; Figure 3b) such that the Alk treatment increased respiration rate over time compared to the NilAlk treatment. Further, there was a temperature × time interaction ($p < 0.001$, respectively; Figure 3b) such that the Heat treatment increased respiration rate over time compared the TN treatment. Moreover, respiration rate was higher in the Alk Heat treatment compared to all other treatments at all times points apart from 8:00 AM ($p < 0.05$), while the Alk TN and NilAlk Heat groups were similar at most time points (apart from 8:00 AM and 3:00 PM).

Back and leg skin temperature increased in the Alk and Heat treatments ($p = 0.004$ and $p < 0.001$ for back, Alk and Heat, respectively and $p = 0.008$ and $p < 0.001$ for leg, Alk and Heat, respectively; Figure 3c,d). There were significant temperature × time interactions for back and leg skin temperature ($p < 0.001$) such that the Heat treatment increased skin temperature over time, while the TN treatment remained lower.

4. DISCUSSION

4.1. Production Response

A key finding from the present study was that DMI was decreased by 22% in response to consumption of rye grass seed that contained moderate levels of endophyte alkaloids. A reduction in DMI due to alkaloid consumption has been reported in previous studies by the same authors when thermoneutral and moderate heat conditions were used for the same period of time [14,16]. Furthermore, this finding is consistent with a previous study which found feed intake decreased in cattle fed endophyte infected tall fescue (containing ergovaline) under 32 °C ambient temperature conditions [3]. Liveweight gain was not significantly affected in this study and this is likely due to the short time period used. Liveweight has been found to decrease in a number of grazing based PRGT studies [19,20,21,22], and with controlled alkaloid intake under thermoneutral and moderate heat conditions [14,16]. A recent review has highlighted the clear production effects of ergot alkaloid consumption [8], reporting the clear link between decreased DMI and liveweight. Further, a recent review has documented clear production effects on sheep consuming perennial ryegrass alkaloids in New Zealand, reporting that liveweight can decrease in sheep, prior to the development of staggers [5]. The findings of the current study support this report.

Faecal moisture increased in the AlkTN treatment in week 1, however, DMD was not reduced by alkaloid or heat in this experiment, contrary to the findings in a previous study [14]. The concentration of lolitrem B consumed in this study was lower than previously used [14] and while increased faecal water has been established previously [15], the alkaloid responsible for increased faecal water has not been clearly identified, lolitrem B has been implicated [23]. This raises the possibility that more marked effects on faecal water and, potentially DMD, may be associated with higher concentrations of lolitrem B intake rather than ergovaline.

Water intake relative to DMI was increased in response to alkaloid consumption (+14%) and in response to Heat (+15%) with these effects being additive. A previous study measured water intake when sheep were exposed to wild-type and AR37 perennial ryegrass pasture. This study found water intake to increase by 25% when sheep consumed wild-type pasture and when ambient temperature stayed above 20 °C [24] which is consistent with the current study. Further, the increase in water intake due to heat is consistent with the increased heat load in these animals and similar observations in other studies [25,26,27]. The implications of increased water intake could be extremely detrimental to the Australian sheep industry. Urine output relative to DMI was unaffected by alkaloid consumption and did not increase due to heat exposure, although urine output was highly variable and lack of significant treatment effects is perhaps not surprising. The early increase in urine output may have been in response to the initial consumption of alkaloids, and a result of altered electrolyte and fluid balance [28]. The subsequent decrease in urine output may have been due to the decrease in DMI and therefore total alkaloid consumption. Similarly, the decrease in alkaloid consumption and short period of time used were the likely causes of little change in liveweight, even though DMI decreased significantly.

4.2. Physiological Response

Alkaloid ingestion and heat resulted in increased rectal temperature and respiration rate. This is probably due to the relatively high heat load imposed by the heating regimen. Presumably, the magnitude of this was such that elevation of respiration rate was insufficient for heat to be dissipated adequately during the TN period as indicated by the 9 h physiological profiles presented in Figure 3b. Respiration rate was extremely high in the Alk Heat treatment during the heat period, well above the other treatment groups. This finding indicates the severity of additive effects of alkaloid and heat, and begins to explain the severity of heat stress which may be occurring during summer in southern Australian PRG areas, a novel finding. The mechanism responsible for these increases may be due to the binding of ergovaline to serotonin-2 receptors causing vasoconstriction [29], resulting in increased rectal temperature and therefore respiration rate in an attempt to reduce heat load. Respiration rate was similar in the NilAlk Heat and Alk TN treatment, suggesting that alkaloid ingestion increased heat load similarly to 38 °C heat. Grazing studies have found increases in rectal temperature and respiration rate over summer and autumn [4,12]. However, the amount of PRG alkaloid consumed by individual sheep was not measured due to the grazing nature of the studies. Ambient temperature was not reported in one study [4], however, the other study reported ambient temperatures were within the 30–38 °C range during the days when physiological observations were recorded [13]. The current work has been able to determine the hyperthermic effects of perennial ryegrass toxins under known ambient temperature conditions similar to those experienced during an Australian summer and autumn period, when toxin intake was known.

Leg and back skin temperature increased due to alkaloid consumption and heat. This may have been due to vasoconstriction occurring in the extremities [3] of the sheep and/or the 38 °C heat. The response from sheep was more severe in this study as compared to a previous study by the same authors which used 33 °C heat and a higher alkaloid dose [14], suggesting the higher 38 °C conditions were causative of these effects.

5. CONCLUSIONS

This study demonstrated that feeding alkaloids under extreme heat conditions can compromise sheep productivity in the form of decreased feed intake and may reduce sheep liveweight over a longer period of time. Alkaloid intake, when fed via seed, combined with 38 °C heat resulted in respiration rate increasing to severe levels. This is the first study to elucidate the severity of PRGT heat stress under ambient temperature conditions similar to those experienced during a severe southern Australian summer, when toxin intake was known. Interestingly, the findings also indicate that alkaloid ingestion and 38 °C heat produce a similar heat stress response. Overall, the findings suggest that producer profitability along with sheep health and welfare may be compromised when sheep are consuming alkaloids or during severe weather events, such as extreme heat, and this effect is most severe when alkaloid consumption occurs under hot conditions. Further study investigating longer term effects of alkaloids under natural southern Australian summer conditions, when toxin intake is known, is warranted to further elucidate the effects of alkaloid consumption and heat on animal production and heat stress, reflecting that experienced in the field.

ACKNOWLEDGMENTS

The authors acknowledge the assistance given by farm staff located at the University of Melbourne, Dookie campus throughout the experimental period, along with students and staff within the Animal Physiology group within the Faculty of Veterinary and

REFERENCES

1. Reed, K. Perennial ryegrass toxins in australian pasture. In *Perennial Ryegrass Toxicosis in Australia*; Reed, K.F.M., Page, S.W., Lean, I.J., Eds.; Meat and Livestock Australia: Attwood, Australia, 2005; pp. 11–17.
2. Popay, A.; Bonos, S. Biotic responses in endophytic grasses. In *Neotyphodium in Cool-Season Grasses*; Roberts, C.A., West, C.P., Spiers, D.E., Eds.; Blackwell Publishing: Oxford, UK, 2005; pp. 163–185.
3. Osborn, T.; Schmidt, S.; Marple, D.; Rahe, C.; Steenstra, J. Effect of consuming fungus-infected and fungus-free tall fescue and ergotamine tartrate on selected physiological variables of cattle in environmentally controlled conditions. *J. Anim. Sci.* **1992**, *70*, 2501–2509.
4. Fletcher, L.; Sutherland, B.; Fletcher, C. The impact of endophyte on the health and productivity of sheep grazing ryegrass-based pastures. In *Ryegrass Endophyte: An Essential New Zealand Symbiosis*; Woodfield, D.R., Matthew, C., Eds.; New Zealand Grassland Association: Napier, New Zealand, 1999; pp. 11–18.
5. Di Menna, M.; Finch, S.; Popay, A.; Smith, B. A review of the neotyphodium lolii/lolium perenne symbiosis and its associated effects on animal and plant health, with particular emphasis on ryegrass staggers. *N. Z. Vet. J.* **2012**, *60*, 315–328.
6. Fiorito, I.; Bunting, L.; Davenport, G.; Boling, J. Metabolic and endocrine responses of lambs fed *Acremonium coenophialum*-infected or noninfected tall fescue hay at equivalent nutrient intake. *J. Anim. Sci.* **1991**, *69*, 2108–2114.
7. Browning, R.; Leite-Browning, M. Effect of ergotamine and ergonovine on thermal regulation and cardiovascular function in cattle. *J. Anim. Sci.* **1997**, *75*, 176–181.
8. Klotz, J. Activities and effects of ergot alkaloids on livestock physiology and production. *Toxins* **2015**, *7*, 2801–2821.
9. Fletcher, L.; Harvey, I. An association of a *lolium* endophyte with ryegrass staggers. *N. Z. Vet. J.* **1981**, *32*, 185–186.
10. Fletcher, L.; Harvey, I. Some observations of neurotoxin production in perennial ryegrass. *N. Z. Vet. J.* **1981**, *30*, 203–204.

11. Finch, S.; Imlach, W.; Dunlop, J.; Meredith, A.; Aldrich, R.; Dalziel, J. Mode of action of tremorgens. In Proceedings of the 6th International Symposium on Fungal Endophytes of Grasses; New Zealand Grassland Association Inc.: Dunedin and Christchurch, New Zealand, 2007; pp. 363–364.
12. Reed, K.; Cummins, J.; Moore, D.; Clark, A. Performance of coopworth ewe lambs exposed to low levels of ryegrass endophyte (*Neotyphodium lolii*) alkaloids and allowed access to a mycotoxin deactivator. *Anim. Prod. Sci.* **2011**, *51*, 225–232.
13. Leury, B.; Henry, M.; Kemp, S.; WebbWare, J. *Developing Increased Understanding, Awareness and Potential Mitigation Strategies for Perennial Ryegrass Toxicosis in Sheep Production Systems*; Meat and Livestock Australia: North Sydney, Australia, 2014.
14. Henry, M.; Kemp, S.; Clarke, I.; Dunshea, F.; Leury, B. Production and physiological effects of perennial ryegrass alkaloids in merinos under mild heat conditions. *Small Rumi. Res.* **2016**, submitted.
15. Harrington, K.; Beskow, W.; Hodgson, J. Recovery and viability of seeds ingested by goats. *Pasture Pests Weeds* **2011**, *64*, 75–80.
16. Henry, M.; Kemp, S.; Clarke, I.; Dunshea, F.; Leury, B. Production and physiological effects of perennial ryegrass alkaloids under thermoneutral conditions in merinos. *Anim. Prod. Sci.* **2015**.
17. Hovermale, J.; Morrie Craig, A. Correlation of ergovaline and lolitrem B levels in endophyte-infected perennial ryegrass (*Lolium perenne*). *J. Vet. Diagn. Investig.* **2001**, *13*, 323–327.
18. Reed, K.; Nie, Z.; Walker, L.; Kearney, G. Fluctuations in the concentration of ergovaline and lolitrem B produced by the wild-type endophyte (*Neotyphodium lolii*) in perennial ryegrass (*Lolium perenne*) pasture. *Anim. Prod. Sci.* **2011**, *51*, 1098–1108.
19. Fletcher, L.; Barrell, G. Reduced liveweight gains and serum prolactin levels in hoggets grazing ryegrasses containing *lolium* endophyte. *N. Z. Vet. J.* **1984**, *32*, 139–140.
20. Fletcher, L.; Sutherland, B.; Fletcher, C.; Easton, H. The impact of endophyte toxins on the health of grazing sheep—An evolving story. In Proceedings of the Second Pan Pacific Veterinary Conference, Massey University, Christchurch, New Zealand, 23–28 June 1996; pp. 31–42.
21. Fletcher, L.; Easton, H. The evaluation and use of endophytes for pasture improvement. In *Neotyphodium/Grass Interactions*; Bacon, C.W., Hill, N.S., Eds.; Plenum Press: New York, NY, USA, 1997; pp. 209–227.
22. Cunningham, P.; Foot, J.; Reed, K. Perennial ryegrass (*Lolium perenne*) endophyte (*Acremonium lolli*) relationships: The australian experience. *Agric. Ecosyst. Environ.* **1993**, *44*, 157–168.
23. McLeay, L.; Smith, B. Effects of the mycotoxins lolitrem B and paxilline on gastrointestinal smooth muscel, and cardiovascular and respiratory systems, and temperature in sheep. In *Ryegrass Endophyte: An Essential New Zealand Symbiosis*; Woodfield, D.R., Matthew, C., Eds.; New Zealand Grassland Association: Napier, New Zealand, 1999; pp. 69–76.
24. Sewell, J.; Hill, R.; Hume, D. Preliminary data on the effect of ryegrass endophyte on water intake and flystrike in sheep. In Proceedings of the 50th Annual Conference of the Grassland Society of Southern Australia Inc., Geelong, Australia, 6–7 August 2009; Warn, L., Ed.; Grassland Society of Southern Australia Inc.: Geelong, Australia; pp. 170–174.
25. Dixon, R.; Thomas, R.; Holmes, J. Interactions between heat stress and nutrition in sheep fed roughage diets. *J. Agric. Sci. Camb.* **1999**, *132*, 351–359.
26. Aldrich, C.; Paterson, J.; Tate, J.; Kerley, M. The effects of endophyte-infected tall fescue consumption on diet utilization and thermal regulation in cattle. *J. Anim. Sci.* **1993**, *71*, 164–170.
27. Hannah, S.; Paterson, J.; Williams, J.; Kerley, M.; Miner, J. Effects of increasing dietary levels of endophyte-infected tall fescue seed on diet digestibility and ruminal kinetics in sheep. *J. Anim. Sci.* **1990**, *68*, 1693–1701.
28. Barger, J.; Tannenbaum, M. Consumption of endophyte-infected fescue seeds and osmoregulation in white-footed mice. *J. Mammal.* **1998**, *79*, 464–474.

29. Turner, L.; Loewer, O.; Taul, K.; Gay, N. Stimulation of summer fescue toxicosis in beef cattle by the mechanism of vasoconstriction. *Am. Soc. Agric. Eng. Winter Meet.* **1985**, *85-4509*, 17–20.

CHAPTER 2

Giraffe genome sequence reveals clues to its unique morphology and physiology

Morris Agaba[1,2,3], Edson Ishengoma[1], Webb C. Miller[3], Barbara C. McGrath[3], Chelsea N. Hudson[3], Oscar C. Bedoya Reina[3,4], Aakrosh Ratan[3,5], Rico Burhans[3], Rayan Chikhi[6,7], Paul Medvedev[6,7], Craig A. Praul[8], Lan Wu-Cavener[3], Brendan Wood[3], Heather Robertson[9], Linda Penfold[10]

1 School of Life Sciences and Bioengineering, African Institute of Science and Technology, Arusha 4222, Tanzania.

2Biosciences Eastern and Central Africa,International Livestock Research Institute, Nairobi GPO00100, Kenya.

3 Center for Genomics and Bioinformatics, Department of Biology, Huck Institutes of the Life Sciences, Pennsylvania State University, University Park, Pennsylvania 16802, USA.

4 MRC Functional Genomics Unit, Department of Physiology,Anatomy and Genetics, University of Oxford, Oxford OX1 3PT, UK.

5 Center for Public Health Genomics, Department of Computer Science, University ofVirginia, Charlottesville, Virginia 22908, USA.

6 Center for Genomics and Bioinformatics, Department of Computer Science and Engineering, Huck Institutes of the Life Sciences, Pennsylvania State University, University Park, Pennsylvania 16802, USA.

7 Department of Biochemistry and Molecular Biology, Pennsylvania State University, University Park, Pennsylvania 16802, USA.

8 Huck Institutes of the Life Sciences, Pennsylvania State University, University Park, Pennsylvania 16802, USA.

9 Nashville Zoo at Grassmere, Nashville, Tennessee 37211, USA.

10 White Oak Holdings, Yulee, Florida 32097, USA.

ABSTRACT

The origins of giraffe's imposing stature and associated cardiovascular adaptations are unknown. Okapi, which lacks these unique features, is giraffe's closest relative and provides a useful comparison, to identify genetic variation underlying giraffe's long neck and cardiovascular system. The genomes of giraffe and okapi were sequenced, and through comparative analyses genes and pathways were identified that exhibit unique genetic changes and likely contribute to giraffe's unique features. Some of these genes are in the HOX, NOTCH and FGF signalling pathways, which regulate both skeletal and cardiovascular development, suggesting that giraffe's stature and cardiovascular adaptations evolved in parallel through changes in a small number of genes. Mitochondrial metabolism and volatile fatty acids transport genes are also evolutionarily diverged in giraffe and may be related to its unusual diet that includes toxic plants. Unexpectedly, substantial evolutionary changes have occurred in giraffe and okapi in double-strand break repair and centrosome functions.

1. INTRODUCTION

The origin of giraffe's iconic long neck and legs, which combine to elevate its stature to the tallest terrestrial animal, has intrigued mankind throughout recorded history and became a focal point of conflicting evolutionary theories proposed by Lamarck and Darwin. Giraffe's unique anatomy imposes considerable existential challenges and three systems bear the greatest burden: the cardiovascular system to maintain blood pressure homeostasis[1,] the musculoskeletal system to support a vertically elongated body mass2 and the nervous system to rapidly relay signalling over long neural networks[3,4]. To pump blood vertically 2 m from the heart to the brain giraffe has evolved a turbocharged heart and twofold greater blood pressure than other mammals[1,5]. The blood vessel walls in the lower extremities are greatly thickened to withstand the increased hydrostatic pressure, and the venous and arterial systems are uniquely adapted to dampen the potentially catastrophic changes in blood pressure when giraffe quickly lowers its head to drink water[1,5-11]. To sustain the weight of the long neck and head, the nuchal ligament, which runs down the dorsal surface of the cervical vertebrae and attaches to the anterior thoracic vertebrae, is greatly enlarged and strengthened[2,12].

Okapi (*Okapia johnstoni*), the giraffe's closest relative and the only other extant member of the *Giraffidae* family, provides a useful comparison, because it does not share these unique attributes seen in giraffe[13]. Nine subspecies of giraffe have been identified that can be distinguished by coat colour and pattern, and have been reproductively isolated as long as 2 mya (refs 14, 15). Two giraffe subspecies are nearly extinct and overall the number of giraffes have declined by 40% since 2000, due to poaching and habitat loss[16]. As all giraffe subspecies share the unique anatomical and physiological adaptation of the giraffe genus, they provide an important cross-check for unique patterns of genetic variation.

Here we sequenced the genomes of the Masai giraffe and okapi, and through comparative analysis with other eutherians mammals, 70 genes were identified that exhibit multiple signs of adaptation (MSA) in giraffe. Several of these genes encode well-known regulators of skeletal, cardiovascular and neural development, and are likely to contribute to giraffe's unique characteristics.

2. RESULTS

2.1 Genome sequencing and *de novo* assembly

The whole-genome sequence of two Masai giraffe (*Giraffa camelopardalis. tippelskirchi*) from the Masai Mara (MA1) in Kenya and the Nashville Zoo (NZOO), and one fetal okapi (*O. johnstoni*) from the White Oak Conservatory was determined by constructing paired-end libraries followed by sequencing using an Illumina HiSeq yielding *ca.* 30 × coverage. Mate-paired libraries were also prepared from the MA1 Masai giraffe and okapi, and sequenced to increase coverage and to span repetitive sequence elements. The initial sequence reads from giraffe and okapi were aligned to the 19,030 cattle *(Bos taurus)* references transcripts[17] to predict homologous genes (Supplementary Table 1), which yielded 17,210 giraffe and 17,048 okapi genes. The giraffe and okapi sequence data were also used to generate a draft genome assembly with a total length of 2.9 and 3.3 Gb for giraffe and okapi, respectively (Supplementary Table 2). To verify gene predictions and gene structure in cases where the original gene annotations for giraffe and okapi were incomplete or ambiguous, the draft assembly was aligned to dog or human gene sequences. To determine whether substitutions unique to Masai giraffe were conserved in other giraffe subspecies, we performed targeted sequencing of several genes in Rothschild (*G.c. rothschildi*) and Reticulated (*G.c. reticulata*) giraffes, which diverged from Masai giraffe ~1-2 mya (refs 15, 18).

2.2 Comparative genome analysis

To identify changes that potentially underlie these unique morphological and physiological adaptations, we analysed the coding sequences of orthologous genes in giraffe, okapi and cattle. Giraffe and okapi genes are highly similar overall with 19.4% of proteins being identical (Fig. 1). Giraffe and okapi genes are equally distantly related to cattle, suggesting that giraffe's unique characteristics are not due to an overall faster rate of evolution. The divergence of giraffe and okapi, based on the relative rates of synonymous substitutions, from a common ancestor is estimated to be 11.5 mya (Fig. 1), substantially less than the previous estimate of 16 mya (refs 19, 20), which was based on mitochondrial DNA sequence comparisons.

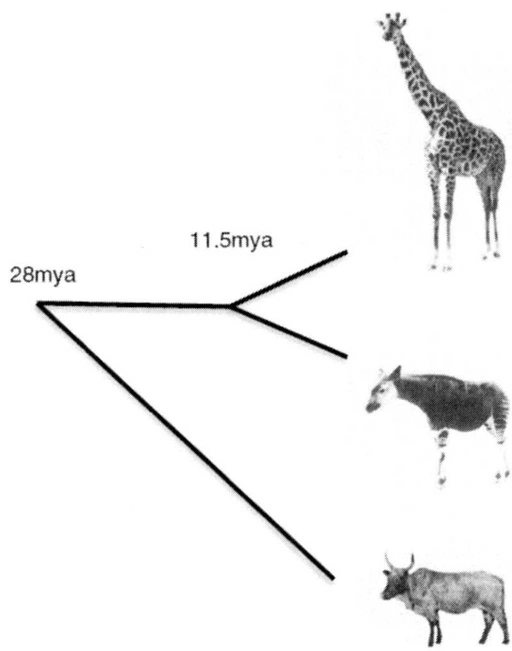

	dN	dS	Proteins identical
Giraffe-okapi	0.0055	0.0255	19.4%
Giraffe-cattle	0.0135	0.0613	9.2%
Okapi-cattle	0.0135	0.0614	9.6%

Figure 1: Divergence of giraffe and okapi from a common ancestor.

Using the average pairwise synonymous substitution divergence (dS) estimates between giraffe, okapi and cattle as calibrated by the pecoran common ancestor (27.6 mya), the divergence of giraffe and okapi from a common ancestor is estimated to be 11.5 mya. Okapi image adapted from a photograph by Raul654.

2.2 Adaptive evolution of giraffe

Adaptive divergence was evaluated by pairwise analysis of 13,581 giraffe, okapi and cattle genes that showed at least 90% coverage by comparing nonsynonymous (dN) changes in protein coding sequences as well as normalized to synonymous (dS) changes (dN/dS, ω). Enrichment analysis based on gene function (gene ontology (GO) biological processes) and pathway relationships Kyoto Encyclopedia of Genes and Genomes (KEGG) revealed elevation of dN or ω for giraffe in genes related to metabolism (tricarboxylic acid cycle, oxidative phosphorylation and butyrate), growth and development (cell proliferation, skeletal development and differentiation), the nervous system and cardiac muscle contraction (Supplementary Table 2). In parallel, we employed Polyphen[2] analysis[21] to identify genes that contain amino acid substitutions that are predicted to cause a significant alteration in function and screened for genes that exhibited evidence for positive selection. Genes exhibiting positive selection in giraffe were enriched in lysosomal transport, natural killer cell activation, immune response, angiogenesis, protein ADP ribosylation, blood circulation and response to pheromones (Supplementary Table 3). Over 400 genes were identified from the giraffe–okapi–cattle analysis that exhibited some degree of genetic differentiation in giraffe by the aforementioned analysis. These selected genes were further compared with orthologues across a large set of mammals, including 14 other cetartiodactyls, to more fully assess evidence of positive selection, relative amino acid sequence divergence and to identify amino acid substitutions unique to giraffe among eutherians. Seventy genes displayed MSA in giraffe by these criteria (Supplementary Table 4 and Supplementary Fig. 1). The unique amino acid substitutions identified in these genes were confirmed in the two unrelated individual Masai giraffe and, in some cases, confirmed in Reticulated and Rothschild giraffe by targeted sequencing. Network analyses based on GO biological process revealed eight functional clusters among the 70 MSA genes including development, cell proliferation, metabolism, blood pressure and circulation, nervous system, double-strand DNA break repair, immunity and centrosome function (Fig. 2). Remarkably, nearly half of these genes are involved in controlling developmental pattern formation and differentiation including homeobox, Notch, Wnt and fibroblast growth factor (FGF) pathway genes, major regulators of growth and cell proliferation including the transcription factors MYC, E2F4, E2F5, ETS2, TGFB1 and CREBBP, and the folate receptor 1 (FOLR1).

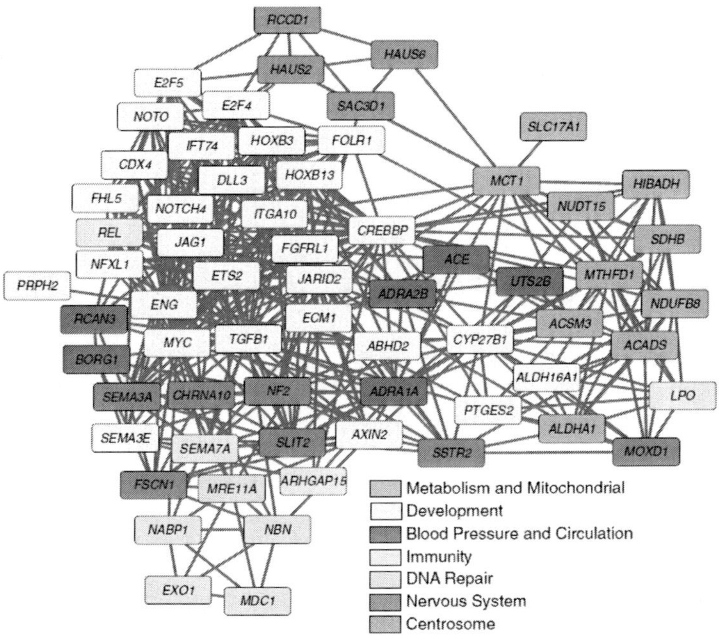

Figure 2: Network analysis of GO biological process of giraffe MSA genes.

Seventy genes were identified that exhibited MSAs based on amino acid sequence divergence as evaluated by neighbour-joining phylogenetic analysis of mammalian orthologous proteins, enrichment of nonsynonymous substitutions, unique amino acid substitutions at sites otherwise fixed in mammals, substitutions predicted to cause functional changes by Polyphen2 analysis and substitutions under positive selection. Cluster analysis was performed on the set of 70 giraffe MSA genes based on GO Biological Process using Cytoscape 3.0 (ref. 68).

2.3 Evolution of regulators of skeletal growth and differentiation

The extraordinarily long neck of giraffe is not due to adding cervical vertebrae as is the case for long-necked birds, but rather to the vertical extension of each of the seven prototypical cervical vertebrae present in mammals[13,22]. The elongation of the cervical vertebrae in giraffe is probably due to the extension of somites, which give rise to the cervical vertebrae during early embryogenesis[22], and is restricted to the cervical region by the combinatorial action of homeobox genes. The major genes and developmental pathways that specify vertebrae differentiation of the axial and appendicular skeleton in giraffe and okapi were compared with other mammals to determine whether unique patterns of amino acid substitutions were found in giraffe (Supplementary Table 5). The homeobox genes *HOXB3*, *CDX4* and *NOTO* exhibit enhanced divergence in giraffe among eutherians and have unique amino acid substitutions predicted to alter protein function. In addition, *HOXB13*, which regulates angiogenic and posterior axial skeletal development, shows high amino acid sequence divergence in giraffe and okapi compared with other mammals (Supplementary Table 4). Modulating the posterior to anterior gradient of fibroblast growth factor signalling or changing the cyclical expression of genes in the *NOTCH* or *WNT* signalling pathways could potentially modulate somite size. We found that *FGFRL1*, a decoy FGF receptor, *AXIN2*, a negative regulator of the WNT pathway, and three genes in the NOTCH pathway including *NOTCH4*, *JAG1* and *DLL3* exhibit amino acid sequence divergence in giraffe and exhibited multiple unique amino acid substitutions compared with other eutherians. The divergence of giraffe *FGFRL1* is particularly striking with a cluster of seven unique substitutions (Fig. 3a) in the domain that interacts with FGF ligands. *FGFRL1* is among nine genes in giraffe that exhibit a significantly higher number of unique amino substitutions at fixed sites in mammals (Supplementary Table 4). FGFRL1 in mammals lacks a tyrosine kinase domain essential for downstream FGF signalling and acts as a competitive inhibitor of the nascent FGF receptors[23]. Interestingly, Badlangana *et al.*[22] speculated that an inhibitor of FGF signalling might be responsible for modulating the size of giraffe cervical vertebrae based on the discovery that chemical inhibition of FGF signalling increased somite size in the chick embryo[24]. Consistent with its hypothesized role in regulating unique features of giraffe, *FGFRL1* mutations in mice and human display severe defects in skeletal and cardiovascular development[25-27]

The Giraffe FOLR1 shows exceptionally strong evidence for adaptive evolution including six positively selected amino acid substitutions of which two are predicted to cause a significant change in function (Fig. 3b). *FOLR1* mutations are embryonically lethal in mice28 and produce hypomyelination and neurological defects in humans[29]. In addition to its role in cellular folate transport, FOLR1 is internalized, processed and transported to the nucleus where it regulates components of the FGF and NOTCH pathways[30]. These changes in giraffe FOLR1 may act in concert with similar changes in *FGFRL1* and *JAG1*, components of the FGF and NOTCH pathways, respectively, to forge major developmental adaptations.

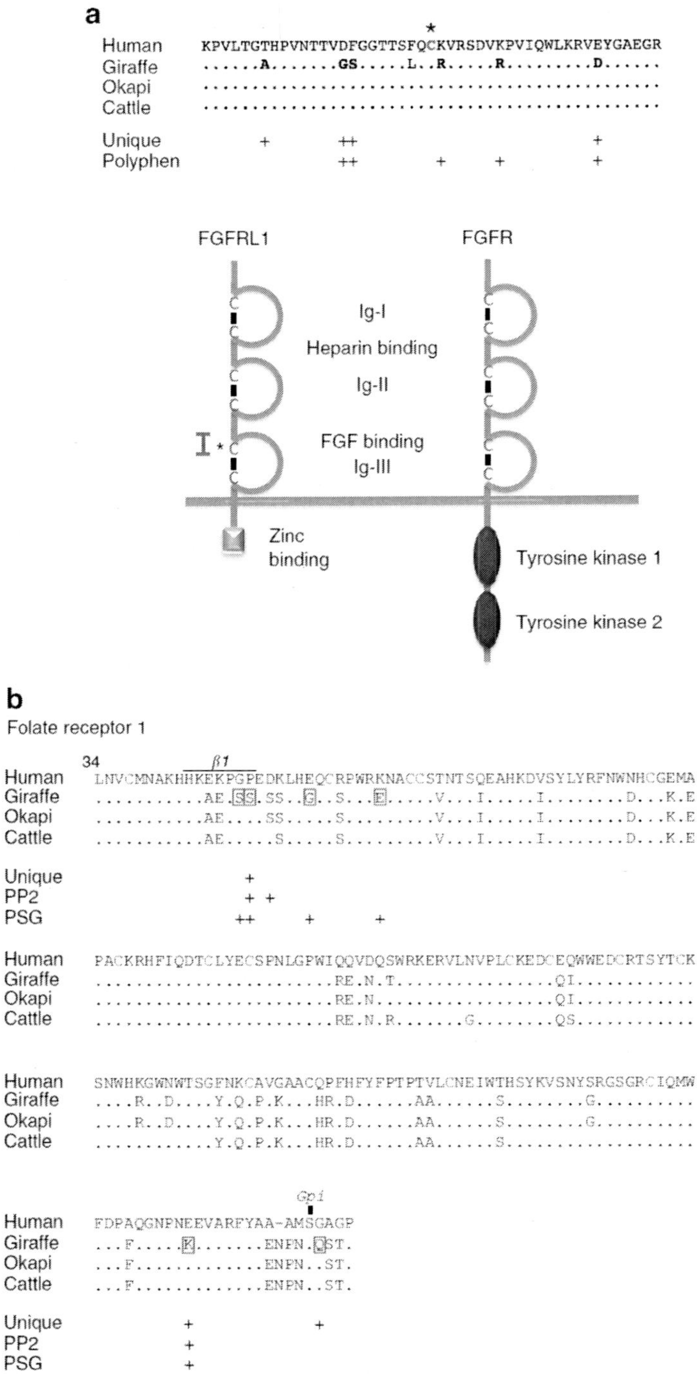
a
*
Human KPVLTGTHPVNTTVDFGGTTSFQCKVRSDVKPVIQWLKRVEYGAEGR
Giraffe ......A.......GS.....L..R.....R.........D......
Okapi ....
Cattle ....
Unique + ++ +
Polyphen ++ + + +
FGFRL1
FGFR
Ig-I
Heparin binding
Ig-II
FGF binding
Ig-III
Zinc binding
Tyrosine kinase 1
Tyrosine kinase 2
b
Folate receptor 1
34
β1
Human LNVCMNAKHHKEKPGPEDKLHEQCRPWRKNACCSTNTSQEAHKDVSYLYRFNWNHCGEMA
Giraffe ...........AE.SS.SS..G..S...E.....V...I.....I........D...K.E
Okapi ...........AE....SS.....S.........V...I.....I........D...K.E
Cattle ...........AE.....S.....S.........V...I.....I........D...K.E
Unique +
PP2 + +
PSG ++ + +
Human PACKRHFIQDTCLYECSPNLGPWIQQVDQSWRKERVLNVPLCKEDCEQWWEDCRTSYTCK
Giraffe ........................RE.N.T.................QI...........
Okapi ........................RE.N...................QI...........
Cattle ........................RE.N.R.......G.........QS...........
Human SNWHKGWNWTSGFNKCAVGAACQPFHFYFPTPTVLCNEIWTHSYKVSNYSRGSGRCIQMW
Giraffe ....R..D....Y.Q.P.K...HR.D......AA......S.........G..........
Okapi ....R..D....Y.Q.P.K...HR.D......AA......S.........G..........
Cattle ............Y.Q.P.K...HR.D......AA......S....................
Gpi
Human FDPAQGNPNEEVARFYAA-AMSGAGP
Giraffe ...F.....K.........ENPN.QST.
Okapi ...F...............ENPN..ST.
Cattle ...F...............ENPN..ST.
Unique + +
PP2 +
PSG +

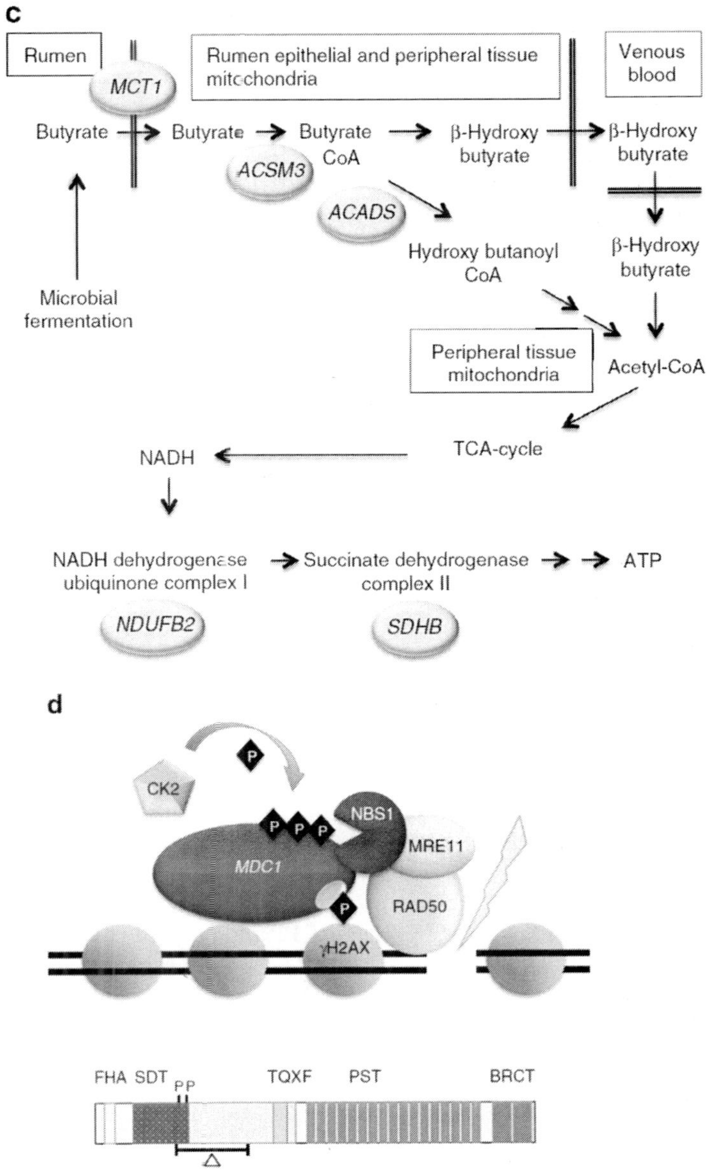

Figure 3: Giraffe genes and pathways exhibiting extraordinary divergence and patterns of amino acid substitutions. (a) Giraffe *FGFRL1* contains seven amino acid substitutions that are unique at fixed sites in other mammals and/or are predicted by Polphen2 analysis to alter function (upper panel). Human reference is shown, which is identical to cattle and okapi in this segment. The unique giraffe substitutions occur in the FGF-

binding domain region flanking the N-terminal cysteine (asterisk) of the Ig-III loop (lower panel). Red bracket in lower panel corresponds to the sequence in the upper panel. The extracellular structure of FGFRL1 (left) is the same as a prototypical FGF receptor (FGFR, right) but lacks the cytoplasmic C-terminal tyrosine kinase domains seen in FGFR and instead contains a zinc-binding domain. (b) Giraffe *FOLR1* contains seven substitutions that each show evidence of positive selection ($P<0.05$) by the branch-site model. Two of the positive selected sites (PSG), P48S and E222K, are also unique substitutions at fixed sites and Polyphen2 (PP2) analysis predicts them to alter function. P48S is within β-sheet-1 that forms part of the folic acid-binding pocket. The FOLR1 protein forms a globular structure maintained by overlapping disulfide bridges between 16 cysteine residues (red) and tethered to the plasma membrane at S233 by a Gpi anchor. The unique substitution in giraffe, G234Q, immediately adjacent to the Gpi anchor site may alter the anchor site or the rate of its formation. (c) Genes encoding key enzymes in butyrate metabolism and downstream mitochondrial oxidative phosphorylation pathways have diverged in giraffe including the monocarboxylate transporter (*MCT1*), acyl-coenzyme A synthetase-3 (*ACSM3*), short-chain specific acyl-CoA dehydrogenase (*ACADS*), NADH dehydrogenase (ubiquinone) 1β subcomplex subunit 2 (*NDUFB2*) and succinate dehydrogenase [ubiquinone] iron-sulfur subunit (*SDHB*). *ACSM3* and *ACADS* are located in the mitochondrial matrix where as *NDUFA2*, *NDUFB2* and *SDHB* are located in the mitochondrial inner membrane. In addition to being present in the rumen epithelial cells, MCT1 is highly expressed in the heart, skeletal muscle and the nervous system where it acts to transport volatile fatty acids (VFAs) and lactate. (d) Double-strand break repair genes exhibit divergence in giraffe and/or okapi. The mediator of DNA-damage check point 1 (MDC1) binds phosphorylated H2AX, which mark DNA double-strand break, and serves as scaffold to recruit the MRN DNA repair complex composed of NBS1, MRE11 and RAD50 (upper panel). The giraffe and okapi *MDC1* gene exhibits a 264 amino acid deletion that removes part of the SDT region that harbours two critical CK2 phosphorylation sites (lower panel). These two phosphorylation sites are among multiple sites that regulate the interaction of MDC1 and NBS1 essential for the recruitment of the MRN complex to double-strand breaks.

2.4 Cardiovascular and metabolic gene evolution

The giraffe cardiovascular system is adapted to regulate blood pressure over a height of 6 m and to maintain cardiovascular homeostasis associated with rapid changes in the relative position of the brain to the heart. The blood pressure of giraffe is 2.5 × higher than man, the left ventricle of the heart is enlarged and the blood vessel walls of the lower extremities are greatly thickened[1,31]. Giraffe exhibits evidence for adaptive evolution of eight genes that regulate blood pressure or cardiovascular function including two of the major adrenergic receptors α1 and β-2, urotensin-2b and angiotensin-converting enzyme (Supplementary Table 4). BCRG1 and RCAN3, which are highly expressed in the heart and purported to have important functions related to cell shape and cardiac muscle contraction, respectively, are also significantly diverged in giraffe[32,33]. The observed distinctive changes in these genes may provide clues as to the evolutionary origins of giraffe's high blood pressure, increased cardiac output and modified vasculature.

Giraffe's elevated stature enables it to feed on acacia leaves and seedpods that are highly nutritious but also contain toxic alkaloids. As with other ruminants, giraffes' gut microbes ferment plants to generate volatile fatty acids that are transported through the gut epithelium and serve as the main energy source[34,35]. Included among the MSA genes in giraffe are those involved in the catabolism of volatile fatty acids such as butyrate (MCT1, ACSM3 and ACADS) or downstream oxidative phosphorylation that generate ATP (NDUB2 and SDHB) (Fig. 3c). In addition, these proteins are essential for lactate transport and metabolism that is particularly important for cardiovascular functions[36].

2.5 Evolutionary changes in DNA and chromosome repair genes

The mediator of damage checkpoint-1 (MDC1) acts as a key scaffold for proteins participating in double-strand DNA break repair, homologous recombination, nonhomologous end-joining and telomere maintenance[37-43], and its sequence exhibits the most radical evolutionary change in giraffe and okapi compared with all other vertebrates. The giraffe and okapi *MDC1* gene contains an in-frame termination substitution in exon 5, suggesting either premature termination or alternative splicing to remove the offending termination codons. The complementary DNAs from both giraffe and okapi liver tissue were truncated in exon 5, indicating the use of a cryptic 5′-splice site resulting in a 264-amino acid internal deletion not seen in any other vertebrate. The deleted region corresponds to the ST/Q domain that contains numerous phosphorylation sites that have an impact on important regulatory protein–protein interactions[44]. Perhaps, not surprisingly, the amino acid sequence of NIBRIN, MRE11 and SOSB2, and BAZB1, which interact with MDC1 (ref. 45) are diverged in giraffe and/or okapi (Fig. 3d). We speculate that the divergence of these genes and those involved in centromeric functions may underlie the unusual degree of chromosomal fusions that occurred in the giraffe lineage[46,47].

The pecoran ancestor that gave rise to the horned, even-toed ungulates is purported to have had a karyotype of $2n$=58–60 as exemplified by cattle[46]. However, giraffe and okapi have unusual karyotypes among pecorans exhibiting reduced chromosome number of $2n$=30 and $2n$=44–46, respectively, due to Robertsonian centric fusions of acrocentric chromosomes.

3. DISCUSSION

Genes regulating fundamental aspects of development and physiology are highly conserved among major mammalian taxa[48,49]. However, we found that two-thirds of the genes most diverged in giraffe have specific roles in regulating skeletal, cardiovascular and/or neural development, or physiology (Fig. 4). In addition, several identified genes functionally intersect metabolism, growth and cardiovascular function, suggesting that giraffe's unique features may have co-evolved to elevate its stature, adapt its metabolism for more toxic food sources and adapt its cardiovascular and nervous system to the increased demands imposed by its unique morphology. The camel's neck is relatively long among mammals and intermediate in length between giraffe and okapi[22]. However, unlike the giraffe, the camel's long neck does not function to increase its stature and we did not detect similar patterns of unique amino acid substitutions between giraffe and camel among the 70 giraffe MSA genes including those that are known to regulate skeletal development. Okapi shares some of the same genetic changes seen in giraffe, which for some genes might underlie shared adaptive traits, whereas in other cases might represent evolutionary remnants of a common *Giraffidae* ancestor that is purported to have had a shorter neck than giraffe but longer than that of okapi[50].

Developmental and physiological regulatory genes in giraffe that exhibit adaptive evolution are enriched in skeletal, cardiovascular and neural functions. The MSA genes that are not known to be related to the regulation of skeletal, cardiovascular, or neural development are listed (right box).

Among the 70 genes exhibiting MSA in giraffe, *FGFRL1* is the strongest candidate for directly having an impact on the unique growth of the axial and appendicular skeleton and the cardiovascular system. FGFRL1 is known to be essential for normal skeletal and cardiovascular development in humans and mice[25-27], and the FGF pathway regulates somite size[51]. Other genes are required to restrict differential growth to the cervical vertebrae and legs, and the homeotic genes, which specify the identity of different regions of the body, probably play that role. We identified three homeobox genes—*HOXB3*, *CDX4* and *NOTO*—which exhibit significant changes in giraffe compared with other mammals. The advent of gene-editing methods provide a means of testing these hypotheses by introducing the unique amino acid substitutions seen in giraffe into the homologous genes of model organisms and determining the functional consequences. Among mammals, giraffe has some of the most challenging physiological and structural problems imposed by its towering height. The solutions to these challenges, in particular related to its turbocharged circulatory system, may be instructive for treatment of cardiovascular disease and hypertension in humans.

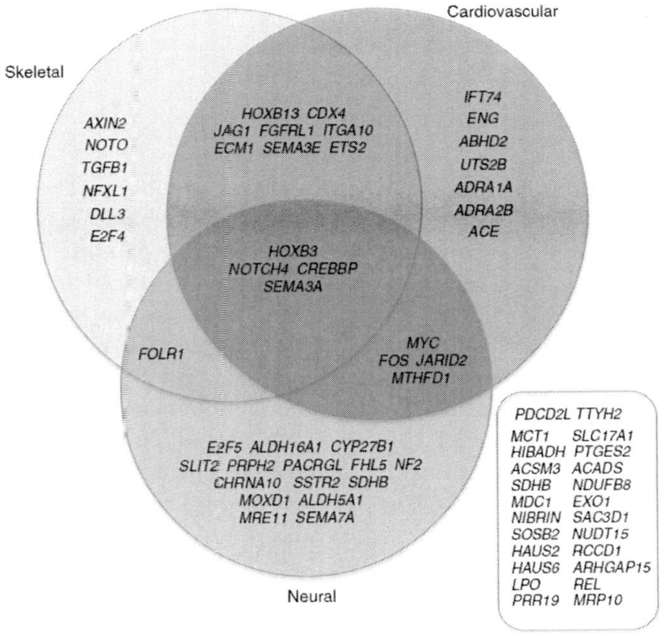

Figure 4: Gene cluster analysis of genes that exhibit evidence of adaptive evolution in giraffe.

4. METHODS

4.1 Genome sequencing

The Illumina TruSeq DNA PCR-Free Library Preparation Kit was used to construct paired-end libraries from liver samples of two female Masai giraffe (*G.c. tippelskirchi*) from the MA1 in Kenya and the Nashville Zoo (NZOO), and one fetal male okapi (*O. johnstoni*) from the White Oak Holdings. Libraries were prepared according to the manufacturer's protocol using 2 µg of input and the 550 bp insert size workflow. The Nextera Mate Pair Sample Preparation Kit was used to construct mate pair libraries from the same three samples using the manufacturer's 'Gel Plus' protocol with 4–8 kb size selection. Libraries were sequenced on an Illumina HiSeq 2500 in Rapid Run mode using 2 × 150-bp paired-end sequencing. All libraries were prepared and sequenced by the Penn State Genomics Core Facility at University Park, PA. Targeted sequencing of specific genes in Rothschild (*G.c. rothschildi)* and Reticulated (*G.c. retulata)* giraffe used genomic DNA that we isolated from primary fibroblast cell cultures obtained from Dr Oliver Ryder at the San Diego Zoo Institute for Conservation Research.

4.2 Quality control and genome coverage

Interspecies variant nucleotides were identified as follows. The sequences that aligned to the reference genome as described above were sorted by the start position of their alignment to the reference genome. These were then assembled using a reference-based approach[52], requiring at least 2-fold and at most 80-fold coverage of the region to be considered for assembly. The sequences from the okapi samples were aligned to the giraffe consensus sequence using BWA[53]version 0.5.9 with default arguments and differences between giraffe and okapi were then identified using SAMtools[54] version 0.1.19 with default arguments and the mpileup command. In-house scripts (available on request) were used to determine the position of variants relative to the (cow or dog) reference sequence.

Reads were discarded if the above process revealed evidence of insufficient read quality or instability of the genomic region, using three criteria. First, reads were required to have a best alignment to the reference assembly with at least 3% more identical nucleotides than the second-best alignment. Second, reference contigs were ignored if the depth of coverage was too high or too low according to the Lander–Waterman statistic. Third, regions with an unusually high putative rate of interspecies differences were ignored, to lessen the impact of duplications and low-complexity regions. The average depth of read coverage for the nucleotide differences identified using the dog reference assembly and applied in subsequent analyses were 20.0 for the giraffe from MA1, 21.6 for the Nashville Zoo (NZOO) giraffe and 16.8 for the okapi.

Approximately 300 genes that displayed relative high dN/dS ratios in giraffe compared with cow and okapi were lacking complete coverage relative to cattle or other orthologues of other mammals. In most cases, incomplete coverage of these genes was due to the fact that the reference cattle gene model that was used was incomplete relative to other mammals. To complete the annotation for these genes, the giraffe and okapi scaffolds containing these genes were identified. The appropriate scaffolds were analysed by the Genewise55 annotation programme using complete reference coding sequences from cattle or human. Ensembl reference transcripts with the highest degree of confidence and information (TSL:1, GENECODE basic, APPRIS P1) were used.

4.3 *De novo* assembly

First, TruSeq adapters from mate-pair data were removed using Nesoni default parameters (v0.115) (https://github.com/Victorian-Bioinformatics-Consortium/nesoni). Then, KmerGenie (v1.6269)[56] was executed with default parameters on both data sets, to determine best k-mer sizes for assembly. Scaffolds were assembled using SOAPdenovo2 (v2.04)[57], setting k-mer size to 91 for the giraffe data set and 81 for the okapi data set, and enabling repeat resolution (-R parameter). Finally, gaps in scaffolds were filled using GapCloser (v1.12) with default parameters.

The same paired-end and mate-pair reads that were used to assemble were mapped back to the giraffe and okapi assemblies. The BWA-MEM programme was executed with default parameters and statistics were extracted using the 'samtools stats' tool. It is noteworthy that the percentage of properly mapping mate pairs was lower than for paired ends, as the larger span of a mate pair makes it more likely to map across different scaffolds.

4.4 Alignments and gene trees

Before aligning sequences, tblastn was run on each sequence against corresponding cow protein RefSeq sequence (downloaded from Ensembl). This ensured correction for frame shifts indels, as it was noted that some sequences were of draft quality and may have some sequencing errors. Sequences were aligned using MUSCLE release 3.8 (ref. 58) and phylogenetic trees were constructed using PhyML Version 3.0 (ref. 59). PhyML uses a likelihood-based tree-searching algorithm to find an optimal phylogeny. Bootstrapping (n=100) was used to test the robustness of the resulting phylogenies.

4.5 Positive selection analyses

To test for signatures of positive selection acting on giraffe lineage for each of the genes, we compared the likelihood scores of selection models implemented in CODEML in the PAML package, version 4.7 (ref. 60), using likelihood ratio tests (LRTs). Branch-site models were used to identify positive selection acting on giraffe versus cattle, okapi and gerenuk. The revised branch-site model A was used, which attempts to detect positive selection acting on a few sites on particular specified lineages, that is, 'foreground branches'[61]. Four classes of sites are assumed in the model and codons are categorized into these site classes based on foreground and background estimates of ω. The alternative hypothesis that positive selection occurs on the foreground branches (ω>1) is compared with the null hypothesis, where ω=1 is fixed, using an LRT[62]. All genes whose LRT χ^2- analysis yielded *P*-values<0.05 were considered significant and these were selected as initial positive selection gene (PSG) candidates. As maximum likelihood methods designed to detect episodes of positive selection are sensitive to taxa sample size[63], we re-analysed the initial PSG candidates list by including the orthologues of all mammals for which high-quality sequence data were available (10–45 species). In addition, genes identified by other means to have shown evidence of selection/divergence in giraffe were subjected to PSG analyses using all the available high-sequence quality mammalian orthologues. The results of the PSG analysis are given for the 70 MSA genes in Supplementary Table 4. Bayesian empirical Bayes values64 were used to identify sites under significant positive selection. Functional classification of positively selected genes was achieved using PANTHER classification of Biological Process ontology terms[65].

4.6 Evaluation of nucleotide and amino acid substitutions

The mappings between giraffe–okapi nucleotide difference and the reference assembly allowed us to predict amino-acid difference (in the case of nonsynonymous protein-coding differences) as follows. Ensembl gene annotations identified protein-coding regions in the reference assembly, which were inferred to map to coding regions in giraffe and okapi, as well as revealing the transcription orientation and phase. These data were analysed extensively on the Galaxy platform[66,67] to determine enrichment of dN and dN/dS (ω) in giraffe–cattle as compared with okapi–cattle. Genes that exhibit higher dN or dN/dS values in the giraffe–cattle dyad were subjected to (a) KEGG pathway analysis and biological function analysis. Approximately 400 genes exhibiting exceptionally higher dN or dN/dS values in giraffe–cattle dyad were further analysed in detail including (a) Polyphen2 analysis[21]to identify amino acid substitutions predicted to be 'probably damaging'; (b) Unique Substitution Analysis to identify unique amino acid substitutions in giraffe at fixed sites in eutherians, and to determine which genes have a statistically significant excess of unique substitutions at fixed sites, unique substitutions were manually curated from BLAST alignments; and (c) protein phylogenetic tree analysis using neighbour-joining method to identify genes that exhibit a high degree of divergence in giraffe as assessed by relative branch lengths. In assessing unique substitutions and constructing phylogenetic trees, all available mammalian orthologues of sufficient sequence quality were used. These data were combined with global analysis of positive selection analysis to identify genes that exhibit MSA in giraffe. This aggregate analysis led to the identification of 70 MSA genes. For these 70 genes, the amino acid substitutions unique to giraffe were confirmed in 2 individual Masai giraffes (MA1 and NZOO) and confirmed in an individual Rothschild and Reticulated giraffe including *FGFRL1*, *FOLR1*, *RCAN3*, *AXIN2* and *HOXD9*.

ACKNOWLEDGEMENTS

This work was supported by the Eberly College of Science and Huck Institutes of Life Sciences, Penn State University; Nelson Mandela African Institute of Science and Technology, Tanzania; Biosciences Eastern and Central Africa–International Livestock Research Institute; Nashville Zoo, Nashville, TN; and White Oak Holding and SEZARC. E.I. was supported by the Tanzania Commission of Science and Technology, COSTECH, Tanzania. We thank the Kenya Wildlife Service for providing the giraffe tissue from the MA1. We thank David Hunter, Penn State University, for advice on the statistical analysis of unique substitutions. We thank Carly Driebelbis and Michael Potter for constructing Giraffe Genome website (https://giraffegenome.science.psu.edu).

REFERENCES

1. Mitchell, G. & Skinner, J. D. *An allometric analysis of the giraffe cardiovascular system*. Comp. Biochem. Physiol. A. Mol. Integr. Physiol. 154, 523–529 (2009).
2. Endo, H. et al. *Modified neck muscular system of the giraffe (Giraffa camelopardalis)*. Ann. Anat. 179, 481–485 (1997).
3. Badlangana, N. L., Bhagwandin, A., Fuxe, K. & Manger, P. R.*Observations on the giraffe central nervous system related to the corticospinal tract, motor cortex and spinal cord: what difference does a long neck make?* Neuroscience 148, 522–534 (2007).
4. More, H. L. et al. *Sensorimotor responsiveness and resolution in the giraffe*. J. Exp. Biol. 216, (Pt 6): 1003–1011 (2013).

5. Hargens, A. R., Millard, R. W., Pettersson, K. & Johansen, K.*Gravitational haemodynamics and oedema prevention in the giraffe*. Nature 329, 59–60 (1987).
6. Brondum, E. et al. *Jugular venous pooling during lowering of the head affects blood pressure of the anesthetized giraffe*. Am. J. Physiol. Regul. Integr. Comp. Physiol. 297, R1058–R1065 (2009).
7. Mitchell, G., Bobbitt, J. P. & Devries, S. *Cerebral perfusion pressure in giraffe: modelling the effects of head-raising and -lowering*. J. Theor. Biol. 252, 98–108 (2008).
8. Ostergaard, K. H. et al. *Left ventricular morphology of the giraffe heart examined by stereological methods*. Anat. Rec. 296, 611–621 (2013).
9. Ostergaard, K. H. et al. *Pressure profile and morphology of the arteries along the giraffe limb*. J. Comp. Physiol. B 181, 691–698 (2011).
10. Paton, J. F., Dickinson, C. J. & Mitchell, G. *Harvey Cushing and the regulation of blood pressure in giraffe, rat and man: introducing 'Cushing's mechanism'*. Exp. Physiol. 94, 11–17 (2009).
11. Petersen, K. K. et al. *Protection against high intravascular pressure in giraffe legs*. Am. J. Physiol. Regul. Integr. Comp. Physiol.305, R1021–R1030 (2013).
12. Solounias, N. *The remarkable anatomy of the giraffe's neck*. J. Zool. 247, 257–268 (1999).
13. Lankester, R. *On certain points in the structure of the cervical vertebrae of the okapi and the giraffe*. Proc. Zool. Soc. Lond. 1908, 320–334 (1908).
14. Bock, F. et al. *Mitochondrial sequences reveal a clear separation between Angolan and South*
15. Brown, D. M. et al. *Extensive population genetic structure in the giraffe*. BMC Biol. 5, 57 (2007).
16. Fennessy, J. Giraffe–The Facts. http://www.giraffeconservation.org/programmes/giraffe-conservation-status-2/ (2014).
17. Bovine HapMap, C. et al. *Genome-wide survey of SNP variation uncovers the genetic structure of cattle breeds*. Science 324, 528–532 (2009).
18. Stanton, D. W. et al. *Distinct and diverse: range-wide phylogeography reveals ancient lineages and high genetic variation in the endangered okapi (Okapia johnstoni)*. PLoS ONE 9, e101081 (2014).
19. Hassanin, A. et al. *Pattern and timing of diversification of Cetartiodactyla (Mammalia, Laurasiatheria), as revealed by a comprehensive analysis of mitochondrial genomes*. C. R. Biol. 335, 32–50 (2012).
20. Hernandez Fernandez, M. & Vrba, E. S. *A complete estimate of the phylogenetic relationships in Ruminantia: a dated species-level supertree of the extant ruminants*. Biol. Rev. Camb. Philos. Soc. 80, 269–302 (2005).
21. Adzhubei, I., Jordan, D. M. & Sunyaev, S. R. *Predicting functional effect of human missense mutations using PolyPhen-2*. Curr. Protoc. Hum. Genet. Chapter 7: Unit 7, 20 . Editorial board, J. L. Haines et al. (2013).
22. Badlangana, N. L., Adams, J. W. & Manger, P. R. *The giraffe (Giraffa camelopardalis) cervical vertebral column: a heuristic example in understanding evolutionary processes?* Zool. J. Linn. Soc. 155, 736–757 (2009).
23. Trueb, B. *Biology of FGFRL1, the fifth fibroblast growth factor receptor*. Cell. Mol. Life Sci. 68, 951–964 (2011).
24. Dubrulle, J., McGrew, M. J. & Pourquie, O. *FGF signaling controls somite boundary position and regulates segmentation clock control of spatiotemporal Hox gene activation*. Cell 106, 219–232 (2001).
25. Catela, C. et al. *Multiple congenital malformations of Wolf-Hirschhorn syndrome are recapitulated in Fgfrl1 null mice*. Dis. Model. Mech. 2, 283–294 (2009).
26. Rieckmann, T., Zhuang, L., Fluck, C. E. & Trueb, B.*Characterization of the first FGFRL1 mutation identified in a craniosynostosis patient*. Biochim. Biophys. Acta 1792, 112–121 (2009).
27. Engbers, H. et al. *Wolf-Hirschhorn syndrome facial dysmorphic features in a patient with a terminal 4p16.3 deletion telomeric to the WHSCR and WHSCR 2 regions*. Eur. J. Hum. Genet. 17, 129–132 (2009).

28. Piedrahita, J. A. et al. *Mice lacking the folic acid-binding protein Folbp1 are defective in early embryonic development.* Nat. Genet.23, 228–232 (1999).
29. Steinfeld, R. et al. *Folate receptor alpha defect causes cerebral folate transport deficiency: a treatable neurodegenerative disorder associated with disturbed myelin metabolism.* Am. J. Hum. Genet. 85, 354–363 (2009).
30. Boshnjaku, V. et al. *Nuclear localization of folate receptor alpha: a new role as a transcription factor.* Sci. Rep. 2, 980 (2012).
31. Goetz, R. H. & Keen, E. N. *Some aspects of the cardiovascular system in the giraffe.* Angiology 8, 542–564 (1957).
32. Joberty, G. et al. *Borg proteins control septin organization and are negatively regulated by Cdc42.* Nat. Cell Biol. 3, 861–866 (2001).
33. Facchin, F. et al. *Identification and analysis of human RCAN3 (DSCR1L2) mRNA and protein isoforms.* Gene 407, 159–168 (2008).
34. Clemens, E. T., Maloiy, G. M. & Sutton, J. D. *Molar proportions of volatile fatty acids in the gastrointestinal tract of East African wild ruminants.* Comp. Biochem. Physiol. A. 76, 217–224 (1983).
35. Aluwong, T., Kobo, P. T. & Abdullahi, A. *Volatile fatty acids production in ruminants and the role of monocarboxylate transporters: a review.* African J. Biotechnol. 9, 6229–6232 (2010).
36. Johannsson, E. et al. *Upregulation of the cardiac monocarboxylate transporter MCT1 in a rat model of congestive heart failure.* Circulation 104, 729–734 (2001).
37. Lukas, C. et al. *Mdc1 couples DNA double-strand break recognition by Nbs1 with its H2AX-dependent chromatin retention.* EMBO J. 23, 2674–2683 (2004).
38. Townsend, K. et al. *Mediator of DNA damage checkpoint 1 (MDC1) regulates mitotic progression.* J. Biol. Chem. 284 33939–33948 (2009).
39. Coster, G. et al. *The DNA damage response mediator MDC1 directly interacts with the anaphase-promoting complex/cyclosome.* J. Biol. Chem. 282, 32053–32064 (2007).
40. Dimitrova, N. & de Lange T. *MDC1 accelerates nonhomologous end-joining of dysfunctional telomeres.* Genes Dev. 20, 3238–3243 (2006).
41. Stucki, M. & Jackson, S. P. *MDC1/NFBD1: a key regulator of the DNA damage response in higher eukaryotes.* DNA Repair (Amst) 3, 953–957 (2004).
42. Lou, Z., Minter-Dykhouse, K., Wu, X. & Chen, J. *MDC1 is coupled to activated CHK2 in mammalian DNA damage response pathways.* Nature 421, 957–961 (2003).
43. Goldberg, M. et al. *MDC1 is required for the intra-S-phase DNA damage checkpoint.* Nature 421, 952–956 (2003).
44. Stewart, G. S., Wang, B., Bignell, C. R., Taylor, A. M. & Elledge, S. J.*MDC1 is a mediator of the mammalian DNA damage checkpoint.* Nature 421, 961–966 (2003).
45. Spycher, C. et al. *Constitutive phosphorylation of MDC1 physically links the MRE11-RAD50-NBS1 complex to damaged chromatin.* J. Cell Biol. 181, 227–240 (2008).
46. Cernohorska, H. et al. *Molecular cytogenetic insights to the phylogenetic affinities of the giraffe (Giraffa camelopardalis) and pronghorn (Antilocapra americana).* Chromosome Res. 21, 447–460 (2013).
47. Huang, L. et al. *Karyotype evolution of giraffes (Giraffa camelopardalis) revealed by cross-species chromosome painting with Chinese muntjac (Muntiacus reevesi) and human (Homo sapiens) paints.* Cytogenet. Genome. Res. 122, 132–138 (2008).
48. Jiang, Y. et al. *The sheep genome illuminates biology of the rumen and lipid metabolism.* Science 344, 1168–1173 (2014).
49. Qiu, Q. et al. *The yak genome and adaptation to life at high altitude.* Nat. Genet. 44, 946–949 (2012).
50. Danowitz, M., Vasilyev, A., Kortlandt, V. & Solounias, N. *Fossil evidence and stages of elongation of the neck.* R. Soc. Open Sci. 2, 150393 (2015).
51. Dubrulle, J. & Pourquie, O. *fgf8 mRNA decay establishes a gradient that couples axial elongation to patterning in the vertebrate embryo.* Nature 427, 419–422 (2004).
52. Ratan, A. *Assembly algorithms for next-generation sequence data.* Pennsylvania State Univ. Thesis (2009).

53. Li, H. & Durbin, R. *Fast and accurate short read alignment with Burrows-Wheeler transform*. Bioinformatics 25, 1754–1760 (2009).
54. Li, H. et al. *The sequence alignment/map format and SAMtools*. Bioinformatics 25, 2078–2079 (2009).
55. Birney, E., Clamp, M. & Durbin, R. *GeneWise and Genomewise*. Genome Res. 14, 988–995 (2004).
56. Chikhi, R. & Medvedev, P. *Informed and automated k-mer size selection for genome assembly*. Bioinformatics 30, 31–37 (2014).
57. Luo, R. et al. *SOAPdenovo2: an empirically improved memory-efficient short-read de novo assembler*. Gigascience 1, 18 (2012).
58. Edgar, R. C. *MUSCLE: multiple sequence alignment with high accuracy and high throughput*. Nucleic Acids Res. 32, 1792–1797 (2004).
59. Guindon, S. & Gascuel, O. *A simple, fast, and accurate algorithm to estimate large phylogenies by maximum likelihood*. Syst. Biol.52, 696–704 (2003).
60. Yang, Z. *PAML 4: phylogenetic analysis by maximum likelihood*. Mol. Biol. Evol. 24, 1586–1591 (2007).
61. Zhang, J., Nielsen, R. & Yang, Z. *Evaluation of an improved branch-site likelihood method for detecting positive selection at the molecular level*. Mol. Biol. Evol. 22, 2472–2479 (2005).
62. Anisimova, M. & Yang, Z. *Multiple hypothesis testing to detect lineages under positive selection that affects only a few sites*. Mol. Biol. Evol. 24, 1219–1228 (2007).
63. Anisimova, M., Bielawski, J. P. & Yang, Z. *Accuracy and power of bayes prediction of amino acid sites under positive selection*. Mol. Biol. Evol. 19, 950–958 (2002).
64. Yang, Z., Wong, W. S. & Nielsen, R. *Bayes empirical bayes inference of amino acid sites under positive selection*. Mol. Biol. Evol. 22, 1107–1118 (2005).
65. Thomas, P. D. et al. *PANTHER: a browsable database of gene products organized by biological function, using curated protein family and subfamily classification*. Nucleic Acids Res. 31, 334–341 (2003).
66. Bedoya-Reina, O. C. et al. *Galaxy tools to study genome diversity*. Gigascience 2, 17 (2013).
67. Blankenberg, D. et al. *Galaxy: a web-based genome analysis tool for experimentalists*. Curr. Protoc. Mol. Biol. Chapter 19: Unit 19, 1–21 (2010).
68. Shannon, P. et al. *Cytoscape: a software environment for integrated models of biomolecular interaction networks*. Genome Res. 13, 2498–2504 (2003).

CHAPTER 3

Equine Welfare during Exercise: An Evaluation of Breathing, Breathlessness and Bridles

David J. Mellor * and Ngaio J. Beausoleil

Animal Welfare Science and Bioethics Centre, Institute of Veterinary, Animal and Biomedical Science, College of Sciences, Massey University, Palmerston North 4442, New Zealand

SIMPLE SUMMARY

Horses have superior athletic capabilities due largely to their exceptional cardiorespiratory responses during exercise. This has particular relevance to horses' potential to experience breathlessness, especially when their athletic performance is reduced by impaired respiratory function. Breathlessness, incorporating three types of unpleasant experiences, has been noted as of significant animal welfare concern in other mammals. However, the potential for breathlessness to occur in horses as usually ridden wearing bitted bridles has not yet been evaluated in detail. Accordingly, key physiological responses to exercise and the consequences of impaired respiratory function are outlined. Then the physiological control of breathing and the generation of the aversive experiences of breathlessness are explained. Finally, the potential for horses with unimpaired and impaired respiratory function to experience the different types of breathlessness is evaluated. This information provides a basis for considering the circumstances in which breathlessness may have significant negative welfare impacts on horses as currently ridden wearing bitted bridles. Potential beneficial impacts on respiratory function of using bitless bridles are then discussed with emphasis on the underlying mechanisms and their relevance to breathlessness. It is noted that direct comparisons of cardiorespiratory responses to exercise in horses wearing bitless and bitted bridles are not available and it is recommended that such studies be undertaken.

ABSTRACT

Horses engaged in strenuous exercise display physiological responses that approach the upper functional limits of key organ systems, in particular their cardiorespiratory systems. Maximum athletic performance is therefore vulnerable to factors that diminish these functional capacities, and such impairment might also lead to horses experiencing unpleasant respiratory sensations, i.e., breathlessness. The aim of this review is to use existing literature on equine cardiorespiratory physiology and athletic performance to evaluate the potential for various types of breathlessness to occur in exercising horses. In addition, we investigate the influence of management factors such as rein and bit use and of respiratory pathology on the likelihood and intensity of equine breathlessness occurring during exercise. In ridden horses, rein use that reduces the jowl angle, sometimes markedly, and conditions that partially obstruct the nasopharynx and/or larynx, impair airflow in the upper respiratory tract and lead to increased flow resistance. The associated upper airway pressure changes, transmitted to the lower airways, may have pathophysiological sequelae in the alveolae, which, in their turn, may increase airflow resistance in the lower airways and impede respiratory gas exchange. Other sequelae include decreases in respiratory minute volume and worsening of the hypoxaemia, hypercapnia and acidaemia commonly observed in healthy horses during strenuous exercise. These and other factors are implicated in the potential for ridden horses to experience three forms of breathlessness—"unpleasant respiratory effort", "air hunger" and "chest tightness"—which arise when there is a mismatch between a heightened ventilatory drive and the adequacy of the respiratory response. It is not known to what extent, if at all, such mismatches would occur in strenuously exercising horses unhampered by low jowl angles or by pathophysiological changes at any level of the respiratory tract. However, different combinations of the three types of breathlessness seem much more likely to occur when pathophysiological conditions significantly reduce maximal athletic performance. Finally, most horses exhibit clear behavioural evidence of aversion to a bit in their mouths, varying from the bit being a mild irritant to very painful. This in itself is a significant animal welfare issue that should be addressed. A further major point is the potential for bits to disrupt the maintenance of negative pressure in the oropharynx, which apparently acts to prevent the soft palate from rising and obstructing the nasopharynx. The untoward respiratory outcomes and poor athletic performance due to this and other obstructions are well established, and suggest the potential for affected animals to experience significant intensities of breathlessness. Bitless bridle use may reduce or eliminate such effects. However, direct comparisons of the cardiorespiratory dynamics and the extent of any respiratory pathophysiology in horses wearing bitted and bitless bridles have not been conducted. Such studies would be helpful in confirming, or otherwise, the claimed potential benefits of bitless bridle use.

Keywords: horse; equine welfare; breathing during exercise; jowl angle; soft palate displacement; lower airway pathophysiology; breathlessness; dyspnoea; bit problems; bitless bridle impacts

1. INTRODUCTION

There is something especially engaging about seeing healthy horses running free, alert, ears forward, nostrils flared, moving across the ground with ease, displaying exceptional musculoskeletal agility, control and power. Through domestication and breeding, horses have been utilised in a wide range of human activities including competitive sporting or recreational pursuits because, compared to many other mammals, horses have superior athletic capabilities [1,2,3]. Knowledge of the integrated dynamics of whole body function underlying these capacities is extensive, and the contribution of key organ systems is well documented (e.g., [4]). Virtually every organ system is involved, directly or indirectly, but of particular interest for the present evaluation is homeostatic cardiorespiratory participation in speedy locomotion. The possibility that specific elements of physiological and pathophysiological responses to exercise may predispose horses to experience "breathlessness", representing unpleasant respiratory sensations identified recently as a significant welfare issue in other mammals [5], is also considered here.

Note at the outset that consideration of affects, i.e., what animals experience subjectively, has a key role in contemporary animal welfare science thinking, the affects of welfare significance being those that are consciously experienced as unpleasant or pleasant rather than as hedonically neutral [6,7,8,9,10,11,12,13]. Note that animal welfare may be characterised by reference to its major features [14], which include: (1) that it is a state which is experienced subjectively by an animal; (2) that welfare-relevant experiences arise as the integrated outcomes of sensory and other neural inputs from within the animal's body and from its environment; (3) that these sensory inputs, which reflect the animal's internal functional states and external circumstances, are processed and interpreted by the brain according to the animal's species-specific and individual nature, and past experience; (4) that the integrated subjective outcome of this neural processing represents the animal's current experience (i.e., its welfare status), and this changes as the balance and character of the inputs change; and (5) that an animal's welfare status at any one time may vary on a continuum from very bad to very good.

In short, welfare thinking focuses on what the animal experiences, the affects, considered in relation to the internal states and/or external conditions responsible for generating them [14]. Accordingly, every evaluation of an animal's general welfare status, or specific features of it, is hypothetical to the extent that it involves making inferences about what affects may be experienced. Those inferences derive credibility from validated knowledge of the underlying systems physiology, neurophysiology and affective neuroscience, as also from the caution exercised when inferring the presence of particular affects, for example, breathlessness, thirst and pain (e.g., [5,15,16,17]). Thus, the process involves cautiously exercising scientifically informed best judgement. Here it entails using physiological knowledge to clarify whether or not particular internal states and/or external circumstances might predispose horses to experience different forms of breathlessness.

To date, no examination of breathlessness per se has been found for horses. Five questions are addressed here. First, what is the likelihood that impediments to airflow in the upper respiratory tract of exercising horses [18,19,20,21,22,23] would lead them to experience a form of breathlessness characterised as "unpleasant respiratory effort" [5]? Second, to what extent might the well-established occurrence of exercise-induced hypoxaemia, hypercapnia and/or acidaemia in horses [24,25,26,27,28,29] contribute to another form of breathlessness characterised as "air hunger" [5]? Third, might a third form of breathlessness, "chest tightness", which is associated with inflammatory processes [5,30,31], arise independently or in association with other pathophysiological conditions [3,32,33,34,35]? Fourth, what, if any, impact might use of bitless bridles have on the ease of airflow through the upper respiratory tract [22,36,37,38,39]; the degrees of exercise-induced hypoxaemia, hypercapnia and/or acidaemia [24,25,26,27,28,29]; and the occurrence and severity of lower respiratory tract pathophysiology [3,38,40,41], and thus breathlessness? Fifth, in light of these observations, how significant might breathlessness be as an animal welfare issue in exercising horses?

This paper begins by describing key features of equine physiology and pathophysiology that are relevant to inferences about whether or not one or more of the three different forms of breathlessness may be experienced during exercise. Thus, information is provided about the horse as an obligate nasal breather, airflow capacity and resistance in the upper respiratory tract, oxygen demands in exercising horses, cardiorespiratory responses to exercise, respiratory-gas homeostasis during exercise and some pathophysiological impediments to respiratory function. The paper then continues with a brief general outline of breathlessness in mammals and an evaluation of respiratory effort, air hunger and chest tightness as they might apply to horses engaged in competitive exercise. There follows a commentary on bitted and bitless bridle use with particular reference to possible consequences for respiratory dynamics related to airflow resistance, respiratory gas homeostasis and lower airway pathophysiology in exercising horses. Finally, some implications of these and other related observations for equine welfare are discussed.

2. SOME KEY ATTRIBUTES OF BREATHING-RELATED FUNCTIONS IN THE HORSE

2.1. Upper Airway Anatomy and Obligate Nasal Breathing

The upper airway of the horse includes the nostrils, nasal passages separated by the nasal septum, paired paranasal sinuses and guttural pouches, and the nasopharynx which extends from the nasal passages to the trachea [42]. The nasopharynx is located dorsal to the soft palate, which separates it from the oropharynx; the soft palate is the anatomical extension of the roof of the mouth (the hard palate) and extends from the end of the hard palate to the larynx (Figure 1 and Figure 2) [42]. It is tightly apposed to the base of the larynx, so that there is usually no communication between the nasopharynx and oropharynx, thereby obliging the horse to breathe through its nose; i.e., it is an obligate nasal breather [43]. Oral breathing can occur, but usually only when anatomical abnormalities or disruptive conditions such as dorsal displacement of the soft palate are present. The horse cannot breathe or exercise normally under these circumstances [42]. Swallowing disengages the larynx and the soft palate, allowing food and water to pass from the oropharynx into the oesophagus (Figure 2).

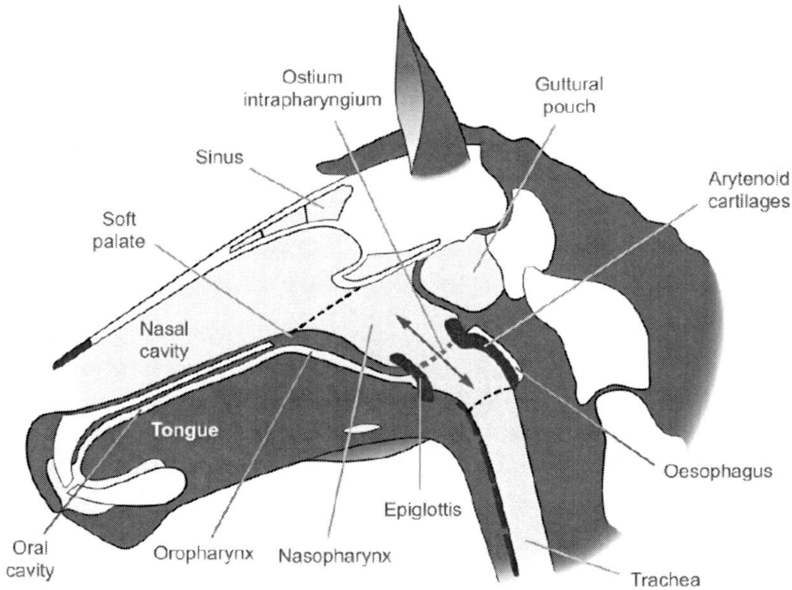

Figure 1. Diagram of the relationship of the soft palate and the larynx of the horse while breathing with its mouth closed (modified from Cook [22] with permission). The larynx (the "button") fits tightly into the ostium intrapharyngium (the "buttonhole") of the soft palate, creating an airtight seal so that air cannot enter the oropharynx (see also Figure 2). This, and closed lips, would enable a negative pressure to be maintained in the oropharynx, which would hold the soft palate against the root of the tongue thereby widening the nasopharyngeal airway [22,44]. Disengagement of the soft palate and larynx and/or loss of the lip seal would dissipate the negative pressure in the oropharynx, which would allow the soft palate to rise, vibrate with each breath, and impede nasopharyngeal airflow. The double-headed arrow indicates the directions of airflow.

2.2. Upper Respiratory Tract Airflow Capacity and Resistance

The maximum O_2 consumption of Thoroughbred and Standardbred racehorses at peak exertion is about 40 times greater than values at rest [42]. This is far greater than the 6–8-fold increase observed in endurance trained human athletes and the 10-fold increase seen in some other mammals [45,46].

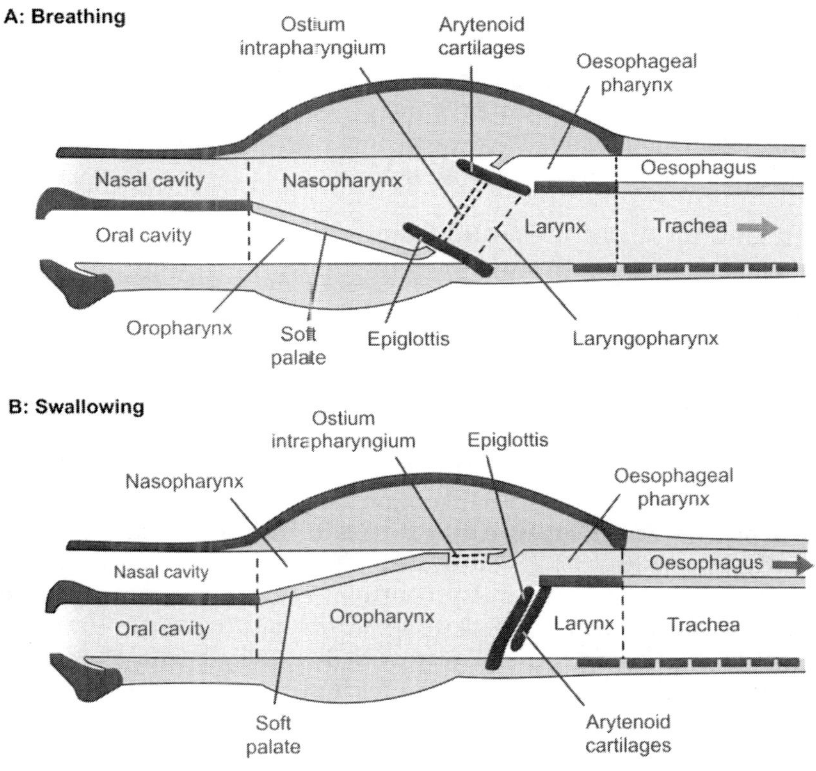

Figure 2. A schematic diagram of changes in pharyngeal function between the states of breathing and swallowing; derived from Cook [22] with permission. For clarity, the mouth is depicted as open and the oropharynx and oesophagus are shown as actual spaces, whereas, except during eating and drinking, the mouth is usually closed and the oropharynx and oesophagus are potential spaces only. (**A**) Breathing: Lowering of the soft palate seals off the oropharynx and enlarges the nasopharynx. The raised arytenoid cartilages close the oesophagus and open the larynx. The lowered epiglottis forms a seal with the soft palate. In this state, the larynx (the "button") now fits snugly into the ostium intrapharyngium (the "buttonhole" of the pharynx). (**B**) Swallowing: The raised soft palate closes off the nasal cavity to prevent food and water from entering it. The arytenoid cartilages rotate down to open the oesophagus and prevent food and water from entering the larynx and trachea. Finally, the epiglottis rotates back over the arytenoid cartilages.

The nasopharyngeal airflow, i.e., respiratory minute volume, required to meet these peak O_2 demands in Thoroughbreds is approximately 1800–2000 L/min, being 25–27 fold greater than resting values of about 65–80 L/min [19,26,47]. Achieving such high airflows poses a significant physiological challenge, exacerbated when narrowing of the nasopharynx, larynx and/or trachea increases flow resistance thereby necessitating the generation of higher negative inspiratory pressures [18,20,22,48,49]. In exercising horses, the nasopharynx, larynx and trachea contribute about 95% to airflow resistance during inspiration and about 45% to it during expiration [18,29].

2.2.1. Jowl Angle and Airflow Resistance

Head-neck position, or head carriage, is an important determinant of airflow resistance [18,22,41,49,50,51,52]. This may be expressed in terms of the jowl angle (Figure 3), i.e., the angle of intersection of the leading edge of the neck and the line of the lower jaw [22,50,51]. It has been established that the cross-sectional area of the nasopharynx decreases the closer the jowl angle approaches 33°, equivalent to the nasal bone being nearly vertical to the ground [22,23,53,54,55]. At rest the horse usually holds its head at a jowl angle of about 90°, whereas, if free to do so, when galloping it may extend its head and neck out towards jowl angles of 120–130° [22,50,51,54]. This straightens and widens the nasopharynx [22,23,53] and disproportionately reduces nasopharyngeal airflow resistance [20,51]; i.e., it *disproportionately* increases the ease of airflow in approximate compliance with Poiseuille's Law of pressure-flow characteristics in a tube [56,57]. Such neck extension also stretches and straightens the extrathoracic trachea, making it less compliant and therefore less subject to dynamic narrowing during inspiration [18,21, 22, 50,54].

In contrast, when tightly reined in during exercise the horse's nasal bone may be nearly vertical to the ground (jowl angle of about 33° or less), thereby increasing the angulation of the nasopharyngeal airway, decreasing its cross-sectional area and markedly impeding airflow at that point [22,23,55,58]. In the extreme example of the hyperflexion of the Rollkur position the cross-sectional area of the laryngeal opening is also significantly reduced [59,60]. As expected, both such reductions disproportionately increase inspiratory airflow resistance [19,48,51, 52,61].

Note that various studies provide evidence that maintaining low jowl angles by rein tension can also induce or exacerbate a range of dynamic upper respiratory tract disorders during strenuous exercise (e.g., [52,54, 64,65,66].

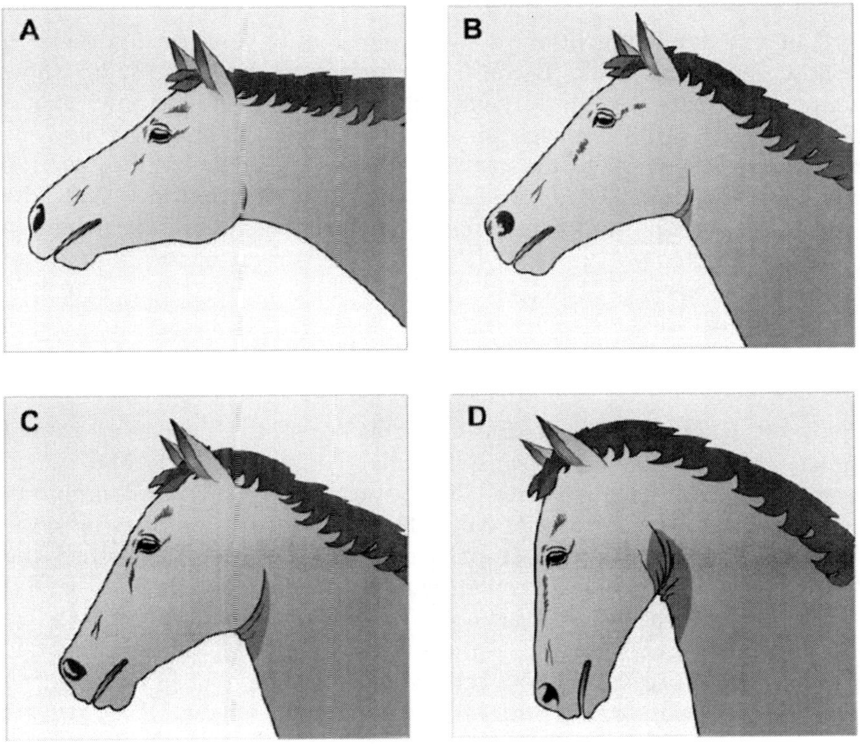

Figure 3. Diagrams based on photographs of horses engaged in different activities whilst wearing bitted bridles (not shown) and displaying different jowl angles: (**A**) derived from Gerring [50]; and (**B–D**) derived from Cook [22] who quantified nasopharyngeal areas from radiographs of the same conscious horse at different jowl angles, and compared them with the area at an extended angle. (**A**) Head-neck position of a *galloping racehorse with low rein tension* showing an extended jowl angle of about 125° observed in some animals when strenuous exercise has continued for several minutes. This jowl angle straightens and widens the diameter of the upper airway, resulting in a smooth (creaseless) "throatlatch". However, the slightly open mouth, often seen when bitted bridles are worn [36,37,62,63], breaks the lip seal. In the remaining three examples traction was applied to the interdental space (diastema) via rein tension; this often leads to the mouth being wide open, although here it is shown as slightly open. (**B**) *A galloping racehorse.* The jowl angle is about 87°, close to the 90° resting neutral position, yet the nasopharyngeal area is reduced to about 90% of that at an extended head-neck position (e.g., A), potentially representing a significant impediment to maximal airflow. Note that the "throatlatch" is creased. (**C**) *A show jumper.* The jowl angle is about 75°, leading to a reduction in the nasopharyngeal area to about

55% of an extended position (e.g., A), representing a major hindrance to airflow. The "throatlatch" is considerably creased. Note that many show jumpers are expected to perform with significantly lower jowl angles than this [22]. (**D**) *A dressage horse*. The jowl angle is about 33° and the nasal bone is vertical to the ground. The degree of airway obstruction will be proportionately much greater than that depicted in C. Note the multiple creases in the "throatlatch". Many dressage horses perform in a more hyperflexed position than this where the airway obstruction would be even more severe [22,59].

2.2.2. Airflow Resistance and Disorders of the Soft Palate and Nasopharyngeal Walls

Two soft palate disorders cause narrowing of the nasopharynx and impede airflow. They are transient palatal instability (PI) and dorsal displacement of the soft palate (DDSP), the causes of which continue to be discussed [38,39,64,67,68,69,70]. One view emphasises the importance during exercise of an airtight seal between the larynx (the "button") and the *ostium intrapharyngium* (the "buttonhole") in the soft palate so that, when sealed, air cannot enter and dissipate the negative pressure in the oropharynx (Figure 1 and Figure 2) [22,44]. Negative oropharyngeal pressure, likely generated by swallowing [44], is said to hold the soft palate against the root of the tongue, whereas, it is argued, its dissipation allows the soft palate to bulge or balloon dorsally thereby partially obstructing the nasopharynx and larynx [44]. Such bulging, observed by dynamic endoscopy, is a distinguishing feature of PI and, moreover, appears to precede complete palatal-laryngeal disengagement and the onset of DDSP [44,71,72]. One way this seal may be broken during vigorous exercise is when the jowl angle is lower than is concordant with exercise-appropriate neck extension (Figure 3A) as a result of the horse being reined in to a marked extent (e.g., Figure 3C or D) [44,71,73]. Another proposed way is via the buccal route when aversive bit sensations are said to cause mouth opening sufficient to break the airtight lip seal [22,62,74]. A dynamic endoscopy study provides indirect support for either or both of these routes, because the wearing of a bitted bridle, as opposed to a halter, to reproduce significant neck flexion by the pressure of the bit in the mouth, was considered to be important for predisposing some horses to DDSP [71]. Potential negative impacts of the bit on respiratory function will be considered in more detail below (Section 5).

Dynamic pharyngeal collapse may lead to partial-to-complete obstruction of the airway rostral to the glottis and marked transient increases in negative pressure with each inspiration [75]. Characterised by ventral displacement of the dorsal pharyngeal wall and/or axial displacement of the lateral walls of the pharynx, the obstruction increases in severity as the number of walls involved rises from one to three, and finally to four if PI/DDSP also occurs [75]. Likely predisposing factors for pharyngeal collapse during high-speed treadmill exercise are considered to include low jowl angles due to reign use, high negative inspiratory pressure and weakness of the stylopharyngeus muscles, normal contractions of which act to dilate the dorsal nasopharynx [51,70,71,75,76,77].

2.2.3. Other Pathophysiological Impediments to Airflow in the Upper Respiratory Tract

In addition, glottal-laryngeal dysfunction can restrict airflow and increase negative inspiratory pressure, and may also be exacerbated or elicited by rein-induced low jowl angles [54,64,65,78]. Such disorders include, but are not limited to, dynamic laryngeal collapse, which may include one or more of laryngeal hemiplegia, laryngeal hemiparesis, vocal cord collapse and axial deviation of the aryepiglottic folds; other disorders include epiglottal entrapment and flaccid epiglottis [44,54,70,73,77,78]. Finally, there is potential for partial dynamic collapse of the extrathoracic trachea when negative inspiratory pressures increase in response to more rostral airway obstructions, and of the intrathoracic trachea due to compressive transmural pressure during forced expiration [18].

2.2.4. Clustering of Multiple Upper Respiratory Tract Airflow Impediments

Clustering of multiple disorders has been reported in horses submitted for dynamic endoscopic examination of upper respiratory tract disorders because of suboptimal athletic performance (i.e., exercise intolerance). For example, in 471 Thoroughbred racehorses exhibiting dynamic collapse of the nasopharynx or larynx, DDSP occurred in 50% and PI in 33%, and clustered forms of dynamic collapse were apparent in 30% of the horses [44]. In 99 harness racehorses, strong associations were demonstrated between PI and intermittent DDSP, between flaccid epiglottis and dynamic laryngeal collapse due to rein-induced low jowl angles, and between axial deviation of aryepiglottic folds, PI and flaccid epiglottis [77]. The overall negative impacts of such clusters on airflow resistance might be expected to be greater than the impact of each disorder alone.

2.3. Oxygen, Carbon Dioxide and Other Key Features of Respiratory Function in Galloping Horses

As already noted, the 40-fold increase in O_2 consumption between the values at rest and at the gallop is achieved, in part, by a 25–27-fold increase in respiratory minute volume [19,26,47]. This itself is due to the combined effects of increases in respiratory frequency from 10–15 to 110–130 breaths/min and in tidal volume from 5–6 to 12–15 L [29]. Equally important for O_2 delivery to tissues are related cardiovascular changes [46]. These include increases in heart rate from resting values of 30 beats/min, or less, to peak values of 210–230 beats/min. At a mean stroke volume of about 1.35 L, this represents a 7–8-fold increase in cardiac output from about 40 L/min or less to about 285–310 L/min.

Oxygen delivery is further enhanced by other changes. These include the effects of splenic contraction increasing the haematocrit from about 40% at rest to about 60% at the gallop and a corresponding increase in arterial blood O_2 content from about 20 to about 30 mL O_2/100 mL [25,46]. In addition, the O_2 extraction by tissues increases by five-fold, or more, indicated by an equivalent widening of the carotid-arterial-to-mixed-venous difference in blood O_2 content [25,46]. Of course, CO_2 disposal is equally important and is referred to below.

2.3.1. Oxygen and Carbon Dioxide Partial Pressures, pH and Chemoreceptor Function

The partial pressures of respiratory gases in carotid arterial blood reflect the extent to which O_2 uptake via the lungs matches its utilisation within the body and, conversely, the extent of disposal of the CO_2 produced by working muscle and transported to the lungs [29]. Carotid arterial pH primarily reflects whole body production of carbonic acid from CO_2(respiratory acidosis) and lactic acid production via anaerobic glycolysis within contracting skeletal muscle (metabolic acidosis) [25,29,79].

Changes in the arterial partial pressure of CO_2 ($PaCO_2$) and/or arterial pH, which are detected by both central and peripheral chemoreceptors, are more potent regulators of ventilation than are changes in the arterial partial pressure of oxygen (PaO_2), which are detected only by peripheral chemoreceptors [29]. The central chemoreceptors, located in the ventral aspect of the medulla, respond to changes in the pH of cerebrospinal fluid (CSF). Rapid responses occur when changes in CSF pH are due mainly to changes in $PaCO_2$ because CO_2 diffuses freely across the blood–brain barrier [29]. In contrast, responses to changes in arterial blood pH are delayed because the barrier is relatively impermeable to H^+ ions [29].

Accordingly, acute increases in circulating H^+ concentrations are first detected by the peripheral chemoreceptors.

The activity of peripheral chemoreceptors, which are located in the carotid bodies and aortic arch of the horse [29], is increased by complex interactions between different degrees of hypoxaemia, hypercapnia and acidaemia [80]. Although peripheral chemoreceptor responses to CO_2 and pH are nearly linear, their response to O_2 evaluated separately is not, as they show enhanced activity only when the PaO_2 decreases below about 60 mm Hg [29]. However, peripheral chemoreceptors exhibit greater sensitivity to rises in $PaCO_2$ the greater the degree of the coexistent hypoxaemia [80].

2.3.2. Hypoxaemia, Hypercapnia and Acidaemia in Galloping Horses

At rest, healthy high performance horses exhibit carotid PaO_2 of 85–100 mm Hg and $PaCO_2$ of 35–45 mm Hg [24,27,29,81]. Both O_2 consumption (VO_2) and CO_2 production (VCO_2) increase progressively as exercise intensity increases from resting values to the maximum [25,81]. In strenuously exercising racehorses, but not in ponies [82,83,84], hypoxaemia (PaO_2 50–70 mm Hg) and haemoglobin desaturation are apparent when exercise intensity exceeds 60% of the maximum O_2consumption capacity (VO_{2max}) [25,81], whereas hypercapnia ($PaCO_2$ 48–51 mm Hg) appears only when CO_2 production is greater than 85–92% of VCO_{2max} [25]. In contrast, there is a near linear inverse relationship between carotid arterial pH and exercise intensity between about 45% and 100% of VCO_{2max} [25]. These changes in blood gas partial pressures and pH in strenuously exercising racehorses are considered to reflect limits on the exchange capacity between alveolar air and blood due to several interacting factors, including: (1) increased cardiac output reducing alveolar capillary transit time and thus impairing gas diffusion; (2) a very low PO_2 in mixed venous blood entering the lungs, representing marked degrees of O_2desaturation that cannot be corrected in a single transit; and (3) the normal maximal ventilation capacity (i.e., respiratory minute volume) being inadequate due to limitations on peak tidal volume and because breathing rate is locked 1:1 with stride frequency in galloping horses [29,35,85].

Exercise-induced arterial hypoxaemia is restored to normoxaemia and hypercapnia is reversed to hypocapnia well within 5 min of healthy strenuously exercised racehorses coming to a standstill [25,29,81]. Moderate-frequency/high-tidal-volume breathing during at least the first 40–60 s after cessation of exercise restores the normoxaemia and causes a respiratory "washout" of CO_2 [86] thereby accounting for the early appearance of hypocapnia. The rapid post-exercise changes in $PaO_2$and $PaCO_2$ show that, in the face of marked reductions in VO_2 and VCO_2, the gas exchange capacity of the lungs greatly exceeds that required for homeostatic restoration of normoxaemia and normocapnia once exercise ceases. In contrast, there is little change in the arterial acidaemia during this period (blood pH ~7.40 at rest and ~7.13 at the gallop and shortly afterwards) [24,27]. However, immediately after exercise and for several minutes, although tidal volume progressively approaches resting values, respiratory frequency remains elevated [86,87]. This suggests that acidaemia might continue to influence respiratory activity during this period. In addition, an evaporative respiratory contribution to dissipating the exercise-induced thermal load is also likely to be involved [88,89,90]. Further salient aspects of the control of breathing are referred to below (Section 3).

Compared to healthy horses, pathophysiological conditions involving the *upper airway* that increase respiratory airflow resistance, for example, PI, DDSP, pharyngeal collapse, epiglottal entrapment and laryngeal hemiplegia [27,28,91], and experimental obstruction of the upper airway [48,91], are associated with an earlier onset of greater degrees of hypoxaemia, hypercapnia and acidaemia during exercise

2.4. Lower Respiratory Tract Pathophysiology and Exercise

Three *lower airway* pathophysiological conditions are considered here, namely exercise-induced pulmonary haemorrhage (EIPH), negative pressure pulmonary oedema (NPPO) and inflammatory airway disease (IAD). EIPH and NPPO represent *lower airway* conditions that are likely to be exacerbated or precipitated by increases in *upper airway* flow resistance, and IAD is considered to be equivalent to equine asthma.

2.4.1. Exercise-Induced Pulmonary Haemorrhage (EIPH)

The subject of EIPH has received attention for over three decades (e.g., [3,35,40,41,48,92,93,94]. This is due to the high prevalence of the condition (70–90%) in strenuously exercising horses [3,32,33,34,35] and its negative impact on the peak athletic performance of significantly affected animals [3,94]. Although primarily occurring in Thoroughbreds, Quarter Horses and Standardbreds and during sprint racing, EIPH is also apparent among high-performance horses in cutting, barrel, roping, polo, cross-country and three-day event competitions, as well as in show jumping, hunter-jumper, steeplechase, dressage and draft horse events [35].

At present, no consensus is apparent regarding the ultimate cause of EIPH [35,38,94]. With regard to the proximate cause, markedly negative alveolar inspiratory pressure and high alveolar capillary blood pressure have been emphasised, and treatments are based on one or both of these factors [3,35]. Thus, during strenuous exercise, intra-alveolar air pressure becomes more negative and alveolar capillary blood pressure increases; this results in marked increases in the transmural pressure difference [35,48,91,95] at a time when pulmonary blood flow, equal to cardiac output, would approach maximum levels of 285–310 L/min [29,41]. The proximate cause of EIPH is therefore considered to be rupture of alveolar capillary membranes leading to extravasation of blood into the interstitial and alveolar spaces, with such ruptures occurring as a secondary consequence of the exercise-induced increase in transmural pressure exceeding the tensile strength of the capillary wall [3,35,94,96]. Note, however, that upper airway factors may be implicated in clarification of the ultimate cause of EIPH [35] because a greater alveolar transmural pressure difference is apparent in exercising horses that have partially obstructed upper airways and associated increases in upper airway resistance [48,91,95].

Grading EIPH by tracheobronchoscopic examination differentiates five levels of severity from 0 (none) to 4 (the highest) based on the amount and distribution of blood in the trachea and/or mainstream bronchi visible from the tracheal bifurcation. Although inter-observer reliability of scoring is high and it is assumed that the assigned scores reflect the severity of haemorrhage, this has not yet been established [3,94]. Nevertheless, the association of EIPH with reduced athletic performance (i.e., exercise intolerance) suggests a significant degree of pulmonary functional impairment, especially in badly affected horses (EIPH grades 3 and 4) [94]. Such severely impaired functionality is associated with filling of a proportion of alveolar air spaces with a blood/interstitial fluid mixture during a race, followed after the race by oedematous, fibrotic and angiogenic responses to haemorrhage into the alveolar interstitium [35,41,96]. The changes that occur during a race would likely act to reduce the surface area available for unimpeded respiratory gas exchange, and those that occur after the race would probably also reduce the effective surface area for gas exchange at that time and represent an impediment likely to be worsened by repeated episodes of EIPH. Although this would be expected to exacerbate the usually observed exercise-induced hypoxaemia, hypercapnia and acidaemia [24,25,27, 28,29,81], there is insufficient information to clarify the extent of such effects [94]. Note, however, that the absence of evidence does not constitute evidence of absence. In this case, difficulties in prospectively identifying and acquiring horses that would be badly affected by EIPH, thereby enabling insertion of arterial catheters only into susceptible animals before exercise testing, would hinder recruitment of a sufficient number of affected animals to provide a credible basis for conclusions about impacts of EIPH on blood gas tensions and pH.

Moderate-to-strong evidence supports the following consequences of EIPH grades 3 and 4 [94]: affected horses exhibit extensive and characteristic pulmonary lesions; the condition is progressive and related to the load of racing; and affected racehorses have shorter careers, inferior competitive race performance and/or the worst race performance.

2.4.2. Negative Pressure Pulmonary Oedema (NPPO)

It is speculated here that NPPO-like effects might manifest as extreme lower respiratory tract outcomes of marked upper airway obstruction due to very severe DDSP and/or pharyngeal collapse when they occur (Section 2.2.2). NPPO is a rare life threatening condition in human beings [97,98], and has also been observed in horses [99,100,101,102,103]; it develops rapidly after the onset of complete or very severe *upper airway obstruction*, for example, due to post-surgical laryngospasm following tracheal extubation. The oedema is precipitated by obstruction-induced marked increases in negative intrapleural pressure being transmitted to the alveolar interstitium, which, when combined with elevated pulmonary capillary pressure, substantially increases the transmural pressure difference [97,98,103,104]. With one reported exception [103], significant alveolar capillary bleeding is less apparent with NPPO [97,98,99,100,101,102,104] than with EIPH grades 3 and 4 [3,35,94,105], possibly because pulmonary blood flow, i.e., cardiac output, would be much lower in the absence of exercise [46]. Moderately obstructing the upper airways experimentally has partly or completely replicated EIPH pressure changes, but at lower levels [48,95,106]. This suggests that very marked nasopharyngeal obstruction may precipitate worse oedema than would less severe nasopharyngeal obstruction. Thus, in strenuously exercising horses with very severe DDSP and/or pharyngeal collapse grades 3 and 4 (Section 2.2.2), such obstruction likely occurs repeatedly with each inspiration. If so, obstruction-induced severe hypoxaemia and hypercapnia would occur immediately due to markedly impeded airflow through the upper respiratory tract, and such oedema, which hinders respiratory gas exchange via deleterious bronchiolar and alveolar effects and also reduces compliance in the *lower respiratory tract*, would likely become apparent within 1 h and persist for 24–48 h after the obstruction has been relieved [97,98,104]. There would be merit in exploring this possibility.

2.4.3. Inflammatory Airway Disease (IAD)

Resembling human asthma, IAD in horses embodies a spectrum of chronic inflammatory dysfunctional effects of the lower respiratory tract [107,108] and is associated with riders' or trainers' subjective impressions of horses' reduced race performance or willingness to perform in show jumping and dressage events [31]. Clinical signs include coughing and difficulty in breathing (i.e., shortness of breath), which may vary in occurrence and intensity, and are associated with variable increases in *expiratory airflow resistance* [31]. The proximate cause is considered to be irritant hypersensitivity giving rise to bronchoconstriction, bronchial oedema and mucus accumulation in bronchi [31,35]. These effects in their turn impede pulmonary gas exchange, which is evidenced by strenuously exercising horses with IAD exhibiting greater degrees of arterial hypoxaemia and a more protracted post-exercise recovery of blood gas levels than do healthy controls [109,110,111]. Clearly, this respiratory impairment will contribute to the exercise intolerance observed in IAD-affected horses [31]. Finally, in a study of 122 horses with impaired athletic performance that were diagnosed with respiratory disease, ~23% had IAD, ~16% EIPH, and ~10% had both IAD and EIPH [110].

3. BREATHING MECHANISMS THAT UNDERLIE BREATHLESSNESS

Much of what follows in this section is derived from a detailed account of breathlessness in which human experience was cautiously extrapolated to other mammals [5]. The bases for this extrapolation include respiratory control mechanisms and breathing responses that appear to be common to most mammals, as well as displays of withdrawal, escape, struggling, and other aversion behaviours by non-human mammals in response to respiratory challenges that human beings find intensely unpleasant.

During normal breathing at rest, human beings usually remain unaware of sensations relating to ventilation. However, in situations where respiration is stimulated, challenged, obstructed or attended to, sensations associated with breathing rise to consciousness. Included among these is breathlessness, which is a well-recognised negative affective experience [112]. In fact, it embodies at least three qualitatively distinct sensations—respiratory effort, air hunger and chest tightness—and each of these reflects comparison by cerebral cortical processing of some combination of heightened ventilatory drive and impaired respiratory function [5]. All three major forms of breathlessness are considered to be relevant to the present evaluation.

Breathing is unusual in that its control is both automatic, via the brainstem, and voluntary, via activity of the motor cortex [113]. Automatic or reflex control of breathing is due to the collective activity of various nuclei in the medulla that stimulate pools of inspiratory motor neurons, thus influencing the frequency, strength and timing of contractions of respiratory muscles. The activity of these neurons generates the normal spontaneous pattern of breathing [114], which, in turn, is influenced by afferent inputs of four major types: (1) inputs from central and peripheral chemoreceptors responding to the PO_2, PCO_2, pH and other factors in arterial blood and cerebrospinal fluid; (2) afferent feedback from various respiratory structures (briefly outlined below); (3) inputs from working locomotor muscles; and (4) afferents from various higher brain centres [30,115,116]. On the other hand, voluntary control of breathing depends on efferent signals from the primary motor cortex which act to control ventilation during increased exertion and also to interrupt the normal rhythm of breathing during vocalisation and breath-holding [116,117,118].

Importantly, afferent feedback, which signals the response of the respiratory apparatus to central motor command, is also integral to both the control of breathing and the conscious awareness of respiratory sensations. Information from various receptors in the airways, lungs, respiratory muscles and chest wall, transmitted to the cortex and limbic structures, plays a key role in generating or modulating sensations of breathlessness [30,112]. The effect of such feedback on the quality and intensity of breathlessness depends on which receptors are stimulated or not stimulated, and how strongly, and on whether the source of central motor command to which the respiratory response is compared is automatic or voluntary [119].

4. LIKELY FORMS OF BREATHLESSNESS IN EXERCISING HORSES

In healthy horses, the voluntary and automatic components of central command to breathe would be expected to exhibit the following pattern, as is usual in healthy human beings [120,121,122]. After an initial phase of exercise dominated by voluntary command, the automatic drive to breathe would progressively increase through activation of peripheral chemoreceptors and muscle metaboreceptors [120,122]. Then, as exercise intensity approaches the aerobic threshold, inputs to the automatic drive to breathe would increase markedly, respiratory minute volume would rise steeply with increasing skeletal muscular work and the sense of the effort or work of breathing (non-aversive respiratory effort) would become increasingly intense [121,122]. Provided that a mismatch does not develop between total central command (sum of voluntary and automatic command) and the respiratory responses during exertion, unpleasant experiences of breathlessness would not be expected to occur in healthy mammals [121]. There follows a brief description of key features of *unpleasant* respiratory effort, air hunger and chest tightness and some circumstances when command-response mismatches that give rise to them may occur in exercising horses.

4.1. "Respiratory Effort" and High Airflow Resistance

Respiratory effort is the conscious experience of the respiratory muscle force required to achieve a necessary or desired level of ventilation, and is described in terms of the work, effort or heaviness of breathing [112,116,118]. This is usually experienced when voluntary motor command to the respiratory muscles needs to be increased to achieve the desired level of ventilation. There are two particular situations when this occurs. The first is when an increase in the depth and frequency of breathing is required during normal exercise; as noted above, this is not usually experienced as unpleasant as long as ventilation is matched with command [117]. The second, which is commonly associated with *unpleasant* respiratory effort, occurs in the presence of several pathological states when the motor command needed to elicit a given level of ventilation is greater than normal [119,121]. Thus, sensations of respiratory effort become unpleasant when the response of the ventilatory apparatus, as reported via afferent sensory inputs, especially from the respiratory muscles, is lower than expected at a given level of motor command, i.e., command and response are mismatched [116,121]. This may arise when the respiratory muscles must generate greater inspiratory or expiratory airway pressures as occurs when airflow is impeded, or when the maximum pressure-generating capacity of the respiratory system is diminished [119,121].

It is apparent from details provided above (Section 2.2) that several factors commonly observed in ridden horses would increase the likelihood that they would experience unpleasant respiratory effort of increased intensity, due primarily to increases in airflow resistance in the upper airway. The first factor is rein use that maintains jowl angles of significantly less than the resting values of about 90° (Section 2.2.1; Figure 3). Such low jowl angles very markedly restrict pharyngeal cross-sectional area near the larynx [22,23,55], and, with the extreme neck hyperflexion of the Rollkur position, reduce the cross-sectional area of the larynx as well [59].

The second factor is pathophysiological conditions that decrease the cross-sectional areas of the nasopharynx, larynx and trachea (Section 2.2.2, Section 2.2.3 and Section 2.2.4), as these structures are responsible for about 95% of inspiratory airflow resistance in healthy exercising horses [18,29]. For example, PI, DDSP, pharyngeal collapse, epiglottal entrapment and laryngeal hemiplegia [27,28] all increase airflow resistance above what are taken to be unimpeded values [48,52,91,95], and this leads to reductions in the maximum respiratory minute volume [70]. Breathlessness in the form of unpleasant respiratory effort is therefore likely to occur under these circumstances.

It is obvious that the greater the levels of exercise demanded of horses in which the cross-sectional area of the upper airway is reduced by rein use or pathophysiology, the greater would be the likelihood that they would experience unpleasant sensations of respiratory effort, and, moreover, the greater would be the likely intensity of those experiences. In addition, if both impediments occurred together, airflow resistance would be magnified, as would the anticipated intensity of the unpleasant experience of respiratory effort.

A third factor may contribute to unpleasant respiratory effort elicited by one or both of the above two factors. It relates to pathophysiological sequelae of *partial obstruction of the upper airway*, which likely increase *inspiratory* airflow resistance in the *lower respiratory tract* (Section 2.4.1 and Section 2.4.2). Identified in association with EIPH (Section 2.4.1), and, if the upper airway obstruction is very severe, possibly with NPPO-like effects (Section 2.4.2), these sequelae are: (1) physical impediments to airflow in bronchioles and bronchi due to partial or complete blockage with blood/oedema fluid; and (2) decreases in alveolar elasticity due to interstitial oedema. In addition, in horses with IAD, bronchoconstriction due to direct irritant inflammatory effects in the lower respiratory tract, including oedema, together with bronchial mucus accumulation would increase airflow resistance leading to increased *expiratory effort*, at least partly because the greater airway pressure generated during expiration tends to collapse the intrathoracic airways [31].

4.2. "Air Hunger" and the Chemoreceptor-Induced Drive to Breathe

Air hunger is the sensation experienced at the end of a long breath hold. Often described as "increased urge to breathe", "shortness of breath", "smothering" or "suffocation", it is always reported to be unpleasant, and even moderate air hunger is more unpleasant than maximal respiratory effort [123].

Unlike respiratory effort, where the motor command is voluntary, air hunger arises mainly from a mismatch between automatic motor command and the degree of lung inflation, i.e., tidal volume [121,123,124]. Automatic drive to breathe is increased by any condition that raises $PaCO_2$ and/or which reduces PaO_2 [122]. As noted above (Section 2.3.1, Section 2.3.2 and Section 3), such changes, detected by central and peripheral chemoreceptors, provide afferent stimulation to brainstem respiratory nuclei that then alter their efferent output to respiratory muscles. During exercise, additional afferent input from metaboreceptors, which sense metabolic activity in working skeletal muscles, is transmitted to the medullary respiratory nuclei and thereby also influences ventilation [116,122]. Further details about the mechanisms underlying air hunger are available from Beausoleil and Mellor [5]. It is sufficient to re-emphasise here that air hunger occurs when there is a mismatch between automatic command and the ventilatory response to that command (i.e., limitations on tidal volume), and that the greater the mismatch the more intense will be the unpleasantness of the sensation [112]. The possible cause of tidal volume limitation is intractably high airway resistance [112] but not girth tightness, as thoracic girth circumference does not change significantly in cantering or galloping horses [85].

It is well established that healthy racehorses exhibit various combinations and degrees of arterial hypoxaemia, hypercapnia and acidaemia at different levels of exercise (Section 2.3.2), with all three conditions apparent when exercise and respiratory function are maximal [24,27,28,29,81]. The chemical drive to breathe therefore becomes progressively more intense as sustained exercise and respiratory function approach maximum levels, potentially eliciting sensations of air hunger if a mismatch develops between chemical command and ventilatory response. However, it is not known if such mismatch-induced air hunger does occur in healthy racehorses that perform athletically without apparent impediment during strenuous exercise.

In contrast, as in exercise-intolerant human beings with lung pathology [125,126], intense air hunger may contribute significantly to the impaired athletic performance often reported in horses that have pathophysiological conditions of the respiratory tract (Section 2.2.2, Section 2.2.3, Section 2.2.4, Section 2.3.2 and Section 2.4) (e.g., [3,64,94]). Likely examples include exercise-intolerant horses with partial upper airway obstruction, such as occurs with PI, DDSP, pharyngeal collapse, epiglottal entrapment and laryngeal hemiplegia [27,28,91]. In such cases, exercise induces earlier and more extreme degrees of hypoxaemia, hypercapnia and acidaemia [28,70]. Poorly performing horses having lower airway pathophysiology, for example, those affected by severe EIPH [94] and/or IAD [31] in which partial or complete blockage of bronchi as well as alveolar oedema would significantly impair pulmonary gas exchange [3,35,41,109,110,111] and likely limit tidal volume, would also be included.

4.3. Potential for Unpleasant "Respiratory Effort" and "Air Hunger" to Occur Simultaneously

In healthy horses, reining-in to maintain low jowl angles may lead to the occurrence of both types of breathlessness at relatively low exercise levels. As already noted, unpleasant respiratory effort may occur when the jowl angle is markedly reduced (Section 2.2.1) due to a disproportionate increase in inspiratory airflow resistance related to decreases in upper airway cross-sectional area [22,23,51,55,59]. Moreover, if any associated reduction in respiratory minute volume [70] were sufficient to impede alveolar gas exchange to an extent that caused significant hypoxaemia, hypercapnia and/or acidaemia [27,28], air hunger may be experienced as well. The likelihood of such air hunger occurring, and its intensity, would increase the greater the level of any accompanying exercise. Moreover, pathophysiological conditions, such as PI, DDSP, pharyngeal collapse, epiglottal entrapment and laryngeal hemiplegia in exercise intolerant horses [27,28,91], would also likely be accompanied both by a heightened airflow resistance as well as an increased chemical drive to breathe, giving rise, respectively, to unpleasant respiratory effort and air hunger. Thus, whether primarily due to maintenance of lower jowl angles or the presence of pathophysiological conditions, or both, both forms of breathlessness would likely be mutually reinforcing [5], and this would probably intensify the unpleasantness of the associated negative experiences especially at high levels of exercise.

4.4. "Chest Tightness" and Lower Respriatory Tract Inflammatroy Processes

Tightness relates specifically to the bronchoconstriction that occurs in inflammatory airway diseases such as asthma [31,80,112] or allergic bronchitis [127,128]. Tightness is understood to reflect afferent input from irritant receptors in the airways and lungs ascending to the cerebral cortex. Thus, inflammatory substances activate airway and lung C fibres and this leads to both parasympathetically controlled airway constriction and the sensation of chest tightness [129].

Such tightness is usually the first symptom described during an asthma attack. It often progresses to unpleasant respiratory effort, which is required to overcome airflow resistance, and because lung hyperinflation due to trapped air puts inspiratory muscles at a mechanical disadvantage [119]. In addition, air hunger may occur if this process restricts tidal volume, and/or if gas exchange is impaired [109,110,111]. It is apparent that conditions in the lower respiratory tract of horses with IAD would duplicate those that elicit chest tightness [31], and would also likely lead to simultaneous experiences of unpleasant respiratory effort and air hunger (Section 4.3).

4.5. Assessment of the Potential for Breathlessness to Occur in Freely Running Horses

A problem arises when considering cardiorespiratory responses to exercise and their potential relevance to unpleasant experiences of breathlessness in horses. None of the detailed information available refers to freely galloping horses unencumbered by a bridle, saddle, rider and/or equipment for making physiological measurements. It is not known, for example, whether healthy feral horses galloping voluntarily, and which are not being chased by predators or humans, would exhibit the marked hypoxaemia, hypercapnia and acidaemia observed in animals studied on treadmills during maximal exercise, nor whether they would exhibit such states when chased for extended periods. The questions of whether or not breathlessness occurs in free-running horses at full gallop, and, if so, the extent to which such animals might voluntarily modulate their exercise levels to avoid it, therefore remain unanswered.

5. SOME IMPLICATIONS OF BITTED AND BITLESS BRIDLE USE

5.1. Bits, Control of Behaviour, and Pain

Bits, fitted within the mouth, were introduced in about 2300 BC to help control horses so that they could be ridden or driven safely [1,37]. Various designs of bitted bridles enable pressure to be applied to different points on the horse's head, especially the lips, interdental space (diastema), hard palate, tongue and lower mandible [130,131]. The bit is a potential source of considerable discomfort for the horse, as the named tissues of the mouth are extremely sensitive to mechanical stimulation. This statement is supported by clear behavioural evidence that horses find bits aversive [62,132] (see Section 5.2), as well as by convincing skeletal evidence of long-term bit-inflicted mouth injuries that would induce pain [133,134,135], features which were absent or had a much lower prevalence in feral equidae [133,134].

Experienced equestrians are well aware of mouth sensitivity as evidenced, for example, by recommendations to gently and carefully introduce the bridle whilst training young stock horses in order to avoid inducing bit shyness or aversions [136]. Even today, however, the desire to exert control over horses [132] overcomes concerns about the harm that can be done by using some bits that have been designed to apply sufficient pressure to cause significant pain-inducing soft tissue injury [131,137]. Use of such bits is increasingly regarded as inhumane [138], abusive [139] and, if tested in some jurisdictions, would likely be illegal (e.g., [140]). Of course, even simple bits that may be used gently, for example, the snaffle bit [36,37,131], when used inexpertly [132,135] or with malicious intent may cause significant injury and pain.

5.2. Behavioural Signs of Bit-Induced Pain and Discomfort

Distinguishing behavioural signs of bit-induced pain from other unpleasant sensations related to the irritancy of having a foreign body in contact with the tongue, gums, cheeks, lips and teeth is problematic. Nevertheless, all of these behaviours may be taken to indicate that horses displaying them find the bit aversive. Carefully observed, such behaviours have been reported to occur in various combinations in horses at rest or whilst being ridden or driven at a walk, trot, canter and gallop, and include the following: mouth slightly or widely open, persistent jaw movements, fussing with the bit; chewing, teeth grinding or holding the bit between the teeth; tongue rolling or relocation behind or over the bit; the tongue persistently moving or protruding from the mouth; excessive salivation or drooling [36,37,63,141]. Other indicative behaviours may include head shaking or tossing, and tail swishing [63,141]. Accordingly, characteristic features often seen in vigorously exercising horses ridden or driven wearing bitted bridles are an open mouth, jaw movements, dorsal relocation of the tongue and profuse salivation.

Moreover, high performance horses exhibit low, sometimes very low, jowl angles, partly open or wide open mouths, and/or head shaking or tossing at the times when reins are used to control speed, agility and direction during flat racing, steeplechase, harness racing, endurance, cross-country, show jumping, barrel racing, roping, polo and other such competitive events (Google: Youtube plus the named event, then follow all links for numerous filmed records of these equine competitions).

Some of these behaviours may also be apparent during events that primarily focus on deportment, comportment and demeanour at low speed, in particular dressage and some draft horse competitions. Such behaviours likely indicate pain, especially when rein use is abrupt and/or strong. Note, however, that during races and other such events at the stages when higher speed is required, rein tension is usually released, head-neck position becomes extended (Figure 3A), and mouth opening is less apparent, for example, in horses "given their heads" or "off the bit" during the mid-to-late stages of flat races [54]. Nevertheless, direct observation reveals that some horses still have at least partly open mouths during the later stages of races, before rein tension is increased to slow them down once the race ends.

5.3. Mouth Behaviour of Feral Horses and Horses Wearing Bitless Bridles or Halters

The authors have carefully observed the mouth disposition of at least 150 feral horses filmed during roundups in Australia (Brumbies), France (Camargue wild horses), New Zealand (Kaimanawa wild horses) and the USA (Mustangs) (Google: Youtube, roundup, wild horse name, country, e.g., Youtube roundup Brumby Australia, then follow all links). When standing alert and when walking, trotting, cantering and galloping during roundups the mouth is invariably closed. Exceptions include vocalisation, biting, drinking and eating. Ridden horses wearing loosely-but-snugly fitted bitless bridles designed with side-pull or cross-under reins also keep their mouths closed while standing at rest or engaging in exercise ranging from walking to galloping [36,37,141,142]; so do horses wearing halters while standing in stalls or moving freely in turnout paddocks.

These observations raise two main points. First, the marked difference in mouth behaviour between animals wearing bitless bridles or halters and most of those fitted with bitted bridles reinforces conclusions regarding the potentially aversive impacts of bits [36,37,41,62,132]. Second, the striking mouth-closed (bitless) and mouth-open (bitted) difference points to a functional significance of a closed mouth that merits consideration.

5.4. Respiratory Functionality of a Closed Mouth in Exercising Horses

It is predicted that when healthy bit-free horses begin to walk or move at a faster pace they swallow [141]. When combined with a continuing airtight lip-seal and full engagement of the larynx and ostium intrapharyngium [44], this would contribute to generating and maintaining negative pressure in the oropharynx. After swallowing, this negative pressure would hold the soft palate firmly against the immobile root of the tongue (Figure 1) and prevent it from being sucked dorsally into the nasopharyngeal airway during inspiration [41,44], thereby maintaining a larger cross-sectional area and minimising airflow resistance as well as negative inspiratory pressure (see Section 2.2.2). In addition, it would prevent soft palatal instability and an associated increase in airflow turbulence that would otherwise occur during forceful expiratory excursions at high respiratory minute volumes [41,44]. The respiratory benefits of these effects in the exercising horse are obvious.

On the other hand, the presence of the bit itself and the mouth-opening behaviours it often elicits break the airtight lip-seal and could dissipate the negative pressure in the oral compartment [74,143]. This would destabilise the soft palate leading to its dorsal displacement into the nasopharynx during inspiration, especially during exercise [44], decreasing the cross-sectional area and thereby increasing airflow resistance as well as negative inspiratory pressure (Section 2.2.2). Palatal vibration upon expiration would also occur [41]. As noted above (Section 2.2.2), PI and DDSP are well-recognised problems in racehorses [39,41,67,68,69,70]. Interestingly, use of a crossed noseband to prevent visible mouth opening and minimise palatal dysfunction [54,144] apparently has some benefits, but the precise mechanism has not yet been clarified [54]. In addition, bit-induced tongue rolling or relocation behind or above the bit [22,67,141] has occasioned the use of tongue-ties, but with limited success [54,145] possibly because, in many cases, their application would itself precipitate loss of an airtight lip-seal.

Additionally, rein use that markedly reduces the jowl angle exacerbates these effects by decreasing the cross-sectional area of the nasopharynx near the larynx with detrimental impacts on airflow resistance and negative inspiratory pressure (Section 2.2.1) [22,23,59]. Finally, swallowing or attempting to swallow excess saliva during exercise would necessarily disengage the larynx and ostium intrapharyngium, interrupting breathing rhythm and potentially eliciting a gag or cough reflex if saliva contacts the larynx [22]. Note that repeated swallowing during high-speed treadmill studies has been observed in association with PI and/or DDSP [44,67], and that this might represent attempts to reinstate negative pressure in the oropharynx after palatal displacement has occurred [44].

5.5. Potential Impacts of Bitless Bridles on Breathlessness

5.5.1. Bridle Type in Treadmill and Other Studies

It is important to acknowledge that cardiorespiratory responses of horses to all levels of exercise from walking to a full gallop have been extensively and rigorously investigated on treadmills and outdoor training tracks (Section 2). However, reports of treadmill studies do not routinely stipulate whether a bitted bridle or halter was used [146]. Moreover, cardiorespiratory function has not yet been compared in horses when wearing bitted bridles and halters, or bitted and bitless bridles, during the same treadmill or outdoor training track study. This hinders interpretation of results because, as indicated above, a bit in a horse's mouth likely has negative impacts on the dynamics of airflow in the upper respiratory tract as well as functionally significant pathophysiological consequences in the lower respiratory tract. Accordingly, it is not known if such bit-induced effects have confounded some results and led to wider between-study ranges in some measured parameters.

5.5.2. Bridle Type and Potential Impacts on Breathlessness

Bitless bridle use would facilitate lip-seal suction within the oral compartment that can help to maintain contiguity of the tongue and soft palate (Section 5.4), thereby minimising or avoiding palatal obstruction of the nasopharynx, especially at jowl angles of between 90° and 120–130°. This might be expected to reduce the likelihood that unpleasant experiences of respiratory effort due to increased airflow resistance within the upper respiratory tract would occur at all. Alternatively, the intensities of such experiences might be proportionately lower at all exercise levels and/or the threshold for occurrence of such experiences might be located at higher levels of exercise. Although bitless bridle use would not rule out the effects of reining-in on upper airway resistance (Section 2.2.1), users of such bridles normally aim to avoid this (e.g., [142,147]).

Impacts of bitless bridle use on the occurrence and intensity of conditions likely to induce air hunger have not been established definitively. Three possibilities present themselves. First, avoidance of bit-induced increases in nasopharyngeal airflow resistance may improve respiratory minute volume and thereby reduce the degrees of any arterial hypoxaemia, hypercapnia and acidaemia that could lead to air hunger. Such a possibility is supported by the observation that pathophysiological conditions that increase airflow resistance (e.g., see Section 2.2.2, Section 2.2.3 and Section 2.2.4) do exacerbate these three states. Second, an explicit comparison of bit and bitless bridle use may reveal a reduction in conditions likely to induce air hunger with bitless bridles, but only at lower exercise levels. This might occur if wearing a bitless bridle improved the total alveolar air-blood O_2 and CO_2 exchange capacities at low-levels but not at high-levels of exercise. During strenuous exercise air hunger might be more likely to occur because gas exchange capacity between alveolar air and blood is considered to be limited by the following interacting factors: (1) increased cardiac output reducing alveolar capillary transit time and thus impairing gas diffusion; (2) a very low PO_2 in mixed venous blood entering the lungs, representing marked degrees of O_2 desaturation that cannot be corrected in a single transit; and (3) the normal maximal ventilation capacity (i.e., respiratory minute volume) being inadequate due to limitations on peak tidal volume and because breathing rate is locked 1:1 with stride frequency in galloping horses [29,35,85]. The third and final possibility is that bit-bitless differences in circulating PO_2, PCO_2 and H^+ ion concentrations may not be detectable at any exercise level.

Cook [38,41] proposed that minimising airflow resistance in the upper respiratory tract by using bitless bridles may also reduce the risk of developing pathophysiological states (e.g., EIPH) in the lower respiratory tract which are understood to impede gas exchange and reduces athletic performance (Section 2.4.1). As already noted, both unpleasant respiratory effort and air hunger are likely to be associated with the onset of severe EIPH in racehorses engaged in strenuous exercise (Section 4.1 and Section 4.2). Moreover, the combined negative affective consequences of EIPH of sufficient severity to impair athletic performance are likely to be significant. This highlights the value of exploring the extent to which bitless bridle use might beneficially reduce, or even eliminate, these untoward effects that would give rise to potentially very intense breathlessness.

6. SOME ANIMAL WELFARE IMPLICATIONS

Three further questions are considered here. First, might urging racehorses to sustain maximal exercise towards the end of races increase the intensity of breathlessness they may experience? Second, what contributions might breathlessness make to fatigue in vigorously exercising horses that are healthy and fully functional or that are impaired by pathophysiological conditions? Third, how compelling is the behavioural evidence that horses find a bit or bits in their mouths aversive?

6.1. Breathlessness and Urging Racehorses to Sustain Maximal Exercise

Significant breathlessness of all three types can be intensely unpleasant experiences [5]. In healthy ridden or driven horses, sustained strenuous exercise generates levels of hypoxaemia, hypercapnia and/or acidaemia, which would prime the animals to experience air hunger if a mismatch were to occur between total command and the required respiratory response (Section 4.2). In addition, rein-imposed low jowl angles increase respiratory effort (Section 2.2.1) and this would also be experienced as unpleasant if a command-response mismatch were to develop during strenuous exercise (Section 4.1, Section 4.2 and Section 4.3). It is not known whether such mismatches occur in healthy fully functional racehorses that sustain apparently unimpeded high athletic performances throughout races. However, such mismatches and breathlessness seem much more likely to occur in those horses whose performance declines before races end, especially if associated with pathophysiological impediments to respiratory tract airflow and/or to alveolar gas exchange. This raises the question of whether or not whip-mediated urging by their jockeys or drivers [148,149] might lead horses to sustain high exercise levels beyond the point at which they would otherwise voluntarily slow down and thereby exacerbate the intensity of any breathlessness they may experience. The answer to this question is not clear, despite the fact that one obvious intention of whip use is to elicit and sustain maximum athletic performance [150,151].

In the absence of persistent significant pathophysiology, a rapid post-race decline in any air hunger experienced is likely to occur due to the speedy reversal of hypoxaemia and hypercapnia (but not acidaemia) during the first 2–3 min after exercise has ceased [24,25,27,29,81]. Likewise, a decline in any unpleasant respiratory effort would be expected to accompany the somewhat less rapid post-exercise return of respiratory muscle work and minute volume to near resting values [86,87].

The potential for bitless bridles to enhance the ease of breathing and minimise pathophysiological effects during strenuous exercise, as outlined above (Section 5), merits detailed investigation. As a first step, it is recommended that the biophysical dynamics, physiology and pathophysiology of salient features of cardiorespiratory function be rigorously compared in horses wearing bitless and bitted bridles as they exercise at different levels. Of course, a matter of direct practical interest to the horseracing industry is whether bitless bridle use can be convincingly demonstrated to achieve a better athletic performance than occurs with bitted bridle use.

6.2. Fatigue, Exercise Intolerance and Breathlessness

Healthy fully functional horses are said to experience *fatigue* when they can no longer sustain particular intensities of exercise and slow down [81,86], and the higher the intensity of the exercise the shorter is its duration before fatigue supervenes [152]. In the past, explanations of fatigue primarily focused on peripheral factors such as impaired energy transactions in skeletal and diaphragmatic muscles decreasing their contractile performance [66,153,154], detrimental impacts of exercise-induced hyperthermia [89], and/or other factors, all regarded as representing different forms of *homeostatic failure* [155]. In contrast, fatigue has recently been proposed to be a *protective affective experience*, generated by brain processing of exercise-related sensory inputs from the periphery, where the experience of fatigue is directed towards avoiding catastrophic failure of homeostatic mechanisms [155]. Originally referring to human athletes, this proposition may also be applicable to other mammals. If so, it would suggest that the affective experience of fatigue might have evolved to demotivate mammals from exercising at hazardously high levels, the purpose being to *safeguard their homeostatic capacity to recover* before key peripheral mechanisms that underlie strenuous locomotion fail catastrophically. It is interesting to question whether or not several centuries of breeding horses for high athletic performance might, in some respects, have exceeded such a protective role of fatigue, for example with EIPH. This question and that related to whether or not breathlessness in its various forms might reinforce or otherwise contribute to the experience of fatigue in healthy fully functioning horses during strenuous exercise, remain unresolved.

The term *exercise intolerance* is usually applied to horses that exhibit lower than expected athletic performances which are credibly linked to pathophysiological conditions of the upper and/or lower respiratory tract (e.g., [24,27,107,110,111,156]). In light of the increases in upper and/or lower airway resistance, the greater degrees of hypoxaemia, hypercapnia and acidaemia, and/or the evidence of irritant hypersensitivity causing bronchoconstriction in some cases, it seems likely that one or more of the three types of breathlessness may contribute to exercise intolerance in affected animals. However, parallel experiences of fatigue might also be generated by sensory inputs related to homeostatic support for strenuous locomotion before it reaches its limit. If so, the onset of fatigue as an affective experience may be expected to occur earlier, and/or at lower exercise intensities in such horses than in those that are healthy and fully functional.

6.3. Behaviours that Validly Indicate Horses' Aversion to Bits

Observing animals living freely in their natural environments can provide valuable insights into the negative impact that human impositions have had on them (e.g., [10,157,158,159,160,161,162]). Particularly informative here are the clearly identifiable mouth behaviours indicating at least discomfort and at worst pain in most healthy horses wearing bitted bridles (Section 5.2) and their absence in feral horses and horses wearing loosely-but-snugly fitted bitless bridles or halters (Section 5.3). However, it is apparent that these behaviours are not well recognised by the majority of equestrians as indicating that horses find bits aversive. It is well established that persons accustomed to a particular spectrum of behaviours, which are almost always present, often do not recognise some of them as indicating a welfare problem. Two non-equine examples illustrate this. The first relates to dairy farmers who, prior to training, underestimated the prevalence of lameness in their herds because they considered all but the most extreme lameness gaits to be normal [163]. The second example refers to 58% of dog owners whose animals had clear clinical signs of brachycephalic obstructive airway syndrome and who reported that the dogs did not have "a breathing problem" [164]. It is suggested here, as briefly noted elsewhere [39,62], that a similar form of non-recognition of clear behavioural evidence of horses' aversion to bits in their mouths (Section 5.2) arises because the indicative behaviours have been and are observed so commonly that, except in more extreme cases, they are considered to be normal. Once recognised, however, these behaviours cannot go "unseen", nor should their welfare implications be ignored.

7. CONCLUSIONS

It is apparent that horses engaged in strenuous exercise display physiological responses that approach the upper functional limits of key organ systems, in particular their cardiorespiratory systems. Maximum athletic performance is therefore vulnerable to factors that diminish these functional capacities. In ridden horses with bits in their mouths, rein use that reduces the jowl angle, sometimes markedly, and conditions that partially obstruct the nasopharynx and larynx, have been highlighted as factors that impair airflow in the upper respiratory tract and lead to increased flow resistance. The associated upper airway pressure changes, transmitted to the lower airways, may elicit pathophysiological sequelae in the alveolae, which, in their turn, may increase airflow resistance and impede respiratory gas exchange. Additional effects include decreases in respiratory minute volume and worsening of the hypoxaemia, hypercapnia and acidaemia that are usually observed in healthy horses during strenuous exercise.

These experimental observations, made in circumstances that approximate to those of horses ridden outdoors, draw attention to factors that may elicit the three different forms of breathlessness. However, breathlessness of any type would only occur if there were a mismatch between the total command to breathe (voluntary plus automatic) and the required respiratory response. It is not known to what extent, if at all, such mismatches would occur in strenuously exercising horses unhampered by low jowl angles, by pathophysiology at any level of the respiratory tract, or the presence of a bit (see below). However, different combinations of the three types of breathlessness seem much more likely to occur when pathophysiological conditions significantly reduce maximal athletic performance, registered as exercise intolerance. Of course, not all horses will be affected to the same extend and in precisely the same way.

Most horses exhibit clear behavioural evidence of aversion to a bit in their mouths, varying from it being a mild irritant to very painful. This in itself is a significant animal welfare issue that should be addressed. A further major point is the potential detrimental effects of bitted bridle use on the maintenance of negative pressure in the oropharynx due to loss of the airtight lip seal and/or the laryngeal-pharyngeal seal. This can apparently contribute to the development of PI and DDSP, which, when severe, can markedly increase airflow resistance in both the upper and lower respiratory tract. These changes can have wider untoward respiratory outcomes that may lead to poor athletic performance and suggest the potential for affected animals to experience significant intensities of breathlessness. Bitless bridle use may minimise or eliminate such effects. However, direct comparisons of the cardiorespiratory dynamics and extent of any respiratory pathophysiology in horses wearing bitted and bitless bridles have not been conducted. Such studies would be helpful in confirming, or otherwise, the anticipated benefits of bitless bridle use.

ACKNOWLEDGMENTS

The authors wish to thank the numerous colleagues and friends nationally and internationally with whom discussion has helped to hone these ideas, but, of course, the authors accept entire responsibility for the content of this paper. We are also most grateful to Michal Dunn and Briar

Graham for practical library support, and Quentin Roper for preparing the figures.

AUTHOR CONTRIBUTIONS

David J. Mellor instigated the writing of this review and wrote the first draft. He integrated widely distributed information on cardiorespiratory function in exercising horses, the information on bitless bridles and other relevant matters, and, in discussion with Ngaio J. Beausoleil, sought to align this material with the substance of a previously published generic review that introduced breathlessness as a significant animal welfare issue. Ngaio J. Beausoleil led the writing and developed the content of that earlier review with David J. Mellor having secondary roles. Led and largely written by David J. Mellor, both he and Ngaio J. Beausoleil worked together to finalise the present review, including responding to referees' comments, carrying the content and structure through several iterations to reach the final form presented here.

REFERENCES

1. Ramey, D.W. A historical survey of human-equine interactions. In *Equine Welfare*; McIlwraith, C.W., Rollin, B.E., Eds.; Wiley-Blackwell: Chichester, UK, 2011; pp. 22–58.
2. Barrey, E. Genetic basis of equine performance. In *Equine Sports Medicine and Surgery: Basic and Clinical Sciences of the Equine Athlete*, 2nd ed.; Hinchcliff, K.W., Kaneps, A.J., Geor, R.J., Eds.; Elsevier: New York, NY, USA, 2014; pp. 43–58.
3. Hinchcliff, K.W. Exercise-induced pulmonary haemorrhage (EIPH). In *Equine Sports Medicine and Surgery: Basic and Clinical Sciences of the Equine Athlete*, 2nd ed.; Hinchcliff, K.W., Kaneps, A.J., Geor, R.J., Eds.; Elsevier: New York, NY, USA, 2014; pp. 633–647.
4. Hinchcliff, K.W.; Kaneps, A.J.; Geor, R.J. (Eds.) *Equine Sports Medicine and Surgery: Basic and Clinical Sciences of the Equine Athlete*, 2nd ed.; Elsevier: New York, NY, USA, 2014.
5. Beausoleil, N.J.; Mellor, D.J. Introducing breathlessness as a significant animal welfare issue. *N. Z. Vet. J.* **2015**, *63*, 44–51.
6. Fraser, D.; Duncan, I.J.H. "Pleasures", "pains" and animal welfare: Towards a natural history of affect. *Anim. Welfare* **1998**, *7*, 383–396.
7. Boissy, A.; Manteuffel, G.; Jensen, M.B.; Moe, R.O.; Spruijt, B.; Keeling, L.J.; Winckler, C.; Forkman, B.; Dimitrov, I.; Langbein, J.; et al. Assessment of positive emotions in animals to improve their welfare. *Physiol. Behav.* **2007**, *92*, 375–397.
8. Fraser, D. *Understanding Animal Welfare: The Science in Its Cultural Context*; Wiley-Blackwell: Oxford, UK, 2008.
9. Yeates, J.W.; Main, D.C.J. Assessment of positive welfare: A review. *Vet. J.* **2008**, *175*, 293–300.
10. Kiley-Worthington, M. Equine psychological needs and quality of life. In *Equine Welfare*; McIlwraith, C.W., Rollin, B.E., Eds.; Wiley-Blackwell: Chichester, UK, 2011; pp. 94–112.
11. Green, T.C.; Mellor, D.J. Extending ideas about animal welfare assessment to include "quality of life" and related concepts. *N. Z. Vet. J.* **2011**, *59*, 316–324.

12. Boissy, A.; Lee, C. How assessing relationships between emotions and cognition can improve farm animal welfare. *Rev. Sci. Tech. Off. Int. Epizoot.* **2014**, *33*, 103–110.
13. Broom, D.M. *Animal Welfare in the European Union*; Policy Department C Citizens: Rights and Constitutional Affairs, Directorate General for Internal Policies, European Parliament: Brussels, Belgium, 2017; Available online: http://www.europarl.europa.eu/ RegData/etudes/workshop/join/2013/474410/IPOL-FEMM_AT(2013)474410_EN.pdf (accessed on 14 April 2017).
14. Mellor, D.J. Updating animal welfare thinking: Moving beyond the "Five Freedoms" towards "A Life worth Living. *Animals* **2016**, *6*, 21.
15. Gregory, N.G. *Physiology and Behaviour of Animal Suffering*; Blackwell Science: Oxford, UK, 2004.
16. Denton, D.A.; McKinley, M.J.; Farrell, M.; Egan, G.F. The role of primordial emotions in the evolutionary origin of consciousness. *Conscious. Cognit.* **2009**, *18*, 500–514. Available online: http://dx.doi.org/10.1016/j.concog.2008.06.009 (accessed on 22 May 2017).
17. Vinuela-Fernandz, I.; Weary, D.M.; Flecknell, P. Pain. In *Animal Welfare*, 2nd ed.; Appleby, M.C., Mench, J., Olsson, I.A., Hughes, B.O., Eds.; CABI: Oxford, UK, 2011; pp. 64–77.
18. Art, T.; Scrteyn, D.; Lekeux, P. Effect of exercise on the partitioning of equine respiratory resistance. *Equine Vet. J.***1988**, *20*, 268–273.
19. Art, T.; Anderson, I.; Woakes, A.I.; Roberts, C.; West, P.J.; Snow, D.H.; Lekeux, P. Mechanics of breathing during strenuous exercise in thoroughbred horses. *Respir. Physiol.* **1990**, *82*, 279–294.
20. Ducharme, N.G.; Hackett, R.P.; Ainsworth, D.A.; Erb, H.N.; Shannon, K.J. Repeatability and normal values for measurement of pharyngeal and tracheal pressures in exercising horses. *Am. J. Vet. Res.* **1994**, *55*, 369–374.
21. Odeh, M.; Schnall, R.; Gavriely, N.; Oliven, A. Dependency of upper airway patency on head position: The effect of muscle contraction. *Respir. Physiol.* **1995**, *100*, 239–244.
22. Cook, W.R. Pathophysiology of bit control in the horse. *J. Equine Vet. Sci.* **1999**, *19*, 196–204.
23. Cehak, A.; Rohn, K.; Barton, A.; Stadler, P.; Ohnesorge, B. Effect of head and neck position on pharyngeal diameter in horses. *Vet. Radiol. Ultrasound* **2010**, *51*, 491–497.
24. Bayly, W.M.; Grant, B.D.; Breeze, R.G. Arterial blood gas tension and acid base balance during exercise in horses with pharyngeal lymphoid hyperplasia. *Equine Vet. J.* **1984**, *16*, 435–438.
25. Bayly, W.M.; Hodgson, D.R.; Schulz, D.A.; Dempseu, J.A.; Gollnick, P.D. Exercise-induced hypercapnia in the horse. *J. Appl. Physiol.* **1989**, *67*, 1958–1966.
26. Butler, P.J.; Woakes, A.J.; Smale, K.; Roberts, C.A.; Hillidge, C.J.; Snow, D.H.; Marlin, D.J. Respiratory and cardiovascular adjustments during exercise of increasing intensity and during recovery in thoroughbred racehorses. *J. Exp. Biol.* **1993**, *179*, 159–180.
27. Tate, L.P.; Corbett, W.T.; Bishop, B.J.; Foreman, J.H. Blood gas tensions, acid-base status, heart rates and venous profiles in exercising horses with laryngeal hemiplegia before and after corrective surgery. *Vet. Surg.* **1993**, *22*, 177–183.
28. Durando, M.M.; Martin, B.B.; Hammer, E.J.; Langsam, P.P.; Birks, E.K. Dynamic upper airway changes and arterial blood gas parameters during treadmill exercise. Equine Exercise Physiology 6. *Equine Vet. J.* **2002**, *34*, 408–412.
29. Art, T.; Bayly, W. Lower airway function: Responses to exercise and training. In *Equine Sports Medicine and Surgery: Basic and Clinical Sciences of the Equine Athlete*, 2nd ed.; Hinchcliff, K.W., Kaneps, A.J., Geor, R.J., Eds.; Elsevier: New York, NY, USA, 2014; pp. 587–603.
30. Widdicombe, J. Lung afferent activity: Implications for respiratory sensation. *Respir. Physiol. Neurobiol.* **2009**, *167*, 2–8.
31. Couetil, L.L.; Cardwell, J.M.; Gerber, V.; Lavoie, J.-P.; Leguillette, R.; Richard, E.A. Inflammatory airway disease of horses—Revised consensus statement. *J. Vet. Int. Med.* **2016**, *30*, 503–515.
32. Whitwell, K.E.; Greet, T.R.C. Collection and evaluation of tracheobronchial washes in the horse. *Equine Vet. J.* **1984**, *16*, 499–508.

33. Sweeney, C.R.; Soma, L.R.; Maxson, A.D.; Thompson, J.E.; Holcombe, S.J.; Spencer, P.A. Effects of furosemide on the racing times of Thoroughbreds. *Am. J. Vet. Res.* **1990**, *51*, 772–778.
34. Lapointe, J.M.; Vrins, A.; McCarvill, E. A survey of exercise-induced pulmonary haemorrhage in Quebec standardbred racehorses. *Equine Vet. J.* **1994**, *26*, 482–485.
35. Poole, D.C.; Erickson, H.H. Exercise-induced pulmonary hemorrhage: Where are we now? *VMRR* **2016**, *7*, 133–148.
36. Cook, W.R.; Mills, D.S. Preliminary study of jointed snaffle vs. crossunder bitless bridles: Quantified comparison of behaviour in four horses. *Equine Vet. J.* **2009**, *41*, 827–830.
37. Quick, J.S.; Warren-Smith, A.K. Preliminary investigations of horses' (*Equus caballus*) responses to different bridles during foundation training. *J. Vet. Behav.* **2009**, *4*, 169–176.
38. Cook, W.R. An endoscopic test for bit-induced nasopharyngeal asphyxia as a cause of exercise-induced pulmonary haemorrhage in the horse. *Equine Vet. J.* **2014**, *46*, 256–257.
39. Cook, W.R. Hypothesis article: Bit-induced asphyxia in the racehorse as a cause of sudden death. *Equine Vet. J.* **2016**, *28*, 405–409.
40. Lyle, C.H.; Uzal, F.A.; McGorum, B.C.; Aida, H.; Blissitt, K.J.; Case, J.T.; Charles, J.T.; Gardner, I.; Horadagoda, N.; Kusano, K.; et al. Sudden death in racing Thoroughbred horses: An international multicentre study of post mortem findings. *Equine Vet. J.* **2011**, *43*, 324–331.
41. Cook, W.R. A hypothetical, aetiological relationship between the horse's bit, nasopharyngeal oedema and negative pressure pulmonary oedema. *Equine Vet. Educ.* **2014**, *26*, 381–389.
42. Cheetham, J.; Holcombe, S.J.; Ducharme, N.G. Upper airway function of normal horses during exercise. In *Equine Sports Medicine and Surgery: Basic and Clinical Sciences of the Equine Athlete*, 2nd ed.; Hinchcliff, K.W., Kaneps, A.J., Geor, R.J., Eds.; Elsevier: New York, NY, USA, 2014; pp. 529–547.
43. Cook, W.R. Some observations on form and function of the equine upper airway in health and disease, Part I: The pharynx. *Proc. AAEP* **1981**, *27*, 355–392.
44. Lane, J.G.; Blandon, B.; Little, D.R.M.; Naylor, J.R.J.; Franklin, S.H. Dynamic obstruction of the equine upper respiratory tract. Part 1. Observations during high-speed treadmill endoscopy in 600 Thoroughbred racehorses. *Equine Vet. J.* **2006**, *38*, 393–399.
45. Taylor, C.R.; Malony, G.M.; Weibel, E.R.; Langman, V.A.; Kamau, J.M.; Seeherman, H.J.; Heglund, N.C. Design of the mammalian respiratory system. *Respir. Physiol.* **1981**, *44*, 25–37.
46. Weber, J.M.; Dobson, G.P.; Parkhouse, W.S.; Wheeldon, D.; Harman, J.C.; Snow, D.H.; Hochachka, P.W. Cardiac output and oxygen consumption in exercising Thoroughbred horses. *Am. J. Appl. Physiol.* **1987**, *253*, R890–R895.
47. Art, T.; Lekeux, P. Training-induced modification of cardiorespiratory and ventilatory measurements in thoroughbred horses. *Equine Vet. J.* **1993**, *25*, 532–536.
48. Ducharme, N.G.; Hackett, R.P.; Gleed, R.D.; Ainsworth, D.M.; Erickson, B.K.; Erb, H.N.; Soderholm, L.V.; Mitchell, L.M. Pulmonary capillary pressure in horses undergoing alternation to pleural pressure by imposition of upper airway resistive loads. *Equine Vet. J.* **1999**, *30*, 27–33.
49. Wilson, W.D.; Lofstedt, J.; Lakritz, J. Abnormal respiratory noise (stridor). In *Large Animal Internal Medicine*, 5th ed.; Smith, B.P., Ed.; Elsevier: St Louis, MI, USA, 2009; pp. 66–70.
50. Gerring, E.L. Differential diagnosis of equine respiratory noises. *Practce* **1985**, *7*, 109–117.
51. Petsche, V.M.; Derksen, F.J.; Berney, C.E.; Robinson, N.E. Effect of head position on upper airway function in exercising horses. *Equine Vet. J.* **1995**, *18*, 18–22.
52. Strand, E.; Fjordbakk, C.T.; Holcombe, S.J.; Risberg, A.; Chalmers, H.J. Effect of poll flexion and dynamic laryngeal collapse on tracheal pressure in Norwegian Coldblooded Trotter racehorses. *Equine Vet. J.* **2009**, *41*, 59–64.
53. Linford, R.L.; O'Brien, T.R.; Wheat, J.D.; Meagher, D. Radiographic assessment of epiglotic length and pharyngeal and laryngeal diameters in the thoroughbred. *Am. J. Vet. Res.* **1983**, *44*, 1660–1666.

54. Allen, K.J.; Terron-Canedo, N.; Hillyer, M.H.; Franklin, S.H. Equitation and exercise factors affecting dynamic upper respiratory tract function: A review illustrated by case reports. *Equine Vet. Educ.* **2011**, *23*, 361–368.
55. Go, L.-M.; Barton, A.K.; Ohnesorge, B. Objective classification of different head and neck positions and their influence on the radiographic pharyngeal diameter in sport horses. *BMC Vet. Res.* **2014**, *10*, 118.
56. Houghton Mifflin Company. Poiseuille's Law: Dictionary.com "poiseuille's law". In *The American Heritage® Science Dictionary*; Houghton Mifflin Company: Boston, MA, USA, 2017; Available online: http://www.dictionary.com/browse/poiseuille-s-law (accessed on 10 March 2017).
57. Poiseuille's Law: IV Fluids, Open Anaesthesia. 2017. Available online: https://www.openanesthesia.org/poiseuilles_law_iv_fluids/ (accessed on 10 March 2017).
58. Meyer, H. "Rollkur", "Hyperflexion" and "LDR"—The natural position of the head and neck of the horse and the modification by the rider. *Pferdeheilkunde* **2010**, *26*, 388–413.
59. Zebisch, A.; May, A.; Reese, S.; Gehlen, H. Effects of different head-neck positions on the larynges of ridden horses. *J. Anim. Physiol. Anim. Nutr.* **2014**, *98*, 894–900.
60. Zebisch, A.; May, A.; Reese, S.; Gehlen, H. Effect of different head-neck positions on physical and psychological stress parameters in the ridden horse. *J. Anim. Physiol. Anim. Nutr.* **2014**, *98*, 901–907.
61. Sleutjens, J.; Smiet, E.; Van Weeren, R.; Van der Kolk, J.; Back, W.; Wijnberg, I. Effect of head and neck position on intrathoracic pressure and arterial blood gas values in Dutch Warmblood riding horses during moderate exercise. *Am. J. Vet. Res.* **2012**, *73*, 522–528.
62. Cook, W.R. A solution to respiratory and other problems caused by the bit. *Pferdeheilkunde* **2003**, *16*, 333–351.
63. Carey, C.; Moriarty, S.H.; Brennan, R. The impact of bitted and bitless bridles on the Therapeutic Riding Horse. In Proceedings of the 12th International Equitation Science Conference on Understanding Horses to Improve Training and Performance, Wagga Wagga, Australia, 22–25 November 2016; p. 99. Available online: http://www.equitationscience.com/documents/Conferences/2016/Proceedings%20ISES%202016.pdf (accessed on 9 October 2016).
64. Franklin, S.H.; Naylor, J.R.; Lane, J.G. Videoendoscopic evaluation of the upper respiratory tract in 93 sport horses during exercise testing on a high-speed treadmill. *Equine Exerc. Physiol. 7 Equine Vet. J.* **2006**, *36*, 540–545.
65. Fjordbakk, C.T.; Strand, E.; Hanche-Olsen, S. Surgical and conservative management of bilateral dynamic laryngeal collapse associated with poll flexion in harness race horses. *Vet. Surg.* **2008**, *37*, 501–507.
66. Franklin, S.H.; Allen, K.J. Assessment of dynamic upper respiratory tract function in the equine athlete. *Equine Vet. Educ.* **2017**, *29*, 92–103.
67. Parente, E.J.; Martin, B.B.; Tulleners, E.P.; Ros, M.W. Dorsal displacement of the soft palate in 92 horses during high-speed treadmill examination (1993–1998). *Vet. Surg.* **2002**, *31*, 507–512.
68. Allen, K.J.; Christley, R.M.; Birchall, M.A.; Franklin, S.H. A systematic review of the efficacy of interventions for dynamic intermittent dorsal displacement of the soft palate. *Equine Vet. J.* **2012**, *44*, 259–266.
69. Allen, K.J.; Franklin, S. Characteristics of palatal instability in Thoroughbred racehorses and their association with the development of dorsal displacement of the soft palate. *Equine Vet. J.* **2013**, *45*, 454–459.
70. Ducharme, N.G.; Cheetham, J. Abnormalities of the upper airway. In *Equine Sports Medicine and Surgery: Basic and Clinical Sciences of the Equine Athlete*, 2nd ed.; Hinchcliff, K.W., Kaneps, A.J., Geor, R.J., Eds.; Elsevier: New York, NY, USA, 2014; pp. 549–586.
71. Kannegieter, N.J.; Dore, M.L. Endoscopy of the upper respiratory tract during treadmill exercise: A clinical study of 100 horses. *Aust. Vet. J.* **1995**, *72*, 101–107.
72. Franklin, S.H.; Burn, J.F.; Allen, K.J. Clinical trials using a telemetric endoscope for use during over-ground exercise: A preliminary study. *Equine Vet. J.* **2008**, *40*, 712–715.

73. Strand, E.; Fjordbakk, C.T.; Sundberg, L.; Spangen, H.; Lunde, H.; Hanche-Olsen, S. Relative prevalence of upper respiratory tract obstructive disorders in two breeds of harness racehorses (185 cases 1998–2006). *Equine Vet. J.* **2012**, *44*, 518–523.
74. Ahern, T.J. Pharyngeal dysfunction during exercise. *J. Equine Vet. Sci.* **1999**, *19*, 226–231.
75. Boyle, A.G.; Martin, B.B., Jr.; Davidson, E.J.; Durando, M.M.; Birks, E.K. Dynamic pharyngeal collapse in racehorses. *Equine Exerc. Physiol.* **2006**, *36*, 546–550.
76. Smith, C.M.; Taylor, R.J.; Dixon, P.M. Unilateral displacement of the roof of the nasopharynx as a cause of stridor in a poney. *Vet. Rec.* **1994**, *134*, 140–141.
77. Strand, E.; Skjerve, E. Complex dynamic upper airway collapse: Association between abnormalities in 99 harness racehorses with one or more dynamic disorders. *Equine Vet. J.* **2012**, *44*, 524–528.
78. Strand, E.; Hanche-Olsen, S.; Grøvold, A.M.R.; Mellum, C.N. Dynamic bilateral arytenoid and vocal fold collapse associated with head flexion in 5 Norwegian Coldblooded Trotter racehorses. *Equine Vet. Educ.* **2004**, *16*, 242–250.
79. Votion, D. Metabolic responses to exercise and training. In *Equine Sports Medicine and Surgery: Basic and Clinical Sciences of the Equine Athlete*, 2nd ed.; Hinchcliff, K.W., Kaneps, A.J., Geor, R.J., Eds.; Elseier: New York, NY, USA, 2014; pp. 747–767.
80. Duffin, J. The chemoreflex control of breathing. *Can. J. Anaesth.* **1990**, *37*, 933–942.
81. Wagner, P.D.; Gillespie, J.R.; Landgren, G.L.; Fedde, M.R.; Jones, B.W.; DeBowes, R.M.; Pieschl, R.L.; Erickson H.H. Mechanisms of exercise-induced hypoxemia in horses. *J. Appl. Physiol.* **1989**, *66*, 1227–1233.
82. Parks, C.M.; Manohar, M. Blood-gas tensions and acid-base status in ponies during treadmill exercise. *Am. J. Vet. Res.* **1984**, *45*, 15–19.
83. Katz, L.M.; Bayly, W.M.; Roeder, M.J.; Kingston, J.K.; Hines, M.T. Effects of training on maximum oxygen consumption of ponies. *Am. J. Vet. Res.* **2000**, *61*, 986–991.
84. Katz, L.M.; Bayly, W.M.; Hines, M.T.; Sides, R.H. Ventilatory responses of ponies and horses to exercise. *Equine Comp. Exerc. Physiol.* **2005**, *2*, 229–240.
85. Franklin, S.H.; Van Erck-Westergren, E.; Bayly, W.M. Respiratory responses to exercise in the horse. *Equine Vet. J.* **2012**, *44* 726–732.
86. Curtis, R.A.; Kusano, K.; Evans, D.L. Observations on respiratory flow strategies during and after intense treadmill exercise to fatigue in thoroughbred racehorses. *Equine Vet. J.* **2006**, *36*, 567–572.
87. Padilla, D.J.; McDonough, P.; Kindig, C.A.; Erickson, H.H.; Poole, D.C. Ventilatory dynamics and control of blood gases after maximal exercise in the Thoroughbred horse. *J. Appl. Physiol.* **2004**, *96*, 2187–2193.
88. Thiel, M.; Tolkmitt, G.; Hoernicke, H. Body temperature changes in horses during riding: Time course and effects on heart rate and respiratory frequency. In *Equine Exercise Physiology*; 2. ICEEP Publications: Davis, CA, USA, 1987; pp. 183–193.
89. Hodgson, D.R.; Davis, R.E.; McConaghy, F.F. Thermoregulation in the horse in response to exercise. *Br. Vet. J.* **1994**, *150*, 219–235.
90. Foreman, J.H. Thermoregulation in the horse exercising under hot and humid conditions. *Pferdeheilkunde* **1996**, *12*, 405–408.
91. Jackson, J.A.; Ducharme, N.G.; Hackett, R.P.; Rehder, R.S.; Ainsworth, D.M.; Shannon, K.J.; Erickson, B.K.; Erb, H.N.; Jansson, N.; Soderholm, L.V.; et al. Effects of airway obstruction on transmural pulmonary artery pressure in exercising horses. *Am. J. Vet. Res.* **1997**, *58*, 897–903.
92. Cook, W.R.; Williams, R.M.; Kirker-Head, C.A.; Verbridge, D.J. Upper airway obstruction (partial asphyxia) as the possible cause of exercise-induced pulmonary haemorrhage in the horse: An hypothesis. *Equine Vet. Sci.* **1988**, *8*, 11–26.
93. Gunson, D.E.; Sweeney, C.R.; Soma, L.R. Sudden death attributable to exercise-induced pulmonary hemorrhage in racehorses: Nine cases (1981–1983). *JAVMA* **1988**, *193*, 102–106.
94. Hinchcliff, K.W.; Couetil, L.L.; Knight, P.K.; Morley, P.S.; Robinson, N.E.; Sweeney, C.R.; Van Ercj, E. Exercise induced pulmonary haemorrhage in horses: American College of Veterinary Internal Medicine consensus statement. *J. Vet. Intern. Med.* **2015**, *29*, 743–758.

95. Hackett, R.P.; Ducharme, N.G.; Ainsworth, D.M.; Erickson, B.K.; Erb, H.N.; Soderholm, L.; Thorson, L.M. Effects of extrathoracic airway obstruction on intrathoracic pressure and pulmonary artery pressure in exercising horses. *Am. J. Vet. Res.* **1999**, *60*, 485–494.
96. West, J.B.; Mathieu-Costello, O.; Jones, J.H.; Birks, E.K.; Logemann, R.B.; Pascoe, J.R.; Tyler, W.S. Stress failure of pulmonary capillaries in racehorses with exercise-induced pulmonary hemorrhage. *J. Appl. Physiol.* **1993**, *75*, 1097–1109.
97. Deepika, K.; Barrocas, Q.M.; Fonseca, J.J.; Bikasi, G.B. Negative pressure pulmonary edema after acute upper airway obstruction. *J. Clin. Anaesth.* **1997**, *9*, 403–408.
98. Bhaskar, B.; Fraser, J.F. Negative pressure pulmonary oedema revisited: Pathophysiology and review of management. *Saudi J. Anaesth.* **2011**, *5*, 308–313.
99. Abrahamsen, E.J.; Bohanon, T.C.; Bednarski, R.M.; Hubbell, J.A.E.; Muir, W.W. Bilateral arytenoid cartilage paralysis after inhalation anaesthesia in a horse. *J. Am. Vet. Med. Assoc.* **1990**, *197*, 1363–1365.
100. Lang, S.A.; Duncan, P.G.; Shephard, D.A.E.; Ha, H.C. Pulmonary oedema associated with airway obstruction. *Can. J. Anaesth.* **1990**, *37*, 210–218.
101. Dixon, P.M.; Railton, D.I.; McGorum, B.C. Temporary bilateral laryngeal paralysis in a horse associated with general anaesthesia and post anaesthetic myositis. *Vet. Rec.* **1993**, *132*, 29–32.
102. Kollias-Baker, C.A.; Pipers, F.S.; Heard, D.; Seeherman, H. Pulmonary edema associated with transient airway obstruction in three horses. *JAVMA* **1993**, *202*, 1116–1118.
103. Tute, A.S.; Wilkins, P.A.; Gleed, R.D.; Credille, K.M.; Murphy, D.J.; Ducharme, N.G. Negative pressure pulmonary edema as a post-anesthetic complication associated with upper airway obstruction in a horse. *Vet. Surg.* **1996**, *25*, 519–523.
104. Senior, M. Postanaesthetic pulmonary oedema in horses: A review. *Vet. Anaes. Analg.* **2005**, *32*, 193–200.
105. Wilson, R.C.; Jones, P.W. Long-term reproducibility of Borg scale estimates of breathlessness during exercise. *Clin. Sci. Lond* **1991**, *80*, 309–312.
106. Holcombe, S.J.; Derksen, F.J.; Stick, J.A.; Robinson, N.E. Effect of nasal occlusion on tracheal and pharyngeal pressures in horses. *Am. J. Vet. Res.* **1996**, *57*, 1258–1260.
107. Leclere, M.; Lavoie-Lamoureux, A.; Lavoie, J.-P. Heaves, an asthma-like disease of horses. *Respirology* **2011**, *16*, 1027–1046.
108. Bullone, M.; Lavoie, J.-P. Asthma, "of horses and men"–how can equine heaves help us better understand human asthma immunopathology and its functional consequences? *Mol. Immunol.* **2015**, *66*, 97–105.
109. Couetil, L.L.; Denicola, D.B. Blood gas, plasma lactate and bronchoalveolar lavage cytology analyses in racehorses with respiratory disease. *Equine Vet. J. Suppl.* **1999**, *30*, 77–82.
110. Sanchez, A.; Couetil, L.L.; Ward, M.P.; Clark, S.P. Effect of airway disease on blood gas exchange in racehorses. *J. Vet. Intern. Med.* **2005**, *19*, 87–92.
111. Courouce-Malblanc, A.; Deniau, V.; Rossignol, F.; Corde, R.; Leleu, C.; Maillard, K.; Pitel, P.-H.; Pronost, S.; Fortier, G. Physiological measurements and prevalence of lower airway diseases in Trotters with dorsal displacement of the soft palate. *Equine Vet. J. Suppl.* **2010**, *38*, 246–255.
112. Parshall, M.B.; Schwartzstein, R.M.; Adams, L.; Banzett, R.B.; Manning, H.L.; Bourbeau, J.; Calverley, P.M.; Gift, A.G.; Harver, A.; Lareau, S.; et al. An official American Thoracic Society statement: Update on the mechanisms, assessment, and management of dyspnea. *Am. J. Respir. Crit. Care Med.* **2012**, *185*, 435–452.
113. Corfield, D.R.; Murphy, K.; Guz, A. Does the motor cortical control of the diaphragm "bypass" the brain stem respiratory centres in man? *Respir. Physiol.* **1998**, *114*, 109–117.
114. Bianchi, A.L.; Denavit-Saubie, M.; Champagnat, J. Central control of breathing in mammals: Neuronal circuitry, membrane properties, and neurotransmitters. *Physiol. Rev.* **1995**, *75*, 1–45.
115. Buchanan, G.F.; Richerson, G.B. Role of chemoreceptors in mediating dyspnea. *Respir. Physiol. Neurobiol.* **2009**, *167*, 9–19.
116. Jensen, D.; Ofir, D.; O'Donnell, D.E. Effects of pregnancy, obesity and aging on the intensity of perceived breathlessness during exercise in healthy humans. *Respir. Physiol. Neurobiol.* **2009**, *167*, 87–100.

117. El-Manshawi, A.; Killian, K.J.; Summers, E.; Jones, N.L. Breathlessness during exercise with and without resistive loading. *J. Appl. Physiol.* **1986**, *61*, 896–905.
118. Gandevia, S.C. Roles for perceived voluntary motor commands in motor control. *Trends Neurosci.* **1987**, *10*, 81–85.
119. Manning, H.L.; Schwartzstein, R.M. Pathophysiology of dyspnea. *N. Engl. J. Med.* **1995**, *333*, 1547–1553.
120. Prabhakar, N.R.; Peng, Y.-J. Peripheral chemoreceptors in health and disease. *J. Appl. Physiol.* **2004**, *96*, 359–366.
121. O'Donnell, D.E.; Ora, J.; Webb, K.A.; Laveneziana, P.; Jensen, D. Mechanisms of activity-related dyspnea in pulmonary diseases. *Respir. Physiol. Neurobiol.* **2009**, *167*, 116–132.
122. Lindinger, M.I.; Heigenhauser, G J.F. Effects of gas exchange on acid-base balance. *Compr. Physiol.* **2012**, *2*, 2203–2254.
123. Banzett, R.B.; Pedersen, S.H.; Schwartzstein, R.M.; Lansing, R.W. The affective dimension of laboratory dyspnea: Air hunger is more unpleasant than work/effort. *Am. J. Respir. Crit. Care Med.* **2008**, *177*, 1384–1390.
124. Lansing, R.W.; Gracely, R.H.; Banzett, R.B. The multiple dimensions of dyspnea: Review and hypotheses. *Respir. Physiol. Neurobiol.* **2009**, *167*, 53–60.
125. Booth, S.; Adams, L. The shuttle walking test: A reproducible method for evaluating the impact of shortness of breath on functional capacity in patients with advanced cancer. *Thorax* **2001**, *56*, 146–150.
126. Australian Lung Foundation. *Better Living with Chronic Obstructive Pulmonary Disease: A Patient Guide*, 2nd ed.; The Queensland Government: Queensland, Australia, 2012; Available online: http://lungfoundation.com.au/wp-content/uploads/2014/02/Better-Living-with-COPD.pdf (accessed on 15 February 2017).
127. Johnson, B.D.; Saupe, K.W.; Dempsey, J.A. Mechanical constraints on exercise hyperpnea in endurance athletes. *J. Appl. Physiol.* **1992**, *73*, 874–886.
128. Reinero, CR. Advances in the understanding of pathogenesis, and diagnostics and therapeutics for feline allergic asthma. *Vet. Rec.* **2011**, *190*, 28–33.
129. Undem, B.J.; Nassenstein, C. Airway nerves and dyspnea associated with inflammatory airway disease. *Respir. Physiol. Neurobiol.* **2009**, *167*, 36–44.
130. Waran, N.; McGreevy, P.D.; Casey, R. Training methods and horse welfare. In *The Welfare of Horses*; Waran, N., Ed.; Boston Kluwer Academic Publishers: Dordrecht, Netherlands, 2002; pp. 151–180.
131. Duberstein, K.J.; Johnson, E.L. Bits 101. In *University of Georgia Bulletin 1379*; University of Georgia: Athens, GA, USA, 2014; pp. 1–6. Available online: http://extension.uga.edu/publications/files/pdf/B%201379_5.PDF (accessed on 10 August 2016).
132. McGreevy, P.; McLean, A. Behavioral problems with the ridden horse. In *The Domestic Horse: The Origins, Development, and Management of Its Behavior*; Mills, D.S., McDonnell, S.M., Eds.; Cambridge University Press: Cambridge, UK, 2005; pp. 196–211.
133. Van Lancker, S.; Van der Broeck, W.; Simoens, P. Incidence and morphology of bone irregularities of the equine interdental space (bars of the mouth). *Equine Vet. Educ.* **2007**, *19*, 103–106.
134. Cook, W.R. Damage by the bit to the equine interdental space and second lower premolar. *Equine Vet. J.* **2011**, *23*, 355–360.
135. Mata, F.; Johnson, C.; Bishop, C. A cross-sectional epidemiological study of prevalence and severity of bit-induced oral trauma in polo ponies and racehorses. *J. Appl. Anim. Welf. Sci.* **2015**, *18*, 259–268.
136. Black, J.B.; Frisbie, D. Welfare concerns in the training and competition of cutting, reining and reined cow horses. In *Equine Welfare*; McIlwraith, C.W., Rollin, B.E., Eds.; Wiley-Blackwell: Chichester, UK, 2011; pp. 302–317.
137. Fenner, K.; Yoon, S.; White, P.; Starling, M.; McGreevy, P. The effect of noseband tightening on horses' behaviour, eye temperature, and cardiac responses. *PLoS ONE* **2016**, *11*.
138. American Quarter Horse Association Rule Book, Rule 44(h,k). In *Show Rules and Regulations*; AQHA: Amarillo, TX, USA, 2009.
139. Leitch, M. Welfare in the discipline of dressage. In *Equine Welfare*; McIlwraith, C.W., Rollin, B.E., Eds.; Wiley-Blackwell: Chichester, UK, 2011; pp. 332–340.

140. AWA, NZ. *Animal Welfare Act 1999 and Animal Welfare Amendment Act 2015 No. 2*; Government of New Zealand: Wellington, New Zealand, 1999. Available online: http://www.legislation.govt.nz/act/public/1999/0142/latest/DLM49664.html (accessed on 10 August 2016).
141. Cook, W.R. Bit-induced asphyxia: Elevation and dorsal displacement of the soft palate at exercise. *J. Equine Vet. Sci.***2002**, *20*, 7–14.
142. Hanson, F.; Cook, R. The Bedouin bridle rediscovered: A welfare, safety and performance enhancer. *Horse's Hoof***2015**, *60*, 1–8. Available online: http://www.bitlessbridle.com/THEBEDOUBRIDLE.pdf (accessed on 10 August 2016).
143. Cook, W.R. Bit-induced pain: a cause of fear, flight, fight and facial neuralgia in the horse. *Pferdeheilkunde* **2003**, *19*, 1–8.
144. Dugdale, D.; Greenwood, R. Some observations on conservative techniques for treatment of laryngopalatal displacement. *Equine Vet. Educ.* **1993**, *5*, 177–180.
145. Franklin, S.H.; Naylor, J.R.; Lane, J.G. The effect of a tongue-tie in horses with dorsal displacement of the soft palate. *Equine Vet. J.* **2002**, *34*, 430–433.
146. Franklin, S.; Allen, K. Laboratory exercise testing. In *Equine Sports Medicine and Surgery: Basic and Clinical Sciences of the Equine Athlete*, 2nd ed.; Hinchcliff, K.W., Kaneps, A.J., Geor, R.J., Eds.; Elsevier: New York, NY, USA, 2014; pp. 11–24.
147. Tabernaberri, C.; Cooper, C.; Clemence, J. Transition to a bitless bridle. In *Hoofbeats*; October/November; Australian Associated Press (AAP); Equine Veterinarians Australia (EVA): Sydney, Australia, 2011; pp. 62–69. Available online: http://www.whisperingacres.com.au/images/articles/Hoofbeats1.pdf (accessed on 17 February 2017).
148. McGreevy, P.D.; Corken, R.A.; Salvin, H.; Black, C.M. Whip use by jockeys in a sample of Australian thoroughbred races—An observational study. *PLoS ONE* **2012**, *7*, e33398.
149. Hood, J.; McDonald, C.; Wilson, B.; McManus, P.; McGreevy, P. Whip rule breaches in major Australian racing jurisdictions: Welfare and regulatory implications. *Animals* **2017**, *7*.
150. British Horseracing Authority. *Responsible Regulation: A Review of the Use of the Whip in Horseracing*; British Horseracing Authority: London, UK, 2011; Available online: http://www.britishhorseracing.com/wp-content/uploads/2014/03/WhipReview.pdf (accessed on 27 January 2017).
151. Evans, D.; McGreevy, P. An investigation of racing performance and whip use by jockeys in Thoroughbred races. *PLoS ONE* **2011**, *6*, e15622.
152. Travers, C.W.; Guthrie, A.J.; Lund, R.J. Characterization of a standard exercise to fatigue test in Thoroughbred horse. *Pferdeheikunde* **1996**, *12*, 463–465.
153. Flaminio, M.J.B.F.; Gaighan, E.M.; Gillespie, J.R. Exercise intolerance in endurance horses. *Vet. Clin. N. Am. Equine Pract.* **1996**, *12*, 565–580.
154. Valberg, S.J. Muscular causes of exercise intolerance in horses. *Vet. Clin. N. Am. Equine Prac.* **1996**, *12*, 495–515.
155. Noakes, T.D. Fatigue is a brain-derived emotion that regulates the exercise behaviour to ensure the protection of whole body homeostasis. *Front. Physiol.* **2012**, *3*, 82. Available online: https://doi.org/10.3389/fphys.2012.00082 (accessed on 22 May 2017).
156. Parente, E.J. Testing methods for exercise intolerance in horses. *Vet. Clin. N. Am. Equine Pract.* **1996**, *12*, 421–433.
157. Fraser, D.; Weary, D.M.; Pajor, E.A.; Milligan, B.N. A scientific conception of animal welfare that reflects ethical concerns. *Anim. Welf.* **1997**, *6*, 187–205.
158. Fraser, D. Assessing animal welfare at the farm and group level: The interplay of science and values. *Anim. Welf.***2003**, *12*, 433–443.
159. Fraser, D. *Understanding Animal Welfare: The Science in its Cultural Constext*; Wiley-Blackwell: Oxford, UK, 2008.
160. Lund, V. Natural living—A precondition for animal welfare in organic farming. *Livest. Sci.* **2006**, *100*, 71–83.
161. Mellor, D.J. Positive animal welfare states and promoting environment-focused and animal-animal interactive behaviours. *N. Z. Vet. J.* **2015**, *63*, 9–16.
162. Mellor, D.J. Positive welfare states and reference standards for welfare assessment. *N. Z. Vet. J.* **2015**, *63*, 17–23.

163. Mill, J.M.; Ward, W.R. Lameness in dairy cows and farmers' knowledge, training and awareness. *Vet. Rec.* **1994**, *134*, 162–164.
164. Packer, R.M.A.; Hendricks, A.; Burn, C.C. Do dog owners perceive the clinical signs related to conformational inherited disorders as "normal" for the breed? A potential constraint to improving canine welfare. *Anim. Welf.* **2012**, *21*, 81–93.

CHAPTER 4

Cortisol coregulation in fish

Ines Fürtbauer[1] & Michael Heistermann[2]

[1]*Department of Biosciences, College of Science, Swansea University, SA2 8PP Swansea, UK*

[2]*Endocrinology Laboratory, German Primate Center, Kellnerweg 4, 37077 Göttingen, Germany*

ABSTRACT

Cortisol coregulation, which is the up- or down-regulation of partners' physiological stress responses, has been described for individuals with strong attachment bonds, e.g. parents and their children, and romantic relationship partners. Research into moderating effects on cortisol coregulation suggests stronger covariation among distressed partners. Whether cortisol coregulation is unique to humans or can also be found in other species that share universal features of the vertebrate stress response remains unexplored. Using a repeated measures approach and non-invasive waterborne hormone analysis, we test the hypothesis that dyads of three-spined stickleback fish (Gasterosteus aculeatus) coregulate their cortisol levels in shared environments. Dyadic cortisol levels were unrelated when cohabiting (home tank), but significantly covaried when sharing a more stressful (as indicated by higher cortisol levels) environment (open field). Time-lag analysis further revealed that open field cortisol levels were predicted by partner's cortisol levels prior to the shared experience. To our knowledge, this study provides the first evidence for coregulatory processes on cortisol responses in a non-human animal that lacks strong bonds and social attachment relationships, suggesting a shared evolutionary origin of cortisol coregulation in vertebrates. From an adaptive perspective, cortisol coregulation may serve to reduce risk in challenging, potentially threatening situations.

1. INTRODUCTION

Social relationships can have a powerful impact on both mental and physical health[1,2,3,4,5,6,7]. The availability and perception of social support, in particular, have been shown to play an important role in the recovery from stressful experiences, a phenomenon commonly described as 'social buffering'[1,8,9]. More recently, however, research on human subjects has discovered that individuals with strong attachment bonds (typically family members) can influence each other's physiological states or, in other words, they 'coregulate'[10]. Cortisol coregulation, in contrast to social buffering which, by definition, is unidirectional, describes the phenomenon of a bidirectional interdependence of social partners' hypothalamic–pituitary–adrenal (HPA) axis activity, i.e. an up- or down-regulation of physiological stress (cortisol) levels[11,12,13.] Cortisol covariation has been documented between parents and their children[11,14,15,16,17] and in cohabiting intimate couples[11,12,13,18.]

Generally, cortisol coregulation is thought to be a "physiological manifestation of shared emotional and behavioral experiences"[19]. Investigations into coregulatory processes of HPA axis activity revealed, for example, that physical closeness–when individuals share an environment or activities–leads to stronger cortisol covariation[12,13,20]. Furthermore, distress situations, such as domestic violence, appear to be related to stronger cortisol covariation between pair partners, and mothers and their infants[11,21].

Cortisol coregulation between parents and children is thought to be of adaptive value since children's self-regulatory capacities are not yet independent, and thus, might be critical for child development[10,22]. Although cortisol coregulation in adults is unlikely to be a by-product of child-parent relationships, its adaptive function is unclear[10]. To date, studies into the interplay of adult partners' HPA axis activity (and other measures of physiological arousal) have involved human individuals with strong attachment bonds and have focussed on the potential implications for relationship functioning, presuming that coregulation becomes activated due to threats to the bond. However, according to social baseline theory, physiological coregulation may represent the most efficient way to save energy and reduce risk in threatening contexts, in a more general sense[23]. Moreover, the susceptibility of the HPA axis to social modulation is well documented in non-human animals[2,8,9,24]. Thus, cortisol coregulation may represent a general mechanism that is not unique to human attachment relationships and also exist in species that lack human-like bonds. However, to the best of our knowledge, the phenomenon of cortisol coregulation has not yet been studied in non-human animals, and it remains unclear whether cortisol coregulation is unique to humans, or an ancestral trait of vertebrates.

In the present study, we test the cortisol coregulation hypothesis in a basal vertebrate, a freshwater teleost fish, the three-spined stickleback (Gasterosteus aculeatus), using an experimental approach with non-invasive waterborne cortisol analysis and a repeated measures design[25,26]. Sticklebacks provide ideal models for studying the potential interdependence of social companions' cortisol responses. First, they are gregarious and respond to shoal partners. For example, in potentially threatening situations, fishes' time to react to conspecifics decreases and shoal synchrony increases[27,28]. Second, the fish hypothalamic–pituitary–interrenal (HPI) axis (similar to the mammalian HPA axis[29,30]) responds to a stressor within minutes30. Third, sticklebacks (and other fish) are susceptible to social mediation of HPI axis activity31.

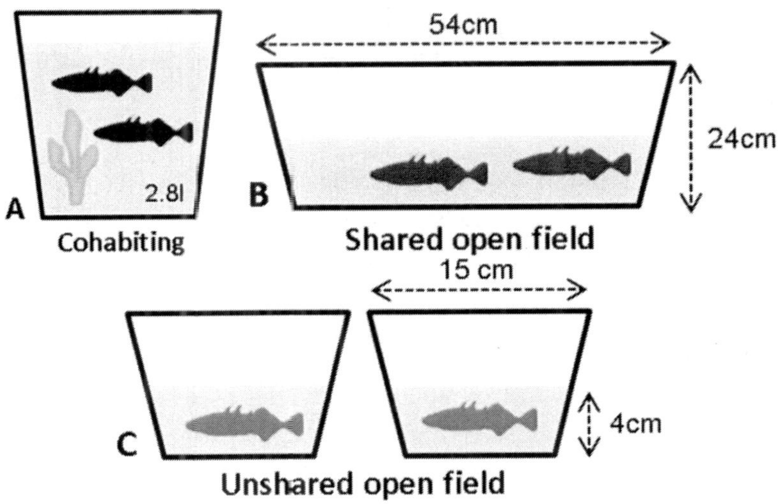

Figure 1: Experimental setup for investigating cortisol coregulation in dyads of three-spined stickleback fish. Fish dyads were sampled in each of three environmental contexts, i.e. when cohabiting (**A**) and after experiencing a shared (**B**) or unshared (**C**) open field environment. Note that test tank sizes (54 × 15 × 24 cm) and water levels (4 cm) were identical in both open field conditions (**B,C**), and that fish had no visual contact in the unshared open field context (**C**). Sampling order was the same for all individuals and for each context, all fish were tested within the same day and the two individuals of a dyad were sampled at the same time. The fish were kept and tested with their partner throughout the study and no re-grouping of dyads took place.

Therefore, inspired by and analogous to human research, we consider different processes potentially affecting cortisol coregulation in sticklebacks. We sample our fish in three different environmental contexts: (i) when cohabiting, which is analogous to the human "at home together" scenario[12,13,20] (Fig. 1A), (ii) when in a shared, potentially stressful open field environment, which is analogous to the human "distress situation"[11,21] (Fig. 1B), and (iii) when in an unshared environment, which is analogous to the human "at work" scenario[12,13,17] (Fig. 1C). In accordance with the cortisol coregulation hypothesis, we first predict that fish dyads' cortisol levels correlate in shared but not in unshared environments, and second, we expect stronger dyadic covariation in a potentially threatening compared to a non-threatening environment.

2. RESULTS

Fish exhibited significantly higher waterborne cortisol concentrations in an open field compared to a cohabiting context (estimate ± se: 0.26 ± 0.06, $t = 4.58$, $p < 0.001$; Fig. 2; full vs. null model: $\chi^2 = 53.2$, $df = 6$, $p < 0.001$, $n = 263$; Table S1), indicating that the open field was perceived as a stressor. Cortisol levels were unrelated to fish sex and weight across contexts (sex: estimate ± se: −0.15 ± 0.08, $t = -1.85$, $p = 0.077$; weight: estimate ± se: 0.19 ± 0.10, $t = 1.93$ $p = 0.065$; Table S1), and were lower in the afternoon than in the morning (estimate ± se: −0.22 ± 0.06, $t = -3.77$, $p < 0.001$; Table S1).

In accordance with our first prediction, we found a significant interaction between partner cortisol and the shared open field (estimate ± se: 0.55 ± 0.14, $t = 3.99$, $p < 0.001$; Fig. 3; Table S1) but not the unshared open field context (estimate ± se: 0.08 ± 0.15, $t = 0.55$, $p = 0.582$; Fig. 3; Table S1), indicating that the dyadic covariation of cortisol levels differed across environmental contexts. Further analysis revealed that cortisol levels of dyad partners were unrelated when cohabiting in their home tank (estimate ± se: − 0.05 ± 0.09, $t = -0.60$, $p = 0.550$; Table S1) and in an unshared open field (estimate ± se: − 0.04 ± 0.11, $t = -0.37$, $p = 0.716$; Table S1). Conversely, dyad partners' cortisol levels were significantly positively correlated when sharing an open field (estimate ± se: − 0.67 ± 0.12, $t = 5.54$, $p < 0.001$; Fig. 4A; Table S1). This indicates dyadic coregulation of HPI axis activity in a shared, more stressful environment, which is in line with our second prediction. Time-lag analysis revealed that partners' cohabiting cortisol levels ('baseline'; prior to the open field test) predicted focal fish cortisol responses in the shared open field environment (estimate ± se: 0.35 ± 0.14, $t = 2.56$, $p = 0.015$; Fig. 4B; Table S1). In other words, individuals' cortisol levels were higher following the shared open field situation when their dyad partner exhibited higher baseline cortisol.

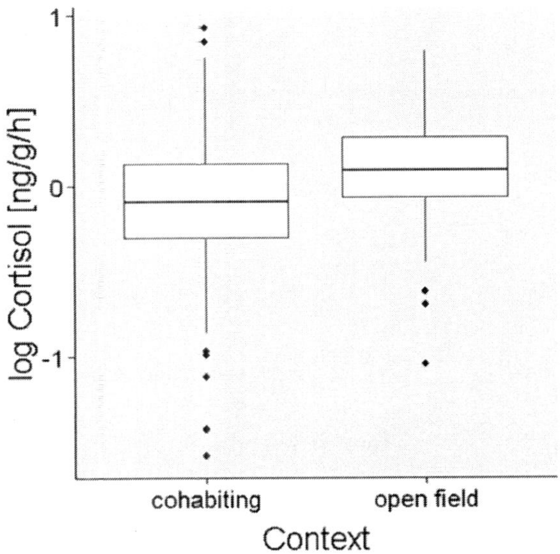

Figure 2: Waterborne cortisol concentration when cohabiting and experiencing an open field environment. Fishes' cortisol levels were significantly higher in an open field context compared to a cohabiting context.

3. DISCUSSION

Research into human cortisol coregulation (or 'linkage') has increased in recent years but many questions still remain unanswered[18,22]. This lack of knowledge may be at least partly attributed to the focus on individuals with strong bonds, for example romantic relationship partners, and parents and their children[18,22]. The findings of this study represent, to the best of our knowledge, the first empirical evidence for comparable coregulatory processes on cortisol responses in a non-human animal species that lacks strong social bonds and attachment relationships. Using fish as an animal model of cortisol coregulation could help to obtain valuable information regarding its mechanisms and functions in humans. Indeed, fish models are becoming increasingly popular due to the intriguing similarities between fish and human socio-endocrinological patterns[32].

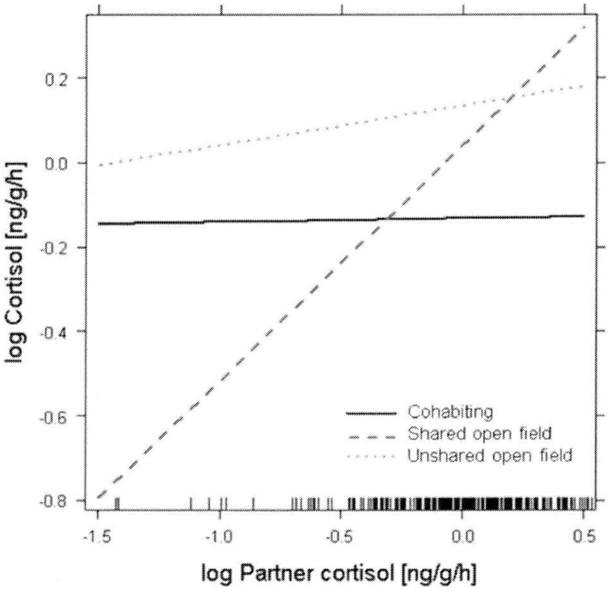

Figure 3: Partner fish cortisol predicting focal fish cortisol, as moderated by environmental context. Dyads' cortisol levels were significantly positively correlated in a shared open field (red dashed line) but not in a shared home tank (black solid line) or an unshared open field (green dotted line). See text for details.

In humans, cortisol levels in socially attached individuals appear to be only associated when individuals are in a shared environment, but not when in an unshared environment[12]. Furthermore, stronger cortisol linkage is found in stressful contexts[11,21]. Similarly, cortisol levels in our stickleback dyads were only associated when fish were in a shared, more stressful environment (as indicated by higher cortisol levels) but not in an unshared environment or when cohabiting. This suggests that dyads' HPI axis activities do not become entrained by cohabiting but that coregulatory processes become activated under stress, and furthermore, can operate at a very fast time scale (i.e. minutes), which is in line with other examples of rapid, socially modulated endocrine plasticity[2,33].

Our findings therefore raise the question: what mechanisms underlie cortisol linkage in a perceived threatening situation? Research into social modulation of mammalian stress responses has shown that, for example, social buffering effects, i.e. the down-regulation of stress responses through the presence/behaviour of a social partner, are mediated by tactile, chemical, and/or visual cues[34,35,36,37]. In fish, both chemical and visual cues are of major importance for inter-individual communication[38], and are likely to also underlie the communication of partners' stress statuses[39]. Given the bidirectional nature of cortisol coregulation, it is possible that different chemicals mediate up- and down-regulation of cortisol responses. For example, in pacus (Piaractus mesopotamicus), behavioural responses to different predator conditions, i.e. alert (predator condition) and attraction (non-predator condition) appear to be mediated by at least two different chemicals, or different proportions of these chemicals[40]. Furthermore, recent work on zebrafish (Danio rerio) revealed that the chemical communication of predation risk is independent of a surge in cortisol levels in 'donor' fish but can still lead to changes in HPI axis activity[41]. Visual cues, such as changes in or the occurrence of certain behaviours in response to unpredictable, potentially threatening environments may also trigger coregulatory physiological processes. If behavioural cues indeed are crucial in the aforementioned context, future research should assess whether familiarity with a social partner is necessary for coregulatory processes on cortisol responses to occur. Partner familiarity has been shown to be a major moderating factor on social buffering effects in birds and mammals[42,43,44] and, in fish, can affect behavioural responses and social interactions in novel situations[45,46].

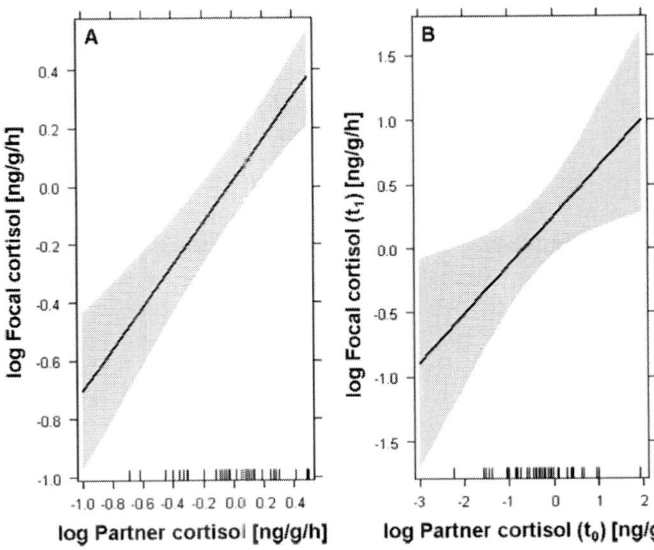

Figure 4: Covariation in dyadic cortisol levels (A) in a shared open field environment and (B) between focal fish open field cortisol (t_1) and partners' baseline cortisol prior to the open field test situation (t_0). Effects shown are predictions from LMMs. The shaded areas indicate upper and lower 95% confidence limits.

Another interesting finding was revealed by our time lag analysis, showing that when paired with a high-stress partner (i.e. an individual with high 'baseline' cortisol levels), fish had higher cortisol levels following the shared open field experience (Fig. 4B) than when paired with a low-stress companion. Generally, our finding suggests that partner's baseline stress reactivity is an important modulator of coregulatory processes of HPI axis activity in stickleback fish and potentially also of HPA axis activity in humans. Future human studies may therefore benefit from investigating whether partners' baseline cortisol levels can be used to predict up- or downregulation of cortisol levels in stressful contexts. This could help to further illuminate if and when cortisol coregulation is "good" or "bad", an issue still under debate[18].

On a functional level, physiological coregulation has been proposed to reduce energy expenditure and risk[23]. In fact, risk dilution is one of the main reasons for group cohesion and behavioural synchrony in social animals, both of which often increase in threatening contexts[47,48]. In fish, shoaling (i.e. the tendency to stick with others) and swimming speed become more synchronous under a perceived threat[27]. Moreover, research into 'stress coping styles', has revealed consistent links between physiological and behavioural stress responses, i.e. reactive (high physiological stress, immobility) versus proactive styles (low physiological stress, high locomotor activity) across and within individuals[26,49]. Stickleback fish, for instance, are shyer on days they exhibit higher cortisol levels[26]. Thus, we hypothesize that cortisol coregulation may underlie synchronized collective behavioural responses, and, ultimately may serve to enhance risk dilution in potentially threatening situations.

Overall, the present study is an important contribution and advances our understanding of cortisol coregulation, a phenomenon which to date was thought to be unique to human attachment relationships. The interplay of fishes' physiological stress responses in the absence of strong social bonds hints at a shared evolutionary origin of cortisol coregulation in vertebrates and suggests that it is not simply a by-product of partners' shared environments. Cortisol coregulation potentially confers important adaptive value across vertebrates, which future studies should look into.

4. METHODS

Subjects and experimental setup N = 22 fish, sourced from a wild population on Swansea University Campus, were initially housed together in a large tank (30 × 29 × 122 cm). Fish were fed daily with defrosted bloodworms (Chironomus sp.). Throughout the study, the fish were kept under non-breeding environmental conditions (16 °C/8L:16D). Four days prior to the experiments, 11 fish dyads (paired randomly with respect to sex; mean weight ± SD: 1.63 ± 0.38 g) were transferred to individual 2.8l gravel-lined aerated tanks (Fig. 1A). The fish were kept and tested with their original partner throughout the study and no re-grouping took place. Fish were tagged with Visible Implant Elastomer (VIE) tags (Northwest Marine Technology, Inc., 2014; www.nmt.us/products/vie/vie.shtml), a harmless method for individual identification of small fish[25]. Each fish was sampled repeatedly in three different contexts: (1) cohabiting (n = 6 samples per fish; Fig. 1A), (2) shared open field (n = 2 samples per fish; Fig. 1B) and (3) unshared open field (n = 4 samples per fish; Fig. 1C). Samples were collected over a 10-day period; sampling order was the same for all individuals and for each condition, all fish were tested within the same day and the two individuals of the dyad were sampled at the same time. The open field test tanks were identical in the shared and unshared contexts (54 × 15 × 24 cm) and were filled with water up to 4 cm, and the water (same source as for home tanks and hormone collection) was changed between each trial to decrease potential olfactory cues (Fig. 1). The fish were exposed to the shared/unshared open field for 30 minutes. All procedures described were carried out in accordance with the guidelines for the use of animals approved by the Association for the Study of Animal Behaviour and were approved by Swansea University's Ethics Committee (IP-1213-3).

Hormone collection, extraction, and analysis A total of n = 263 waterborne hormone samples were collected and analysed for cortisol, following published procedures[25,26]. Briefly, either straight out of the home tank or after experiencing a shared/unshared open field environment, fish were placed individually in 150 ml glass beakers filled with 50 ml water (same source as used for home tanks) for one hour. All subjects were exposed to beaker confinement for habituation twice prior to experiments[26]. Net-filtered water samples were collected in 60 ml polypropylene bottles and stored at −18 °C until extraction. Steroids were extracted using solid phase extraction procedures and using Sep-Pak Plus C18 solid phase extraction Cartridges (Waters #WAT020515), and a Visiprep 12-port vacuum manifold connected to a KNF Laboport vacuum pump. Steroids were eluted from the cartridges with 5 ml methanol, collected in a glass tube and evaporated under nitrogen at 45 °C. Steroids were redissolved in 350 µl assay buffer, and analysed using an enzyme immunoassay for immunoreactive cortisol[50]. All samples were run in duplicate, and samples with a c.v. >7% between duplicates were re-measured. Sensitivity of the assay at 90% binding was 0.5 pg. Intra- and inter-assay c.v., calculated from replicate determinations of high and low-value quality controls, were 7.4% (n = 16) and 7.2% (n = 10) (high) and 8.9 (n = 16) and 12.7% (n = 10) (low).

Statistical analyses All data were analysed using Linear Mixed Models (LMM[51]) fitted in R (R Development Core Team: www.r-project.org), using the package *lme4*[52]. Cortisol data were log-transformed to normalise model residuals. In all models, fish sex, weight, and sampling time (A.M. versus P.M.) were controlled for. To allow for individual- and dyad-specific differences, random intercepts were fitted for "Individual" and "Dyad". The significance of the full models as compared to the null models was established using likelihood ratio tests (R function anova). Co-linearity of fixed effects was controlled for by calculating Variance Inflation Factors[53] for standard linear models excluding the random effects. Model diagnostics were performed using graphical procedures (Q-Q plot and standardized residuals vs. fitted values).

First, in order to test whether the open field environment was perceived as a stressor, we tested for overall differences in cortisol concentrations between cohabiting and open field contexts (model 1; Table S1). The model was fitted with a random slope for context. Second, to simultaneously test for a correlation between partners' cortisol levels and a potential moderating effect of environmental context (cohabiting, shared and unshared open field), we fitted a LMM with cortisol as response and a two-way interaction between partner cortisol and context (cohabiting was used as reference variable) (model 2, Table S1). The model was fitted with a random slope for context. Third, since significant interaction terms make the interpretation of main effects of the interacting predictors unreliable54, we ran separate models for each environmental context (models 3–5; Table S1). Fourth, to further assess coregulatory influences in a shared open field context, we ran a time-lag analysis, and modelled open field cortisol levels (t_1) as a function of partners' 'baseline' cortisol levels prior to the occasion (t_0) (model 6; Table S1). The time interval between t0 and t1 and open field hormone collection was the same for all individuals (1.5 hours between the start of each of the two hormone collections). For all analyses, the level of significance was set at $P < 0.05$.

ACKNOWLEDGEMENTS

We thank Andrea Heistermann for help with the hormone analyses, and Andrew King, Christopher Coates, and three anonymous reviewers for comments on the manuscript. The work was supported by a research grant from the Fisheries Society of the British Isles (FSBI) awarded to IF. IF was funded by the German Research Foundation (DFG; FU-985/1-1).

REFERENCES

1. Cohen, S. Social Relationships and Health. Am. Psychol. 59, 676–684 (2004).
2. DeVries, A., Glasper, E. & Detillion, C. Social modulation of stress responses. Physiol. Behav. 79, 399–407 (2003).
3. Nunn, C. L., Craft, M. E., Gillespie, T. R., Schaller, M. & Kappeler, P. M. The sociality–health–fitness nexus: synthesis, conclusions and future directions. Phil. Trans. R. Soc. B 370 (2015).
4. House, J. S., Landis, K. R. & Umberson, D. Social relationships and health. Science 241, 540–545 (1988).
5. Kappeler, P. M., Cremer, S. & Nunn, C. L. Sociality and health: impacts of sociality on disease susceptibility and transmission in animal and human societies. Phil. Trans. R. Soc. B 370, 20140115 (2015).
6. Sapolsky, R. The influence of social hierarchy on primate health. Science 308, 648–652 (2005).
7. DeVries, A. C., Craft, T. K. S., Glasper, E. R., Neigh, G. N. & Alexander, J. K. 2006 Curt P. Richter award winner: Social influences on stress responses and health. Psychoneuroendocrinology 32, 587–603 (2007).
8. Hennessy, M. B., Kaiser, S. & Sachser, N. Social buffering of the stress response: diversity, mechanisms, and functions. Front. Neuroendocrinol. 30, 470–482 (2009).
9. Kikusui, T., Winslow, J. T. & Mori, Y. Social buffering: relief from stress and anxiety. Phil. Trans. R. Soc. B 361, 2215–2228 (2006).
10. Sbarra, D. A. & Hazan, C. Coregulation, Dysregulation, Self-Regulation: An Integrative Analysis and Empirical Agenda for Understanding Adult Attachment, Separation, Loss, and Recovery. Pers. Soc. Psychol. Rev. 12, 141–167 (2008).
11. Saxbe, D. E. et al. Cortisol covariation within parents of young children: Moderation by relationship aggression. Psychoneuroendocrinology 62, 121–128 (2015).
12. Saxbe, D. E. & Repetti, R. L. For better or worse? Coregulation of couples' cortisol levels and mood states. J. Pers. Soc. Psychol. 98, 92–103 (2010).
13. Papp, L. M., Pendry, P., Simon, C. D. & Adam, E. K. Spouses' Cortisol Associations and Moderators: Testing Physiological Synchrony and Connectedness in Everyday Life. Fam. Process 52, 284–298 (2013).
14. Laurent, H. K., Ablow, J. C. & Measelle, J. Taking stress response out of the box: Stability, discontinuity, and temperament effects on HPA and SNS across social stressors in mother–infant dyads Dev. Psychol. 48, 35–45 (2012).
15. Neu, M., Laudenslager, M. L. & Robinson, J. Coregulation in Salivary Cortisol During Maternal Holding of Premature Infants. Biol. Res. Nurs. 10, 226–240 (2009).
16. Saxbe, D. E. et al. Relative Influences: Patterns of HPA Axis Concordance During Triadic Family Interaction. Health Psychol. 33, 273–281 (2014).
17. Hibel, L. C., Trumbell J. M. & Mercado, E. Work/non-workday differences in mother, child, and mother–child morning cortisol in a sample of working mothers and their children. Early Hum. Dev. 90, 1–7 (2014).
18. Timmons, A. C., Margolin, G. & Saxbe, D. Physiological linkage in couples and its implications for individual and interpersonal functioning: A literature review. J. Fam. Psychol. 29, 720–731 (2015).
19. Hibel, L. C., Granger, D. A., Blair, C. & Finegood, E. D. The Family Life Project Key, I. Maternal-child adrenocortical attunement in early childhood: Continuity and change. Dev. Psychobiol. 57, 83–95 (2015).
20. Papp, L. M., Pendry, P. & Adam, E. K. Mother-Adolescent Physiological Synchrony in Naturalistic Settings: Within-Family Cortisol Associations and Moderators. J. Fam. Psychol. 23, 882 (2009).
21. Hibel, L. C., Granger D. A., Blair, C. & Cox, M. J. Intimate partner violence moderates the association between mother–infant adrenocortical activity across an emotional challenge. J. Fam. Psychol. 23, 615–625 (2009).

22. Feldman, R. Parent–Infant Synchrony: Biological Foundations and Developmental Outcomes. Curr. Dir. Psychol. Sci. 16, 340–345 (2007).
23. Coan, J. A. In Handbook of Attachment: Theory, Research, and Clinical Applications (eds J. Cassidy & P. R. Shaver) 241–265 (The Guilford Press, NY, 2008).
24. Creel, S., Dantzer, B., Goymann, W. & Rubenstein, D. R. The ecology of stress: effects of the social environment. Funct. Ecol. 27, 66–80 (2013).
25. Fürtbauer, I., King, A. J. & Heistermann, M. Visible Implant Elastomer (VIE) tagging and simulated predation risk elicit similar physiological stress responses in three-spined sticklebacks (Gasterosteus aculeatus). J. Fish Biol. 86, 1644–1649 (2015).
26. Fürtbauer, I., Pond, A., Heistermann, M. & King, A. Personality, plasticity, and predation: linking endocrine and behavioural reaction norms in stickleback fish. Funct. Ecol. 29, 931–940 (2015).
27. Bode, N. W. F., Faria, J. J., Franks, D. W., Krause, J. & Wood, A. J. How perceived threat increases synchronization in collectively moving animal groups. Proc. R. Soc. B 277, 3065–3070 (2010).
28. Ward, A. J. W., Sumpter, D. J. T., Couzin, I. D., Hart, P. J. B. & Krause, J. Quorum decision-making facilitates information transfer in fish shoals. Proc. Natl. Acad. Sci. USA 105, 6948–6953 (2008).
29. Galhardo, L. & Oliveira, R. F. Psychological stress and welfare in fish. Annu. Rev. Biomed. Sci. 11, 1–20 (2009).
30. Pankhurst, N. W. The endocrinology of stress in fish: An environmental perspective. Gen. Comp. Endocrinol. 170, 265–275 (2011).
31. Mommer, B. C. & Bell, A. M. A test of maternal programming of offspring stress response to predation risk in threespine sticklebacks. Physiol. Behav. 122, 222–227 (2013).
32. Soares, M. C., Oliveira, R. F., Ros, A. F. H., Grutter, A. S. & Bshary, R. Tactile stimulation lowers stress in fish. Nat. Commun. 2, 534 (2011).
33. Oliveira, R. F. Social behavior in context: Hormonal modulation of behavioral plasticity and social competence. Integr. Comp. Biol. 49, 23–440 (2009).
34. Boissy, A., Terlouw, C. & Le Neindre, P. Presence of Cues from Stressed Conspecifics Increases Reactivity to Aversive Events in Cattle: Evidence for the Existence of Alarm Substances in Urine. Physiol. Behav. 63, 489–495 (1998).
35. Kikusui, T., Takigami, S., Takeuchi, Y. & Mori, Y. Alarm pheromone enhances stress-induced hyperthermia in rats. Physiol. Behav. 72, 45–50 (2001).
36. Kiyokawa, Y., Shimozuru, M., Kikusui, T., Takeuchi, Y. & Mori, Y. Alarm pheromone increases defensive and risk assessment behaviors in male rats. Physiol. Behav. 87, 383–387 (2006).
37. Nakashima, S. F., Ukezono, M., Nishida, H., Sudo, R. & Takano, Y. Receiving of emotional signal of pain from conspecifics in laboratory rats. R. Soc. Open Sci. 2 (2015).
38. Sloman, K., Wilson, R. & Balshine, S. Behaviour and Physiology of Fish. (Elsevier Academic Press, San Diego, California, 2006).
39. Barcellos, L. J. G., Volpato, G. L., Barreto, R. E., Coldebella, I. & Ferreira, D. Chemical communication of handling stress in fish. Phys. Behav. 103, 372–375 (2011).
40. Jordão, L. & Volpato, G. Chemical transfer of warning information in non-injured fish. Behaviour 137, 681–690 (2000).
41. Barcellos, L. J. G., Koakoski, G., da Rosa, J. G. S., Ferreira, D., Barreto, R. E., Giaquinto, P. C. & Volpato, G. L. Chemical communication of predation in zebrafish does not depend on cortisol increase. Sci. Rep. 4, 5076 (2014).
42. Kiyokawa, Y., Honda, A., Takeuchi, Y. & Mori, Y. A familiar conspecific is more effective than an unfamiliar conspecific for social buffering of conditioned fear responses in male rats. Behav. Brain Res. 267, 189–193 (2014).
43. Remage-Healey, L., Adkins-Regan, E. & Romero, L. M. Behavioral and adrenocortical responses to mate separation and reunion in the zebra finch. Horm. Behav. 43, 108–114 (2003).
44. Livia Terranova, M., Cirulli, F. & Laviola, G. Behavioral and hormonal effects of partner familiarity in periadolescent rat pairs upon novelty exposure. Psychoneuroendocrinology 24, 639–656 (1999).

45. Ward, A. J. W. & Hart, P. J. B. The effects of kin and familiarity on interactions between fish. Fish Fish. 4, 348–358 (2003).
46. Galhardo, L., Vitorino, A. & Oliveira, R. F. Social familiarity modulates personality trait in a cichlid fish. Biol. Lett. 8, 936–938 (2012).
47. Krause, J. & Ruxton, G. D. Living in Groups. (Oxford University Press, 2002).
48. King, A. & Fürtbauer, I. Being Social. In The Behavior of Animals: Mechanisms, Function And Evolution. J. Bolhuis & L.-A. Giraldeau eds. (Wiley) (in press).
49. Koolhaas, J. M. et al. Coping styles in animals: current status in behavior and stress-physiology. Neurosci. Biobehav. Rev. 23, 925–935 (1999).
50. Palme, R. & Möstl, E. Measurement of cortisol metabolites in faeces of sheep as a parameter of cortisol concentration in blood. Int. J. Mamm. Biol. 62, 192–197 (1997).
51. Baayen, R. H. Analyzing Linguistic Data. (Cambridge University Press, Cambridge, 2008).
52. Bates, D., Maechler, M., Bolker, B. & Walker, S. lme4: Linear mixed-effects models using Eigen and S4. R package version 1.0–6. http://CRAN.R-project.org/package=lme4 (2014).
53. Field, A. Discovering Statistics Using SPSS (Sage Publications, 2005).
54. Zar, J. H. Biostatistical Analysis (Prentice-Hall, Upper Saddle River, NJ, 1999).

CHAPTER 5

Strontium mineralization of shark vertebrae

Vincent Raoult[1], Victor M. Peddemors[2], David Zahra[3], Nicholas Howell[3], Daryl L. Howard[4], Martin D. de Jonge[4] & Jane E. Williamson[1]

[1]*Department of Biological Sciences, Macquarie University, Sydney NSW 2109, Australia*

[2]*Fisheries NSW, Sydney Institute of Marine Science, 19 Chowder Bay Road, Mosman NSW 2088, Australia.*

[3]*ANSTO Life Sciences, Lucas Heights, NSW 2234, Australia.*

[4]*Australian Synchrotron, 800 Blackburn Road, Clayton, Victoria 3168, Australia.*

ABSTRACT

Determining the age of sharks using vertebral banding is a vital component of management, but the causes of banding are not fully understood. Traditional shark ageing is based on fish otolith ageing methods where growth bands are assumed to result from varied seasonal calcification rates. Here we investigate these assumptions by mapping elemental distribution within the growth bands of vertebrae from six species of sharks representing four different taxonomic orders using scanning x-ray fluorescence microscopy. Traditional visual growth bands, determined with light microscopy, were more closely correlated to strontium than calcium in all species tested. Elemental distributions suggest that vertebral strontium bands may be related to environmental variations in salinity. These results highlight the requirement for a better understanding of shark movements, and their influence on vertebral development, if confidence in age estimates is to be improved. Analysis of shark vertebrae using similar strontium-focused elemental techniques, once validated for a given species, may allow more successful estimations of age on individuals with few or no visible vertebral bands.

1. INTRODUCTION

Humanity's growing demand for protein has led to substantial pressure on both terrestrial and oceanic ecosystems. Harvesting of large predatory fishes by industrial fishing techniques has lowered populations to a fraction of their historic numbers in some areas[1,2], with some species such as tuna or grouper more depleted than others[3,4]. Sharks are especially vulnerable to fishing due to their low reproductive rates and delayed maturity[5,6,7]. Their disappearance has negative effects on ecosystems[8], and considerable resources have been allocated to their management and conservation. Basic information about life history parameters such as reproduction, diet, and age structure are often lacking, however, making assessment and management of shark stocks difficult.

Ageing sharks is problematic due to the large numbers of samples required, which are often difficult to obtain[9]. Ageing also requires cross-validation, and a lack of calcified structures in the many deeper-water species has contributed to discrepancies between growth and age estimates[9]. Inaccurate estimates can lead to poor predictions of population growth and lead to overfishing[10] or, conversely, under-exploitation of resources. Typically, shark ageing has been based on a modification of fish otolith ageing but uses vertebral bands instead of otolith rings[9], which are counted as a proxy for yearly growth patterns under the assumption that banding patterns result from changes in calcification rates over time in a manner similar to that of fish[11,12,13]. Although the seasonality of these bands has been validated for many species[14,15], the processes that determine the formation of these bands are not fully understood. Moreover, published methods of shark ageing vary substantially because different preparation methods must be tested to determine the most effective technique for each newly tested species[9]. Different ageing methods and subjective differences in readability make comparative studies problematic[11,16,17].

To investigate the elemental composition of vertebral bands, and by association, the factors that may influence their deposition, elemental distribution within vertebrae was assessed in five diverse species of sharks using Scanning X-ray Fluorescence Microscopy (SXFM)18. Assessments of SXFM images were compared to traditional microscope ageing9 to determine their use as an alternate ageing technique.

2. RESULTS

Calcium and strontium maps showed clear banding within vertebrae of all species. However, strontium bands were consistently more visible than calcium bands (Figs 1 and 2). Strontium peak counts agreed with, or were slightly under, counts obtained microscopically, while calcium peak counts could be both less than or greater than microscope band counts (Figs 3, 4, 5, 6, 7, 8). In all cases, counts of strontium peaks corresponded more closely to banding counts determined by light microscopy than to counts of calcium peaks (Fig. 9). Neither strontium concentrations ($df = 4$, $F = 6.7$, $R^2 = 0.69$, $p = 0.08$) nor calcium concentrations ($df = 4$, $F = 0.43$, $R^2 = 0.12$, $p = 0.55$) significantly correlated with light microscopy band counts. However, the correlation coefficient (R^2) of the strontium to light microscopy relationship was much higher than the calcium relationship (0.69 vs 0.12).

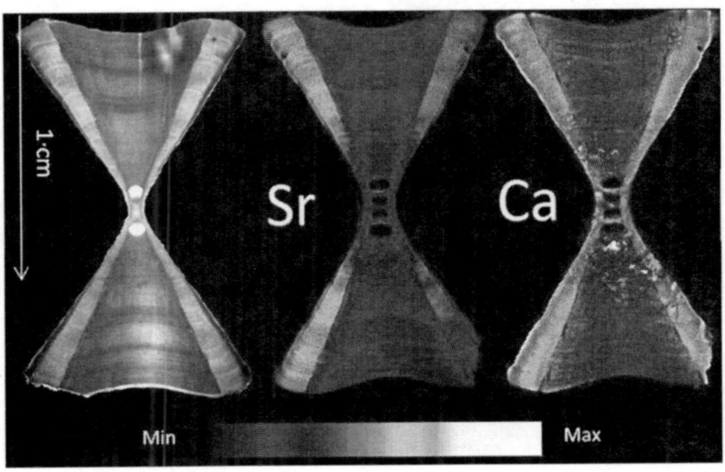

Figure 1: Images of light microscopy (left side) and SXFM (centre, right side) at different elemental spectra (in ppm) of unstained sagittal sections of a 1.7 m female Smooth Hammerhead (*Sphyrna zygaena*) at 25 µm resolution. Banding studies usually focus on the *corpus calcareum*, the denser area on the outside of the vertebrae.

3. DISCUSSION

In some species, shark vertebral growth banding may be more closely related to strontium absorption rather than to calcification. Strontium and calcium peaks did not correlate.

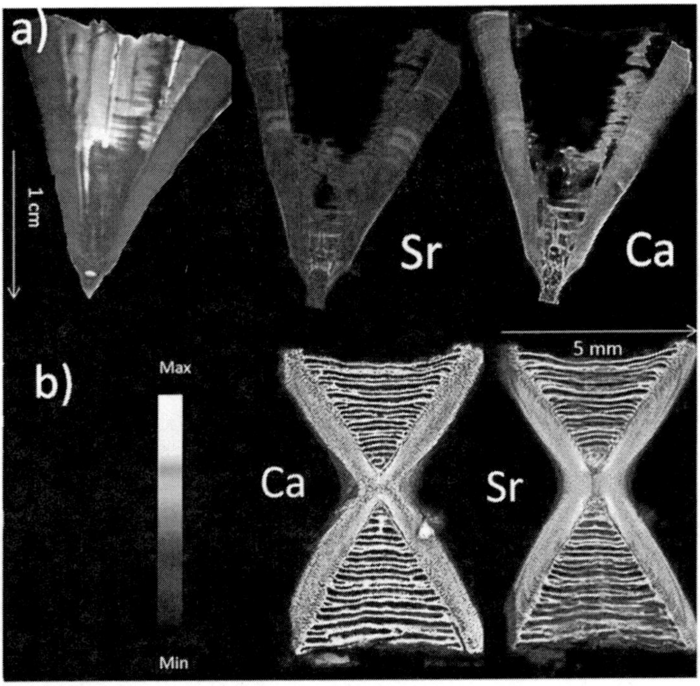

Figure 2 Images of light microscopy (top left) and SXFM (all others) at different elemental spectrums (in ppm) of sectioned vertebrae of (**a**) a 2.7 m White Shark (*Charcharodon carcharias*) vertebra and (**b**) an 0.9 m Eastern angel shark (*Squatina albipunctata)* at 25 and 15 μm resolution, respectively. Light microscopy of angel shark not shown due to lack of visible growth bands.

Rather, strontium peak numbers were more closely related to counts obtained using light microscopy than were calcium peaks.

It is possible that the small sample size in this study prevented the establishment of significant statistical relationships. However, strontium data had a substantially stronger relationship with traditional counts via light microscopy than the calcium data from the same samples. If further individuals were analysed in a similar fashion, it is likely that a statistically significant correlation between strontium and light microscopy bands would exist. In addition, the peaks on the dorsal and ventral face of the vertebrae had symmetrical shapes and positioning (peaks were equidistant on dorsal and ventral sides), whereas calcium peaks were asymmetrical. Since growth deposition is physiological, symmetrical patterns are expected in the banding.

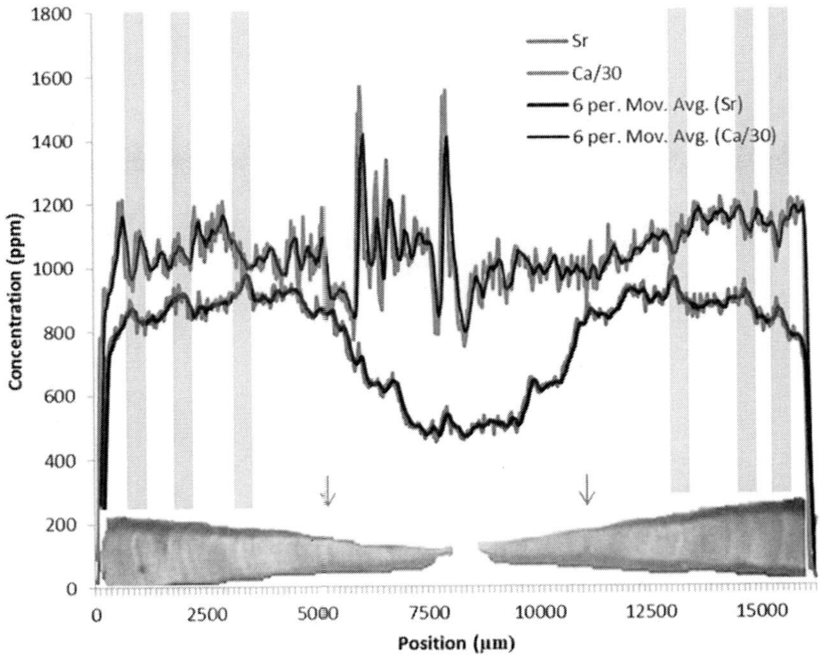

Figure 3: SXFM calcium and strontium concentrations of the *corpus calcarum* from the dorsal to ventral edge of the 1.7 m Smooth Hammerhead (*Sphyrna zygaena*) shown in Fig. 1. Calcium concentrations were reduced by a ratio of 30 to make them comparable to strontium concentrations. Six point moving averages were added to reduce noise. Birth marks are at roughly 5000 and 12000 µm (indicated with red arrows). Data are compared to a flattened horizontal microscope image of the *corpus calcarum* of the vertebra with marked 'traditional' growth bands (grey bands that are lined up with the opaque winter bands throughout the thickness of the vertebra for comparison with strontium and calcium). Note how strontium peaks are consistent on both sides of the vertebra, while calcium concentrations are not.

3.1 Origins of strontium variation

Strontium is absorbed during otolith mineralization in fish, often replacing calcium[19], and otolith absorption rates are dependent on environmental concentrations of the element[20]. Strontium is preferentially absorbed relative to other trace elements present in the marine environment (e.g., magnesium, barium)[21], whereas calcium is absorbed at a relatively constant rate regardless of its environmental availability[20].

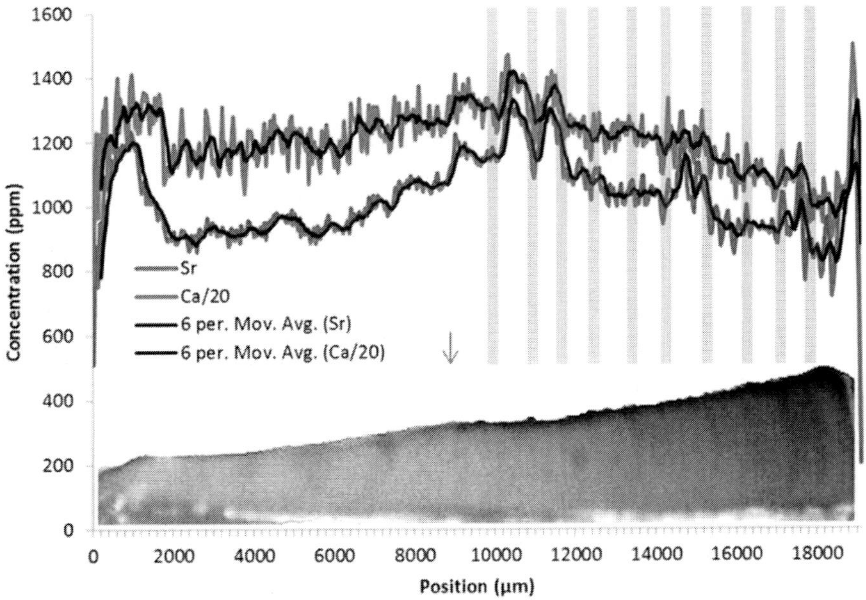

Figure 4: SXFM calcium and strontium concentrations of the *corpus calcarum* from the dorsal to central edge of a 2.7 m female White Shark (*Charcharodon carcharias*). Calcium concentrations were reduced by a ratio of 20 to make them comparable to strontium concentrations. Six point moving averages were added to reduce noise. Birth mark is at 9000 µm (indicated with red arrow). Data were compared to a flattened horizontal microscope image of the *corpus calcarum*, with marked 'traditional' growth bands (grey bands that are lined up with the opaque winter bands throughout the thickness of the vertebra for comparison with strontium and calcium).

Though the processes that govern the absorption of strontium in fish are still debated, the model that best explains these contradicting results is that preferential strontium absorption is linked to the rate of proteinacious matrix formation[21].

Incorporation of strontium into fish otoliths is not immediate: changes in concentrations of ambient water strontium have effects on mineralized concentrations but only after ten days, and sometimes does not become visible until sixty days after exposure[22]. Despite the apparent role of strontium in fish development, the physiological role of this element is still unknown[23].

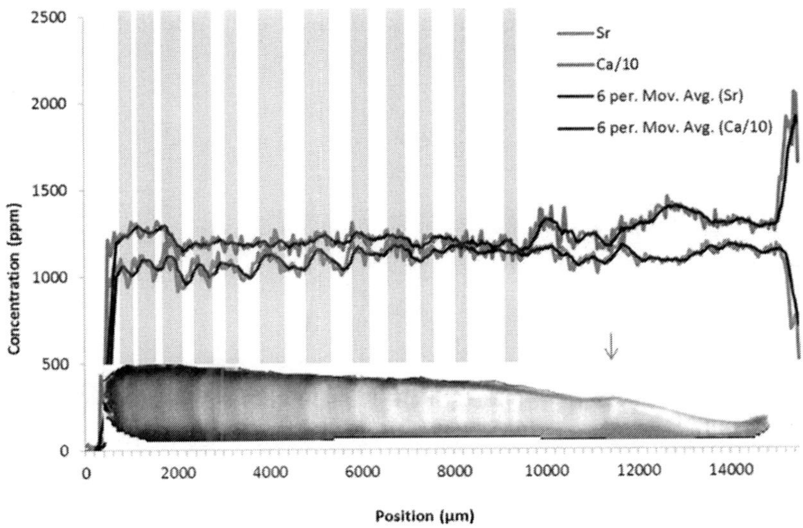

Figure 5: SXFM calcium and strontium concentrations of the *corpus calcarum* from the dorsal to ventral edge of a 2.4 m female Spinner Shark (*Carcharhinus brevipinna*). Calcium concentrations were reduced by a ratio of 10 to make them comparable to strontium concentrations. Six point moving averages were added to reduce noise. Birth mark is roughly at 11500 μm (indicated with red arrow). Data are compared to a flattened microscope image of the *corpus calcarum* of the vertebra with marked 'traditional' growth bands (grey bands that are lined up with the opaque winter bands throughout the thickness of the vertebra for comparison with strontium and calcium).

Strontium does modulate bone metabolism in mammals: higher strontium concentrations increase bone mass[24] or prevent bone loss[25]. In humans, high-strontium-content derivatives significantly decrease vertebral and non-vertebral fracture risk[26,27], offering promise as an anti-osteoporitic and anti-osteoarthritic in medicine. Recent studies that have tried to determine how strontium affects mammalian bone growth have found that nearly 50% of available strontium is incorporated into calcium hydroxyapatite or absorbed into collagen[28]. Strontium effects on mammalian cartilage are less well studied, but evidence suggests that increased strontium levels promote cartilage matrix formation[29]. While mammalian and chondrichthyan bone/cartilage development are unlikely to be completely analogous, mammalian skeletal elements are deposited on cartilaginous models, and the mineral fraction of elasmobranch vertebrae is similar in composition to that of mammals[30]. It is, therefore, probable that strontium also affects the resorption/formation of elasmobranch cartilage and vertebral bands in a similar fashion as in mammalian bones.

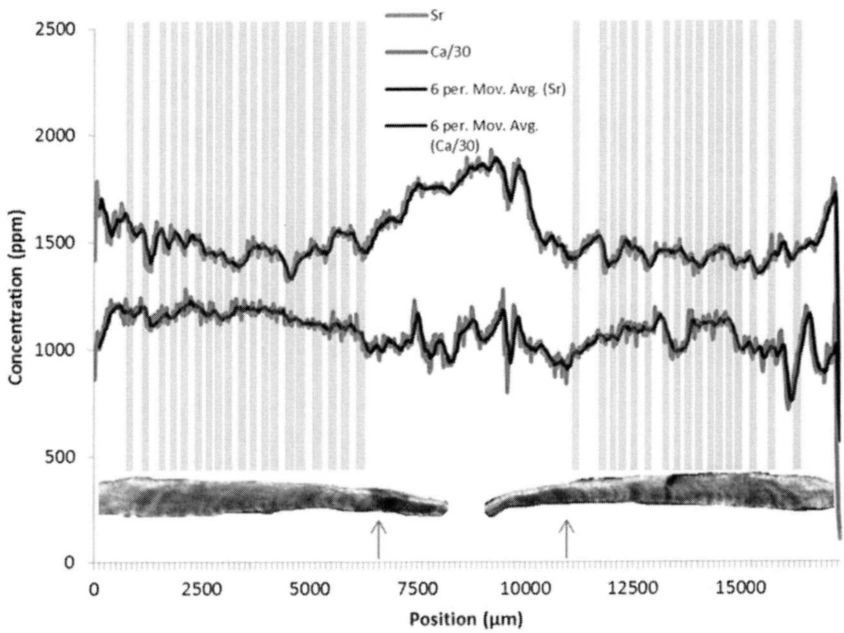

Figure 6: SXFM calcium and strontium concentrations of the *corpus calcarum* from the ventral to dorsal edge of a 1.15 m female Port Jackson Shark (*Heterodontus portusjacksoni*). Calcium concentrations were reduced by a ratio of 30 to make them comparable to strontium concentrations. 6 point moving averages were added to reduce noise. Birth marks are roughly at 7000 and 11000 µm (indicated with red arrow). Data are compared to a flattened microscope image of the *corpus calcarum* of the vertebra.

In this study, strontium peaks sometimes corresponded with thinner, opaque regions known as winter bands[31]. These areas precede areas of higher growth that have lower strontium concentrations. It is possible that the accumulation of strontium allows increased bone development in subsequent months, resulting in thinner, high strontium 'winter' bands and thicker, low strontium 'summer' bands in the sharks. The opacity of vertebral bands observed in this study may also be due to a denser cartilage matrix, similar to those detected in mammals[29].

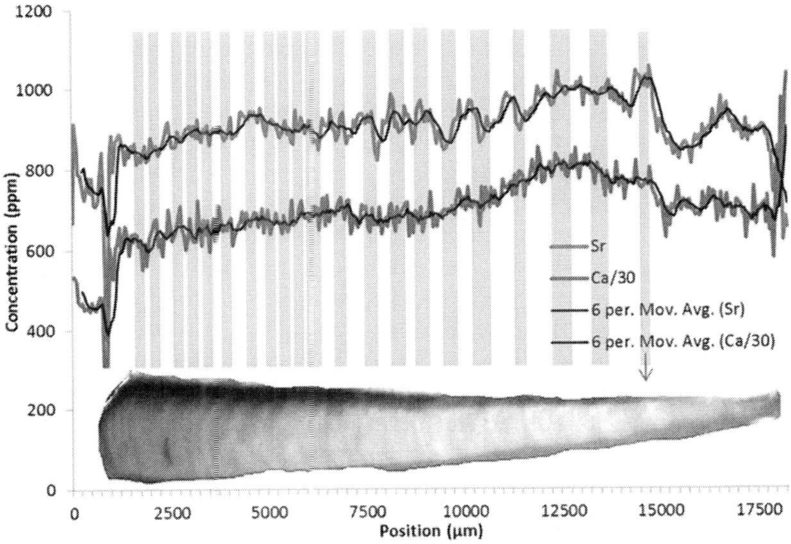

Figure 7: SXFM calcium and strontium concentrations of the *corpus calcarum* from the dorsal to ventral edge of a 2.8 m female Dusky Shark (*Carcharhinus obscurus*). Calcium concentrations were reduced by a ratio of 30 to make them comparable to strontium concentrations. 6 point moving averages were added to reduce noise. Birth mark is roughly at 14500 µm (indicated with red arrow). Data are compared to a flattened microscope image of the *corpus calcarum* of the vertebra with marked 'traditional' growth bands (grey bands that are lined up with the opaque winter bands throughout the thickness of the vertebra for comparison with strontium and calcium).

Previous research assumed that shark growth bands were caused by environmental variables such as temperature and salinity, resulting from changes in environmental conditions[32]. Spectrometry has shown that temperature has no effect on strontium concentrations in elasmobranch vertebrae, although increased temperatures increase rates of vertebral growth in these species[33]. Otolith strontium/calcium ratios in teleost fish are positively linked to environmental strontium concentrations[34], and are directly related to salinity in some species[35,36], but are usually unrelated to temperature[35,37]. Since salinity varies seasonally[38] and increases with depth[39], vertebral growth bands may be partially caused by seasonal or depth salinity gradients rather than by changes in metabolic rates resulting from seasonal variations in temperature. This also raises the possibility of determining the salinity of areas visited by the shark, as estuarine movements have been estimated from chemical signatures of teleost otoliths[40,41,42]. Similar conclusions were obtained from analyses of mass spectrometry on shark vertebral composition, where lower strontium/barium ratios in *Carcharhinus leucas* were suggested to be related to movements into estuarine waters[43]. Smalltooth sawfish (*Pristis pectinata*) also show strontium variations in relation to changes in salinity[44]. One of our model species, *Squatina albipunctata*, is thought to come inshore every year to pup or mate, which may explain the changes in strontium concentrations we observed. While correlations between strontium and visual growth bands for *Heterodontus portjacksoni* were weaker at younger ages, they more closely match strontium peaks in the animals matured. Port Jackson Sharks (*H. portjacksoni*) congregate annually for reproduction in estuaries once sexually mature[45]. It is possible that this post-maturity yearly migratory pattern may be reflected in the increased correlation between visual bands in later years. *Carcharodon carcharias* movement patterns in south-east Australia are not well understood, but tagging of young individuals (1.8–3.6 metres in total length) suggests that they undertake sporadic, large-scale movements[46], which are consistent with our findings.

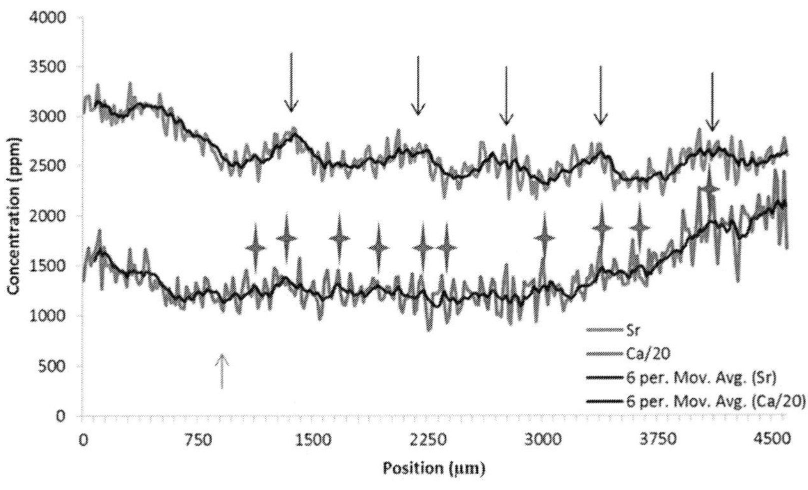

Figure 8: SXFM calcium and strontium concentrations of the *corpus calcareum* from the dorsal to ventral edge of an 839 mm male Eastern angel shark (*Squatina albipunctata*). Calcium concentrations were reduced by a ratio of 30 to make them comparable to strontium concentrations. 6 point moving averages were added to reduce noise. Strontium peaks are marked with black arrows and calcium peaks with blue stars. Birth mark is roughly at 1000 µm (indicated with red arrow). Image of light microscopy not shown as no bands were visible on the *corpus calcareum* in this study or in others52,53.

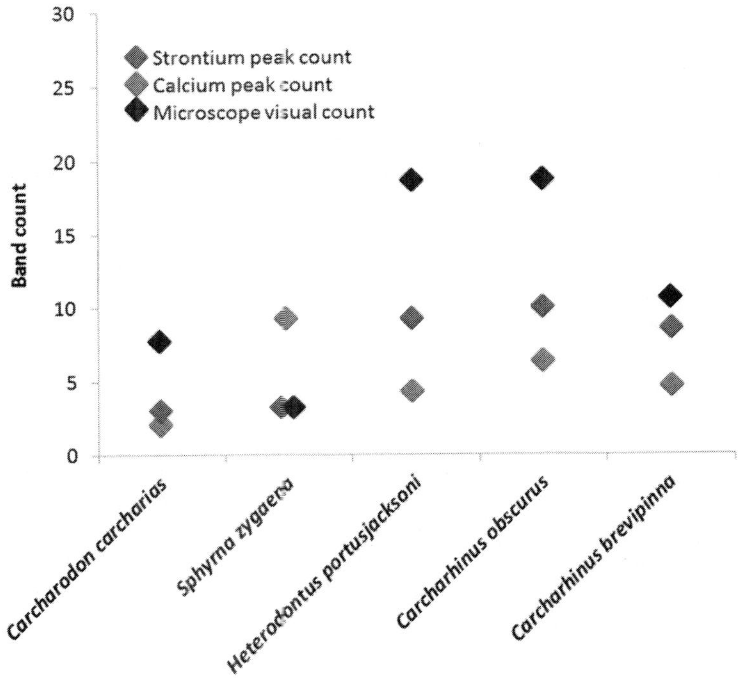

Figure 9: Counts of strontium and calcium peaks, compared to traditional microscope ageing length for (left to right): White Shark (*Carcharodon carcharias)*, Smooth Hammerhead (*Sphyrna zygaena*), Port Jackson Shark (*Heterodontus portjacksoni*), Dusky Shark (*Carcharhinus obscurus*) and Spinner Shark (*Carcharhinus brevipenna*).

Future studies should examine the potential drivers of vertebral banding and their links with environmental variables, with a particular focus on the interactions between temperature, salinity, and strontium availability.

3.2 Strontium maps

Shark ageing can be affected by observer bias[9], however, the use of strontium maps can decrease subjectivity. In conjunction with traditional ageing methods, strontium mapping may increase the accuracy of measurements and/or allow ageing where it was previously unsuccessful (e.g. *Squatina sp.*), but only when species-specific seasonal movement patterns are understood and/or the technique has been validated with individuals of known age. Round stingrays (*Urobatis halleri*) have vertebral strontium concentrations that correlate well with growth bands[47], and they also have seasonal movement patterns. Strontium deposition needs to be further investigated to determine whether it is more reliable for band analysis than methods that analyse calcium concentrations, and how/why strontium is absorbed. If strontium is a better banding indicator than calcium, it would explain why chemical stains historically used in shark ageing (alizarin red, silver nitrate) have varying results, since they are indicators for calcium, strontium, magnesium and iron. The use of directed strontium reporters may, therefore, produce better results.

The effectiveness of strontium maps as an ageing tool could be validated as a technique by using age-validated samples for each species (from tagging programs, bomb-radiocarbon dating specimens, or marking captive sharks with oxytetracycline), but may only be effective if species' migration patterns are known. Movement patterns of elasmobranchs can change with ontogeny[48], can be sex-linked[49], and may show a high degree of intraspecific variation[46]. Consequently, reliable ageing of species using strontium mapping requires highly predictable migratory patterns that occur at a population level, and at consistent intervals, and with known ontogenetic variations. Some species of sharks do not fit these criteria. For example, *C. carcharias* (White Shark) on the east coast of Australia can disperse north and south, with individual preferences[46]. Consequently, strontium maps could not be used to determine the age of *C. carcharias* off Australia. This may explain the disparity between visual band counts and strontium peak counts for the *C. carcharias* sample in this study. However, generalisations should not be made a species level. Female *C. carcharias* in the northeast Pacific have two-year migration intervals between aggregation areas, and male sharks return annually[49,50]. Due to their highly reliable movement patterns, strontium maps could be used to determine the age of *C. carcharias* from the northeast Pacific if the sex of the shark was known. Strontium mapping may thus be useful for elasmobranch vertebrae sourced from populations with known seasonal migrations that cannot be aged using traditional techniques. If population-specific information was unavailable, strontium maps would still be useful to infer movement patterns across salinity gradients, but not for aging.

3.3 SXFM

SXFM allowed novel, high resolution, and rapid analysis of the elemental composition of shark vertebrae. Nevertheless, future use of SXFM for analysis of shark vertebrae could be improved. A moving average, used to reduce noise, reduced the effective spatial resolution to six times the sampling interval (between 90 µm and 150 µm). The lowered resolution would explain why the oldest shark in this study did not display strontium peaks that corresponded to visual bands after nine years of age (Fig. 7). Scanning with a finer sampling interval, longer dwell (time spent exposing the area) and a wider sampling band are all feasible using the SXFM, but those measurement parameters would have increased scanning time. Using growth band counts to age older individuals is known to lead to age under-estimation due to the greater proximity of growth bands[47,51], and so higher resolution scans – of order 1 μm – would be of great interest. While variations in specimen thickness and surface topography can confuse the clear identification of banding, the measurements reported in this study are relatively free of such artefacts. We are currently developing additional methods to reduce the impact of measurement bias that will improve the resolution of this approach.

In summary, the presence of strontium bands in shark vertebrae supports new, comparative, alternative ageing methods once validation studies have been conducted. More precise and better-understood shark ageing will lead to more accurate depictions of life histories and population dynamics, supporting more effective management of shark populations and associated ecosystems exploited by humans.

4. MATERIALS AND METHODS

4.1 Collection of sharks

Six species of shark were analysed in this study: *Sphyrna zygaena* (Smooth Hammerhead), *Charcharodon carcharias* (White Shark), *Carcharhinus brevipinna* (Spinner Shark), *Heterodontus portusjacksoni* (Port Jackson Shark), *Carcharhinus obscurus* (Dusky Shark), and *Squatina albipunctata* (Eastern Angel shark; this genus is known to have no visible growth bands in the corpus calcarum[52,53] and little known of their biology[54]). Sharks were caught by the New South Wales (NSW) Shark Meshing (Bather Protection) Program, with the exception of *S. albipunctata*, which was caught by a commercial fishing trawler near Sydney, Australia. NSW shark nets are set during the summer period along the Newcastle to Wollongong region in New South Wales, Australia[55]. The Animal Ethics Committee at Macquarie University agreed that ethics approval was not required for this study. Tissue samples were retrieved from animals caught by commercial fisheries for sale in local markets and/or the NSW Shark Meshing (bather protection) program, which occasionally catches and kills local wildlife, and thus sharks were obtained for purposes other than research.

4.2 Laboratory preparation of vertebral sections

Vertebral samples were collected by removing the first cervical vertebra from each carcass and freezing it (−20 °C) in a sealed plastic bag until preparation. During preparation, each vertebra was thawed to room temperature, and then manually cleaned of debris with a sterile scalpel before being sectioned. To avoid possible contamination no chemical agents (e.g., sodium hypochlorite, ethanol) were used during cleaning, and the decision was made to avoid using other cleaning techniques (i.e., sonication) because the SXFM is less prone to surface contamination than other elemental techniques. Vertebrae were sectioned using an Isomet circular saw with a single 0.1 mm-increment adjustable diamond-edged blade. Sections were cut dorso-ventrically through the centre of each vertebra. Cuts were 0.6 mm in thickness or thinner, dependent on the degree calcification (more calcified specimens could be cut thinner). Each cut was immediately placed on Kapton film and covered with Kapton adhesive tape. This created an airtight seal that prevented dehydration of the samples that can cause severe tissue warping. Samples were then flattened between two microscope slides and transported to the Australian Synchrotron in Melbourne.

4.3 Scanning x-ray fluorescence imaging of vertebral sections

Samples were attached to polycarbonate frames at the Australian Synchrotron's X-ray Fluorescence Microprobe56 using clear double-sided tape, with roughly 15–20 samples per frame. Smaller samples were scanned at 15 μm intervals (e.g., *S. albipunctata*) while larger samples were scanned at 25 μm intervals (e.g., *Charcarhinid* spp., *C. carcharias, S. zygaena*). Sampling intervals were chosen to balance scanning time and measurement sensitivity with the expected length scale of elemental variations. Per-pixel dwell times were typically 10 ms, and the scanning time for each frame was between 16 and 22 h, depending on the dwell, the samples, and scanned area. Elemental maps were created and viewed using GeoPIXE57, which uses a detailed specimen model to determine first-order depth-independent elemental concentrations. A series of x-ray fluorescence concentration standards (Micromatter) were measured at regular intervals through the experiment to control for experimental drift if present; none was observed, and so all data were processed using a single fit model.

The quantitative accuracy of elemental mapping relies on the uniform illumination of a narrow column through a specimen, and the measurement of characteristic x-ray fluorescence emitted from that column. The high incident x-ray energy (18.5 keV) ensured uniform illumination; however, the lower energy of the characteristic x-ray fluorescence can result in significant absorption of the x-ray fluorescence for various elements. The absorption length of calcium fluorescence (~3.7 keV) in water (a crude model for wet cartilage) is around 110 μm, whereas that of strontium

(~14.1 keV) is 500 μm, leading to a strong dependence of the depth sensitivity on elemental species. The matrix correction applied by GeoPIXE corrects for this differential escape depth, assuming that the elemental distribution is uniform through the thickness. Here we have recorded elemental maps with the illumination closely aligned along the elemental bands so that this assumption is valid. Any sample misalignment would result in a blurring of the strontium distribution relative to calcium due to the increased depth penetration, and such is not observed in our elemental images.

To determine age, elemental lineouts were taken from data from the *corpus calcarum*, from the centre to the outer edge. Data were transformed using a moving six-sample mean to reduce noise. Sample results post-birth mark (defined as an angle change on the centrum face9) were counted for large peaks using a moving framework and a six-point moving average (more than 2.5% variation) in both the calcium and strontium spectrums.

4.4 Sample microscope imaging

Traditional visual ageing was done on 0.6 mm sections of the vertebrae directly adjacent to the vertebra used for the SXFM analysis (thus in total, two vertebrae were used for each shark in this study, for a total of 12 vertebrae and 6 species). Vertebrae were manually cleaned with a scalpel and sectioned using an Isomet diamond saw. Excess tissue remaining after the sectioning was also removed with a scalpel. No chemicals were used when preparing these vertebrae to enable effective comparison with the SXFM samples. Vertebrae were observed in saltwater under a high-contrast binocular microscope immediately after sectioning. Images were taken via USB camera and saved with ImageJ software, where optimum lighting was applied to increase the contrast of growth bands. Images of the samples were then independently aged by two experienced observers after establishing band criteria for each species. A yearly increment consisted of one opaque and one translucent band[9]. Only vertebrae where the growth bands were well defined and visible were scored. Outside experts helped determine the band counts using similar criteria of the *C. carcharias, C. brevipenna* and *C. obscurus*, as the available samples were difficult to age. These experts aged the specimens blindly, with no knowledge of the sharks' age or sex. To compare how well strontium and calcium peak counts matched the numbers of visual growth bands, data were analyzed using a multiple linear regression analysis with an alpha of 5%. All data analyses were done using Microsoft Excel and the Real Statistics data analysis tool pack.

ACKNOWLEDGEMENTS

We thank the NSW Department of Primary Industries and the Department of Biological Sciences at Macquarie University for providing funding and some of the samples for this research. A research grant from AINSE allowed a dialogue between ANSTO and Macquarie University that led to this project. This research was undertaken on the XFM beamline at the Australian Synchrotron, Victoria, Australia. Access to the Australian Synchrotron was supported under the NSW Industry Synchrotron Access Scheme, funded by the NSW Office of Science and Research. Many thanks to those involved in the SXFM: Kathryn M. Spiers, Simon A. James, and David J. Paterson. Thanks to R. Bagnato for his aid in obtaining angel shark samples. Thanks to Macquarie University for scholarship support of V.V.R., and to Michael Gillings, Marie Herberstein, Mark Westoby, and Dorrit Jacob for critiquing an earlier draft of this manuscript. Thanks to Drs. Lisa Natanson, Sabine Wintner, and Greg Skomal for aid in ageing the white shark vertebra, and thanks to Dr. Pascal Geraghty for help ageing the dusky and spinner vertebrae.

REFERENCES

1. Worm, B. & Branch, T. A. The future of fish. *Trends Ecol Evol* **27,** 594–599, doi: 10.1016/j.tree.2012.07.005 (2012).
2. Dulvy, N. K. *et al.* Extinction risk and conservation of the world's sharks and rays. *Elife* **3,** e00590 (2014).
3. Fromentin, J.-M. *et al.* The spectre of uncertainty in management of exploited fish stocks: The illustrative case of Atlantic bluefin tuna. *Mar. Pol.* **47,** 8–14 (2014).
4. Sadovy de Mitcheson, Y. *et al.* Fishing groupers towards extinction: a global assessment of threats and extinction risks in a billion dollar fishery. *Fish fish.* **14.2,** 119–136 (2013).
5. Baum, J. K. *et al.* Collapse and Conservation of Shark Populations in the Northwest Atlantic. *Science* **299,** 389–392, doi: 10.1126/ science.1079777 (2003).
6. Baum, J. K. & Myers, R. A. Shifting baselines and the decline of pelagic sharks in the Gulf of Mexico. *Ecol. Lett.* **7,** 135–145 (2004).
7. Ward-Paige, C. A. *et al.* Large-scale absence of sharks on reefs in the Greater-Caribbean: a footprint of human pressures. *PloS one* **5,** e11968 (2010).
8. Heithaus, M. R., Frid, A., Wirsing, A. J. & Worm, B. Predicting ecological consequences of marine top predator declines. *Trends Ecol. Evol.* **23,** 202–210 (2008).
9. Goldman, K. J., Cailliet, G. M., Andrews, A. H. & Natanson, L. J. In Assessing the Age and Growth of Chondrichthyan Fishes. *Biology of Sharks and Their Relatives* 13, 423 (CRC Press, 2012).
10. Campana, S. E. Accuracy, precision and quality control in age determination, including a review of the use and abuse of age validation methods. *J Fish Biol* **59,** 197–242, doi: 10.1111/j.1095-8649.2001.tb00127.x (2001).
11. Cailliet, G., Smith, W., Mollet, H. & Goldman, K. In *Special Issue: Age and Growth of Chondrichthyan Fishes: New Methods, Techniques and Analysis Vol. 25 Developments in Environmental Biology of Fishes* (ed. JohnK Carlson & KennethJ Goldman) Ch. 2, 211–228 (Springer Netherlands, 2006).

12. Cailliet, G., Martin, L., Kusher, D., Wolf, P. & Welden, B. In Proceedings international workshop on age determination of oceanic pelagic fishes: tunas, billfishes, sharks. *NOAA Tech. Rep. NMFS*. 157–165 (1983).
13. Jones, B. C. & Geen, G. H. Age determination of an elasmobranch (*Squalus acanthias*) by X-ray spectrometry. *J Fish Res Board Can*, **34,** 44–48 (1977).
14. Brown, C. A. & Gruber, S. H. Age assessment of the lemon shark, *Negaprion brevirostris*, using tetracycline validated vertebral centra. *Copeia*, 747–753 (1988).
15. Passerotti, M. S., Carlson, J. K., Piercy, A. N. & Campana, S. E. Age validation of great hammerhead shark (*Sphyrna mokarran*), determined by bomb radiocarbon analysis. *Fish Bull* **108,** 346–351 (2010).
16. Francis, M. P., Campana, S. E. & Jones, C. M. Age under-estimation in New Zealand porbeagle sharks (*Lamna nasus*): is there an upper limit to ages that can be determined from shark vertebrae? *Mar Freshwater Res* **58,** 10–23, doi: http://dx.doi.org/10.1071/MF06069 (2007).
17. Geraghty, P. T. *et al.* Age and growth parameters for three heavily exploited shark species off temperate eastern Australia. *ICES J Mar Sci*, **71,** 559–573 (2013).
18. Lombi, E., Jonge, M. D., Donner, E., Ryan, C. G. & Paterson, D. Trends in hard X-ray fluorescence mapping: environmental applications in the age of fast detectors. *Anal Bioanal Chem* **400,** 1637–1644, doi: 10.1007/s00216-011-4829-2 (2011).
19. Schoenberg, H. P. Extent of strontium substitution for calcium in hydroxyapatite. *Biochim. Biophys. Acta* **75,** 96–103, doi:http://dx.doi.org/10.1016/0006-3002(63)90583-3 (1963).
20. Farrell, J. & Campana, S. E. Regulation of calcium and strontium deposition on the otoliths of juvenile tilapia, *Oreochromis niloticus*. *Comp Biochem Physiol* **115,** 103–109 (1996).
21. Campana, S. E. Chemistry and composition of fish otoliths: pathways, mechanisms and applications. *Mar Ecol-Prog Ser* **188,** 263–297 (1999).
22. Yokouchi, K. *et al.* Time lag of the response on the otolith strontium/calcium ratios of the Japanese eel, *Anguilla japonica* to changes in strontium/calcium ratios of ambient water. *Environ Biol Fish* **92,** 469–478 (2011).
23. Hartl, M. G. Homeostasis and the toxicology of non-essential metals-Edited by C. M. Wood, A. P. Farrell & C. J. Brauner. *J Fish Biol* 83, 1476–1477 (2013).
24. Delannoy, P., Bazot, D. & Marie, P. J. Long-term treatment with strontium ranelate increases vertebral bone mass without deleterious effect in mice. *Metabolis* **51,** 906–911, doi: http://dx.doi.org/10.1053/meta.2002.33360 (2002).
25. Marie, P. J. *et al.* An uncoupling agent containing strontium prevents bone loss by depressing bone resorption and maintaining bone formation in estrogen-deficient rats. *J Bone Miner Res* **20,** 1065–1074 (2005).
26. Seeman, E. *et al.* Five years treatment with strontium ranelate reduces vertebral and nonvertebral fractures and increases the number and quality of remaining life-years in women over 80 years of age. *Bone* **46,** 1038–1042, doi: http://dx.doi.org/10.1016/j.bone.2009.12.006 (2010).
27. Roux, C. *et al.* Strontium ranelate reduces the risk of vertebral fracture in young postmenopausal women with severe osteoporosis. *Ann Rheum Dis* **67,** 1736–1738 (2008).
28. Frankar, C., Raffalt, A. & Stahl, K. Strontium Localization in Bone Tissue Studied by X-Ray Absorption Spectroscopy. *Calcif Tissue Int* **94,** 248–257, doi: 10.1007/s00223-013-9806-7 (2014).
29. Henrotin, Y. *et al.* Strontium ranelate increases cartilage matrix formation. *J Bone Miner Res* **16,** 299–308 (2001).
30. Porter, M. E., Beltran, J. L., Koob, T. J. & Summers, A. P. Material properties and biochemical composition of mineralized vertebral cartilage in seven elasmobranch species (Chondrichthyes). *J Exp Biol* **209,** 2920–2928 (2006).
31. Davenport, S. & Stevens, J. Age and growth of two commercially imported sharks (*Carcharhinus tilstoni* and *C. sorrah*) from Northern Australia. *Mar Freshwater Res* **39,** 417–433 (1988).

32. Branstetter, S. & Stiles, R. Age and growth estimates of the bull shark, *Carcharhinus leucas*, from the northern Gulf of Mexico. *Environ Biol Fish* **20**; 169–181, doi: 10.1007/bf00004952 (1987).
33. Smith, W. D., Miller, J. A. & Heppell, S. S. Elemental Markers in Elasmobranchs: Effects of Environmental History and Growth on Vertebral Chemistry. *PloS one* **8,** e62423 (2013).
34. Kennedy, B. P., Folt, C. L., Blum, J. D. & Chamberlain, C. P. Natural isotope markers in salmon. *Nature* **387,** 766–767 (1997).
35. Kawakami, Y. *et al.* Factors influencing otolith strontium/calcium ratios in *Anguilla japonica* elvers. *Environ Biol Fish* **52,** 299–303, doi: 10.1023/a:1007415420540 (1998).
36. Kraus, R. T. & Secor, D. H. Incorporation of strontium into otoliths of an estuarine fish. *J Exp Mar Biol Ecol* **302,** 85–106 (2004).
37. Clarke, L. M. & Friedland, K. D. Influence of growth and temperature on strontium deposition in the otoliths of Atlantic salmon. *J Fish Biol* **65,** 744–759, doi: 10.1111/j.0022-1112.2004.00480.x (2004).
38. Delcroix, T. & Henin, C. Seasonal and interannual variations of sea surface salinity in the tropical Pacific Ocean. *J Geophys Res- Oceans* **96,** 22135–22150, doi: 10.1029/91jc02124 (1991).
39. Mantyla, A. W. & Reid, J. L. Abyssal characteristics of the World Ocean waters. *Deep-Sea Res, Part A* **30,** 805–833, doi: http://dx.doi.org/10.1016/0198-0149(83)90002-X (1983).
40. Bath, G. E. *et al.* Strontium and barium uptake in aragonitic otoliths of marine fish. *Geochim Cosmochim Ac* **64,** 1705–1714, doi: http://dx.doi.org/10.1016/S0016-7037(99)00419-6 (2000).
41. Walther, B. & Limburg, K. The use of otolith chemistry to characterize diadromous migrations. *J Fish Biol* **81,** 796–825 (2012).
42. Elsdon, T. S. *et al.* Otolith chemistry to describe movements and life-history parameters of fishes: hypotheses, assumptions, limitations and inferences. *Ocean Mar Biol: Ann Rev* **46,** 297–330 (2008).
43. Tillett, B. J. *et al.* Decoding fingerprints: elemental composition of vertebrae correlates to age-related habitat use in two morphologically similar sharks. *Mar Ecol-Prog Ser* **434,** 133–142 (2011).
44. Scharer, R. M., Patterson Iii, W. F., Carlson, J. K. & Poulakis, G. R. Age and Growth of Endangered Smalltooth Sawfish (*Pristis pectinata*) Verified with LA-ICP-MS Analysis of Vertebrae. *PLoS ONE* **7,** e47850, doi: 10.1371/journal.pone.0047850 (2012).
45. Powter, D. & Gladstone, W. The reproductive biology and ecology of the Port Jackson shark *Heterodontus portusjacksoni* in the coastal waters of eastern Australia. *J Fish Biol* **72,** 2615–2633 (2008).
46. Bruce, B., Stevens, J. & Malcolm, H. Movements and swimming behaviour of white sharks (*Carcharodon carcharias*) in Australian waters. *Mar Biol* **150,** 161–172 (2006).
47. Hale, L. F., Dudgeon, J. V., Mason, A. Z. & Lowe, C. G. In *Special Issue: Age and Growth of Chondrichthyan Fishes: New Methods, Techniques and Analysis* 317–325 (Springer, 2006).
48. Andrews, K. S., Williams, G. D. & Levin, P. S. Seasonal and ontogenetic changes in movement patterns of sixgill sharks. *PLoS One* **5,** e12549 (2010).
49. Domeier, M. L. & Nasby-Lucas, N. Two-year migration of adult female white sharks (*Carcharodon carcharias*) reveals widely separated nursery areas and conservation concerns. *Anim Biotel* **1,** 1–10, doi: 10.1186/2050-3385-1-2 (2013).
50. Domeier, M. & Nasby-Lucas, N. Annual re-sightings of photographically identified white sharks (*Carcharodon carcharias*) at an eastern Pacific aggregation site (Guadalupe Island, Mexico). *Mar Biol* 150, 977–984, doi: 10.1007/s00227-006-0380-7 (2007).
51. Andrews, A. H., Natanson, L. J., Kerr, L. A., Burgess, G. H. & Cailliet, G. M. Bomb radiocarbon and tag-recapture dating of sandbar shark (Carcharhinus plumbeus). *Fishery Bulletin* **109,** 454–465 (2011).
52. Baremore, I. E., Andrews, K. I. & Hale, L. F. Difficulties associated with modeling growth in the Atlantic angel shark (*Squatina dumeril*). *Fish Res* **99,** 203–209 (2009).
53. Natanson, L. J. & Cailliet, G. M. Vertebral growth zone deposition in Pacific angel sharks. *Copeia,* 1133–1145 (1990).

54. Raoult, V., Peddemors, V. & Williamson, J. E. Biology of angel sharks (*Squatina* sp.) and sawsharks (*Pristiophorus* sp.) caught in south-eastern Australian trawl fisheries and the New South Wales shark-meshing (bather-protection) program. *Mar Freshwater Res* http://dx.doi.org/10.1071/MF15369 (2016).
55. Reid, D., Robbins, W. & Peddemors, V. Decadal trends in shark catches and effort from the New South Wales, Australia, Shark Meshing Program 1950–2010. *Mar Freshwater Res* **62,** 676–693 (2011).
56. Paterson, D. *et al.* The X-ray Fluorescence Microscopy Beamline at the Australian Synchrotron. Paper presentat at *The 10th international conference on X-ray microscopy, Chicago.* 219–222 (AIP Publishing, 2010).
57. Ryan, C. Quantitative trace element imaging using PIXE and the nuclear microprobe. *Int J Imag Sys Tech* **11,** 219–230 (2000).

CHAPTER 6

Impact of Providing Feed and/or Water on Performance, Physiology, and Behavior of Weaned Pigs during a 32-h Transport

Arlene Garcia [1,2], Mhairi Sutherland [3], Glenna Pirner [1], Guilherme Picinin [1], Matthew May [1], Brittany Backus [1] and John McGlone [1,*]

[1] *Department of Animal and Food Sciences, Texas Tech University, Lubbock, TX 79409, USA*

[2] *Department of Agricultural and Natural Resources, University of Minnesota, Crookston, MN 56716, USA*

[3] *Ruaura Research Centre, AgResearch Ltd., Hamilton 3214, New Zealand*

SIMPLE SUMMARY

Transportation has the potential to negatively affect the health and welfare of weaned pigs, especially those already experiencing weaning stress. Piglets were transported for 32 h, with and without feed and water, and measures of performance, physiology, and behavior were taken to assess piglet welfare. Transportation negatively impacted body weight, Neutrophil to Lymphocyte Ratio (N:L), and post-transport body weight gain, indicating that not providing water during transport can negatively impact the well-being of recently weaned pigs. Provision of water may aid in reducing stress during long distance transport and improve the animals' well-being.

ABSTRACT

Transportation at weaning is a complex stressor made up of many factors, including withdrawal from feed and water, which can potentially negatively affect the health and welfare of pigs, especially those already experiencing weaning stress. The objective of this study was to evaluate the effect of weaning and extended transport durations (up to 32 h), with and without the provision of feed and/or water, on pig welfare. Treatment groups included: pigs neither weaned nor transported, control (CON); weaned pigs transported and provided with feed and water (T+); weaned pigs transported without feed and water (T−); weaned pigs transported with only feed (T+F); and weaned pigs transported with only water provided (TRAN+W). The effect of transport (with and without feed and/or water) on weaned pigs was assessed using behavior, performance, and physiology. After a 32-h transport period, pigs transported without water lost markedly more weight than those transported with water ($p < 0.01$). Furthermore, the neutrophil to lymphocyte ratio was markedly higher in male pigs transported without water ($p < 0.05$). Overall, transportation had a negative effect on pig well-being, especially when water was not provided.

Keywords: animal welfare; behavior; transport; weaned pigs

1. INTRODUCTION

Road transport of weaned piglets from rearing to growing-finishing farms is a common practice in the swine industry [1]. Transportation is a complex stressor made up of many factors including fluctuating temperatures, stocking density, withdrawal from feed and water, mixing with unfamiliar pigs and motion [2]. Therefore, transportation has the potential to affect the health and welfare of pigs, especially in pigs already experiencing weaning stress. In Europe, animals are often transported between different countries. However, current European Union regulations limit transport of animals to 8 h, although pigs can be transported for up to 24 h as long as water is provided [1,3]. Currently in the United States, animals can be transported consecutively for 28 h without the need to be off-loaded, or provided with feed, water, or rest [4].

This 28 h law was enacted in 1873 and applied to rail transportation of cattle, sheep, swine, and other animals and was later amended to include transportation by express or common carriers involving confinement in a "vehicle or vessel" [4]. It is important to provide scientific validation for recommendations to enhance the welfare of animals. However, the 28 h law was not based on science or current common transport methods for weaned pigs. Therefore, research is needed to develop science-based recommendations for transporting weaned pigs that is relevant to modern conditions.

Transportation of pigs at weaning is becoming more common in the U.S. and is an essential aspect of modern swine production [1,5,6]. Weaned piglets are typically transported from short to very long distances from rearing to growing-finishing farms typically to reduce vertical transfer of disease, to enhance production, and overall farm efficacy [6] and the duration of transport can vary from short to very long distances. Transport of weaned pigs has many common factors with transport of finishing pigs, including mixing, fasting, temperature fluctuations, vibration, novelty and noise [7]. Similar to studies in finishing pigs, transport has been shown to cause behavioral and physiological changes indicative of stress in weaned pigs [6]. However, there is limited information regarding the effect of transport duration and the associated effect of feed and water deprivation on the welfare of weaned pigs. Weaning itself is a stressful event for piglets. Piglets are separated from the sow, their diet changes from milk to solid feed, they are mixed with unfamiliar pigs, are exposed to a novel environment, and all these factors can contribute to stress [8]. Weaning and its associated stressors can lead to a period of underfeeding, growth retardation and metabolic changes in pigs until they adapt [8,9]. In addition, weaning can affect the immune system, performance and behavior of pigs [10,11]. Therefore, pigs already experiencing stress due to weaning maybe more vulnerable to additional stressors, such as transport.

Furthermore, there is limited information regarding the effect of transport duration and the associated effect of feed and water deprivation on the welfare of weaned pigs. The current study is a continuation of a previous study conducted by this laboratory, where we concluded that feed and water should be provided to weaned pigs if transport will be over 24 h, based on measures of body weight, physiology, and behavior [12]. Hence, it is necessary for the industry to understand the impact of transport on the welfare of weaned pigs so that standards can be put into place to reduce any detrimental effects associated with transporting pigs at weaning.

The objective of this study was to evaluate the effect of weaning and extended transport durations (up to 32 h), with and without feed and/or water, on pig welfare using measures based on behavior, performance, and physiology. Transport times up to 32 h were assessed to provide scientifically based information to determine if the 28 h law in the USA is appropriate for weaned pigs.

2. EXPERIMENTAL SECTION

Pigs were PIC USA (Hendersonville, TN, USA) genetics using the Camborough-22 sow line. All animals were fed a diet to meet or exceed National Research Council (NRC) nutrient requirements. Feed and water were provided *ad libitum* or according to treatment. All animal procedures were approved by the Texas Tech University Animal Care and Use Committee (13040-06).

Pigs were ear notched, tail docked, needle teeth clipped, and males were surgically castrated at 3 d of age. Pigs were ear tagged 3 d prior to the beginning of the study for identification purposes. At 18 to 22 d of age (the average weaning age of pigs on commercial swine farms in the U.S.), pigs were randomly assigned to one of five treatment groups. Pigs were selected by date of birth ±2 d and weight (in order to get similar weights among groups). Each treatment group weighed 6.0 ± 0.58 kg when selected for the study. Both gilts and barrows were evenly represented in each treatment group (2 sexes × 5 treatments = 10 pigs × 6 replications = 60 pigs). Pigs were randomly selected for one of five treatment groups:

1) Not weaned or transported: pigs remained with the sow (CON); n = 12.
2) Weaned, transported, but had access to feed and water: pigs were removed from the sow and transported for 32 h in pens that had feed and water (T+); n = 12.
3) Weaned, transported but had no access to feed and water: pigs were removed from the sow and transport for 32 h in pens that did not have feed and water (T−); n = 12.
4) Weaned, transported with access to feed: pigs were removed from the sow and transported for 32 h in pens with access to feed only (T+F); n = 12.
5) Weaned and transported with access to water: pigs were removed from the sow and transport for 32 h in pens with access to water only (T+W); n = 12.

The study was conducted in the United States during early fall (avoiding temperatures that were too cold or too hot, to prevent thermal stress). The study began at 07:00 h and ended by 16:00 h the following day. At weaning, blood samples were taken from each individual control (CON) pig and pigs were either left in the farrowing pen with the sow or placed in one of the four treatment groups. All pigs were blood sampled by jugular venipuncture, weighed, checked for lesions or lameness at 0 h and every 8 h thereafter up to 32 h. Feed and/or water intake was measured every 8 h.

For treatment groups two to five, two pigs (one barrow and one gilt) from the same litter were randomly placed in pens (0.76 m × 0.76 m) with wood shavings, approximately 10 cm in depth, in a goose-neck trailer (6 m × 2 m). Space allowance was determined based on the Transport Quality Assurance [13] guidelines for transporting pigs (0.19 m^2/head) and the additional space (0.14 m × 0.14 m) was added to provide space for feed and water. HOBO® (Onset Computer Corporation, Bourne, MA, USA) data loggers were placed on each experimental pen to record temperature and humidity in the trailer at the level of the pig. The average temperature in the trailer was 25.5 °C and relative humidity inside the trailer ranged from 17.9% to 83.5%.

Pigs were transported for 32 h on main highways for the majority of the trip, which were sealed, predominantly straight, and involved minimal stopping/acceleration or inclines/declines. The 32 h of transport were broken down into 8-h phases, in which the truck/trailer would return to the original farm site for sampling purposes. Sampling took approximately 15 to 25 min and involved weighing, bleeding, and assessing pigs for injuries. Feed (starting at 2.5 kg) and water (starting at 5 L) consumption were determined by weighing the feed containers and measuring volume in water jugs every 8 h for the 32 h treatment period. Travel time continued during the sampling periods, so that sampling took place at 0, 8, 16, 24 and 32 h. At the end of the 32-h transport period, all pigs including CON pigs were placed in weaning pens (1.5 m × 1.5 m) with the same pen mate to avoid further stress caused by mixing of unfamiliar pigs.

To assess piglet welfare, measures of performance (body weights, health, and injury), physiology, and behavior (during transport, at the farm, and post-transport) were measured.

Performance: Piglets were weighed by placing them in a plastic tub that was set on top of a digital bench scale (OHAUS, Melrose, MA, USA). The pigs were weighed before (0 h), during (at 8, 16, and 24 h), immediately after (32 h), and at 7- and 14-d post study period. Percent weight change was calculated. Before, during, and after transport each pig was examined for injuries, wounds, abscesses, and lameness.

Physiology: Pigs were placed on their backs in a V-trough with their forelegs and hind legs manually restrained by trained personnel for blood collection, which took on average 2 to 3 min. Two milliliters of blood was collected through jugular venipuncture into vacutainers (BD Vacutainers®, Becton, Dickinson and Company, Franklin Lakes, NJ, USA) containing 5.4 mg of K2 EDTA and an additional 2 mL were collected into vacationers (BD Vacutainers®, Becton, Dickinson and Company, Franklin Lakes, NJ, USA) without additives, for blood analysis. Four mL in total were collected from every piglet every 8 h for 32 h. Whole blood was examined for hematocrit (HCT), and neutrophil to lymphocyte ratio (N:L) within 1 h after collection. Blood samples were centrifuged and plasma and serum were collected and frozen until further analysis for cortisol concentrations. Cortisol concentrations were analyzed using an enzyme immunoassay kit (Enzo® Life Sciences, Farmingdale, NY, USA). The resulting intra- and inter-assay coefficient of variation (CV) were 6% and 10.4% respectively. The sensitivity of the assay was 1 pg/mL.

Behavior: Piglet behavior was recorded for all treatment groups during and 24 h after the study. Lying, standing, sitting, drinking, and eating behaviors were recorded according to Garcia *et al.* [12] (Table 1). To record behaviors of all transported treatment groups during transport, wild life cameras (Moultrie Products, Alabaster, AL, USA) were placed across from each experimental pen. The behavior of four pigs (2 pens) were recorded per frame and 10 min scan samples were done for each individual pen. The cameras were motion activated and instantaneous shots were taken upon activation (movement of the pigs). Digital video recorders (DVRs) (Supercircuits VSS Series, Supercircuits®, Austin, TX, USA) were used to record behaviors of CON pigs in farrowing pens. Post-transport pigs were penned with the same pen mate they had during transport. Video was also taken for 24 h post-transport for all groups and quantified to examine potential treatment effects. Videos from DVRs were analyzed in 10 min scan samples in 2 h intervals.

Table 1. Ethogram, description of behaviors during and post-treatment.

Behavior	Definition
Lying	Animal is recumbent, flat on its side, or ventrally
Standing	Animal is upright on all fours, legs are extended
Sitting	Resting on the caudal part of the body with the forelimbs extended
Drinking	Animal placing its mouth on the water nipple and consuming water
Eating/Nursing	Animal placing its mouth in the feed container and consuming feed or latching on to the sows' nipple and suckling

Statistical Analysis: The study was a Complete Randomized Design. The pen was the experimental unit, consisting of one gilt and one barrow. All data were tested for homogenous variances and departures from normal distribution using the univariate procedure of SAS version 9.1 SAS (SAS, 2010 SAS Inst., Inc., Cary, NC, USA). All data were subjected to analysis of variance using the general linear model procedure of SAS. The statistical model included the main fixed effects of sex, treatment, pen, time, and all possible interactions. Pen within treatment was used as the model's error term. An F-protected Least Significant Difference test was also used within SAS. Behavior was recorded using 10 min scan samples in 2-h intervals.

3. RESULTS AND DISCUSSION

3.1. Performance

3.1.1 Body Weight Loss

The treatment by time interaction was significant for percent change in body weights ($p < 0.01$). All treatments groups were different from CON starting at 8 h of transport (Table 2). All transported treatment groups displayed similar losses in body weight from 0 to 24 h. However, by 32 h of transport the T− and T+F groups had lost more weight compared to T+, T+W and the CON group ($p < 0.05$).

Table 2. Least Squares means ±0.09 for percent change in body weight over 32 h for pigs weaned and transported or neither weaned nor transported ($p < 0.01$). Treatments: neither weaned nor transported (CON), weaned and transported; with feed and water provided (T+), without feed and water provided (T−), only feed provided (T+F), and with only water provided (T+W). $n = 12$ pigs/treatment.

Time, h	Treatments				
	CON	T+	T−	T+F	T+W
0	0	0	0	0	0
8	6.09 ^{a}	−2.43 ^{b}	−2.48 ^{b}	−2.45 ^{b}	−2.86 ^{b}
16	9.04 ^{a}	−3.80 ^{b}	−4.65 ^{b}	−4.58 ^{b}	−3.92 ^{b}
24	11.24 ^{a}	−4.91 ^{b}	−6.11 ^{b}	−5.83 ^{b}	−4.78 ^{b}
32	13.80 ^{a}	−5.66 ^{b}	−7.44 ^{c}	−7.51 ^{c}	−5.51 ^{b}

$^{a-c}$ Means with different superscripts within each time point differ at $p < 0.05$.

These data suggest that water withdrawal during transport has a greater effect on weight loss than feed withdrawal. Weight loss was more than likely due to dehydration, and therefore water withdraw may be more important than feed withdrawal and the effect increases with duration of transport. Similarly, Garcia *et al.* [12] found that pigs transported and/or weaned without the provision of feed and water lost significantly more weight than control pigs and pigs that were weaned and transported and provided with feed and water. These findings are also consistent with results from studies investigating slaughter weight pigs that have been transported and fasted for 25 to 48 h [6,7,8,14,15,16,17,18,19,20,21]. However in the present study, with the exception of CON pigs, percent change in body weight was similar among treatment groups prior to 32 h. These results are in general similar to the findings of Garcia *et al.* [12], who found differences in body weight loss between pigs transported with and without the provision of feed and water within 24 h of transport.

3.2. Physiology

3.2.1. Neutrophil to Lymphocyte Ratio

N:L was not significantly different among treatment groups. However, a significant treatment by sex interaction for N:L was observed ($p < 0.01$). T+F males had higher N:L than T+F females ($p < 0.05$; Figure 1). T− males also exhibited higher N:L than T− females ($p < 0.05$). Whereas, the N:L was similar among male and female pigs transported with water. The current findings suggest that male pigs that did not have access to water or feed during transport may have experienced more stress than females in response to the same stressors. Differences in the stress response between genders could potentially be explained by the differences in their early life experiences. For example, male pigs experience a greater degree of stress at processing due to surgical castration than females. The stress caused by castration may predispose male pigs to be more responsive to future stressors, such as human handling, due to conditioning effects causing pigs to associate handling with acute pain [22]. Literature is limited on whether painful processes, such as castration in pigs, applied early in life increases pain perception later on. It has been documented that circumcision in young boys is associated with greater pain perception at vaccination than uncircumcised boys [23]. Therefore, the negative experience of castration may cause an increase in stress during handling, blood sampling, and transportation compared to females. However, further investigation is needed to understand the relationship between pig gender, castration and the stress response associated with transport (with and without feed/water).

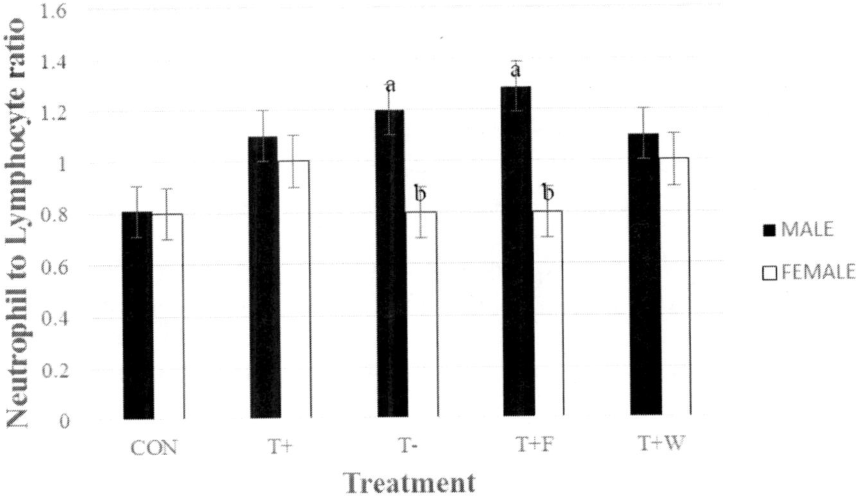

Figure 1. Least Squares means ± SEM for neutrophil to lymphocyte ratio (N:L) for sex by treatment interaction over 32 h for pigs weaned and transported or neither weaned nor transported ($p < 0.01$). Treatments: neither weaned nor transported (CON), weaned and transported; with feed and water provided (T+), without feed and water provided (T−), only feed provided (T+F), and with only water provided (T+W). $n = 12$ pigs/treatment. Superscripts without a common letter within treatment group differ at $p < 0.05$.

Transport and weaning individually have been shown to influence the immune response, performance, physiology, and behavior of young pigs [10,11,24]. Weaning alone, which involves transitioning from milk to solid feed, fasting, and being mixed with unfamiliar pigs, can lead to an increase in the N:L. Studies on the effect of fasting on the N:L have been reported [7,8,18,19,20,21,23,25,26,27] but there are limited studies that describe the effect of providing feed and/or water during transport on the physiology of recently weaned pigs.

3.2.2. Hematocrit

HCT values were not different among the treatment groups ($p > 0.05$). The values for the CON, T+, T−, T+F, and T+W groups were 28.70% ± 1.61%, 30.32% ± 1.70%, 29.23% ± 1.72%, 28.29% ± 2.5%, and 28.85% ± 1.91%, respectively. Previous evaluation of the effects of 6, 12, 18, 24, or 30 h of transport in gilts showed that HCT values did not increase in response to transport [20], but other measure of dehydration such as albumin and total protein concentrations were elevated. In a similar study by Garcia *et al.* 2015 the effect of long distance transport in weaned pigs for up to 32 h showed that both albumin and HCT values did not increase regardless of whether water and feed were provided. These finding are in contrast with previous reports where hematocrit values have been reported to increase during transport of varying durations [1,20,28].

3.2.3. Cortisol

Cortisol concentrations were not different among the treatment groups ($p > 0.05$). The values for the CON, T+, T−, T+F, and T+W groups were 20.32 ± 8.90, 44.72 ± 11.55, 29.09 ± 8.78, 41.97 ± 8.93 and 28.89 ± 8.90 pg/mL, respectively. Cortisol concentrations have been reported to increase in response to transport stress [20,21]. Averos *et al.* [1] did not find any cortisol differences in weaned pigs being transported regardless of short (0.6 h) or long distance transport (8.3 h). In the current study, just as in other reports, it seems that stress associated with handling may have been greater than any potential treatment effects [12,24].

3.3. Behavior

3.3.1. During Transport

During the 32 h transport period, CON pigs spent 58% of their time lying. T+, T−, T+F, and T+W pigs spent 66%, 67%, 73%, and 54% of their time lying, respectively. Lying behavior varied among treatments over time, but T+F displayed more differences than the other groups (Table 3). T+F spent more time lying than CON pigs during 2 h, 6 h, 16 h, and 32 h of transport ($p < 0.05$). T− displayed lying behavior differences compared to the CON at 14 h, 18 h, and 32 h; T+ at 10 h and 32 h; and T+W at 22 h and 32 h of transport ($p < 0.05$).

T+ spent more time lying than T+W at 10 h, 18 h, 20 h, and 22 h of transport, but were similar to the CON except at 10 h of transport, where T+ spent more time lying than both T+W and CON ($p < 0.05$). T+W spent less time lying than T+ but were similar to the CON, except at 22 h, where T+W spent less time lying than both the T+ and CON ($p < 0.05$). T− spent similar amounts of time lying to T+F, except at 2 h of transport, where both T− and CON spent less time lying than T+F ($p < 0.05$). All transported treatment groups spent more time lying compared to CON pigs at 32 h ($p < 0.05$). A diurnal cycle, although difficult to interpret can still be identified from 18 h to 22 h.

Table 3. Least squares means ±2.13 for treatment by time interaction for lying behavior over 32 h for pigs weaned and transported or neither weaned nor transported ($p < 0.05$). Treatments: neither weaned nor transported (CON), weaned and transported; with feed and water provided (T+), without feed and water provided (T−), only feed provided (T+F), and with only water provided (T+W). $n = 12$ pigs/treatment.

Time, h	Treatments				
	CON	T+	T−	T+F	T+W
2	8.67 ^{a}	12.67 ^{a}	13.33 ^{a}	14.67 ^{b}	9.17 ^{a}
4	12.17 ^{a,b}	9.83 ^{a}	16.0 ^{b}	19.17 ^{b}	11.50 ^{a,b}
6	9.67 ^{a}	12.00 ^{a}	15.00 ^{a,b}	18.00 ^{b}	9.50 ^{a}
8	11.33 ^{a,b}	9.57 ^{a,b}	13.00 ^{a,b}	15.17 ^{a}	9.17 ^{b}
10	12.33 ^{a,c}	19.50 ^{b}	15.50 ^{a}	16.00 ^{a}	8.67 ^{c}
12	13.17	14.00	11.67	15.00	12.67
14	15.67 ^{a}	12.50 ^{a,b}	9.50 ^{b}	13.67 ^{a,b}	14.83 ^{a,b}
16	10.33 ^{a}	14.33 ^{a,b}	14.17 ^{a,b}	19.50 ^{b}	12.67 ^{a}
18	16.67 ^{a,c}	19.67 ^{a}	22.50 ^{b}	20.00 ^{a,b}	11.50 ^{c}
20	17.17 ^{a,b}	22.00 ^{a}	21.50 ^{a}	18.50 ^{a,b}	14.17 ^{b}
22	20.83 ^{a}	21.00 ^{a}	22.00 ^{a}	18.00 ^{a,b}	13.17 ^{b}
24	13.00	13.67	15.50	16.00	11.50
26	13.50	15.50	11.50	14.67	13.33
28	17.00	19.50	17.00	19.83	14.83
30	19.00	17.67	18.83	19.50	20.50
32	9.26 ^{a}	18.33 ^{b}	21.67 ^{b}	23.83 ^{b}	19.00 ^{b}

$^{a–c}$ Means with different superscripts within each time point differ at $p < 0.05$.

Transported pigs may spend more time lying due to the unstable conditions caused by transport. Garcia *et al.* [12] similarly found that recently weaned pigs, transported with or without the provision of feed and water, spent more time lying than control (non-transported) pigs. Alternatively, increased lying behavior may be indicative of increased welfare, as these pigs may be less restless or habituated to their environment.

The current study has similar results to other studies where weaned pigs have been previously reported to spend most of their time lying and standing during transport, 75.6% and 21.6%, respectively [7]. It has also been reported that high levels of resting after weaning is common in pigs, which could be associated with the increased lying behavior during transport [27].

CON pigs spent 16% of their time standing during the 32-h study period. T+, T−, T+F, and T+W pigs spent 20%, 25%, 18%, and 21% of their time standing, respectively.

Sitting behavior was not observed in CON pigs but was observed in all other treatment groups. Both T+ and T+F pigs spent 3% of their time sitting during transport while T+W and T− spent 2% of their time sitting.

In addition, as transport time increases, pigs tend to spend more time lying [8,28]. Sitting is a behavior that has been identified as a potential stress indicator [5], and has been reported to be more common during the first 12 h of transport (2.8%) than in the second 12 h (0.3%) [8]. Increased standing or sitting behavior has also been suggested as an indication of competition for space or instability due to too much space [29]. However, neither was a problem in the present study due to the space allowance used. Increased resting and a decrease in sitting behavior later in transport may indicate that piglets have become habituated to some of the elements of transport [7].

It is important to also consider that the type of vehicle used to transport weaning pigs, as they may have a different effect on transport behaviors. In the current study, a 6 m × 2 m livestock trailer was used, which is different to what may typically be used in a commercial setting (in the USA straight deck and potbelly livestock trailers may range from 14.7 m to 16.2 m long and 2.6 m wide). There may be differences between large and small trailers with respect to movement related behavior of the vehicle, just like there are differences in morbidity and mortality associated with large commercial trailers [30]. The main aim of this study was to evaluate the effect of feed/water withdrawal, so, in this instance, the dimensions of the trailer used in the study are less critical. However, the construction of the vehicle and driving style is still very important. Driving style may overrule all other precautions to protect piglet welfare. Additionally, raising the standards of the type of vehicle used in the U.S. to transport piglets, such as special suspensions, heating/air conditioning, feeding and watering for long distance transport like is currently used in Europe can help improve weaned pig welfare. Based on the current findings, larger studies on commercial farms using more commercially applicable trailers are needed to substantiate/confirm these results.

3.3.2. Feed and Water Consumption during Transport

CON pigs were observed to spend 23% of their time nursing during the 32 h study period. T+F pigs spent 1% of their time eating while T+ pigs spent 2% of their time eating. There was no significant treatment effect for feed consumption, both T+ and T+F groups ate similar amounts of feed, 0.19 ± 0.77 kg and 0.29 ± 0.09 kg, respectively. However, a treatment by time interaction was observed for pigs provided with feed ($p < 0.05$; Table 4). T+F pigs spent more time eating at 10 h of transport compared to T+ pigs. Pigs consumed more feed at 24 h of transport ($p <$

0.05) compared to 8, 16, and 32 h of transport. Feed consumption over the 32 h transport period starting at 8 h was 0.06 kg, 0.06 kg, 0.64 kg, and 0.20 kg, respectively. Drinking behaviors were not observed during transport, or may have been missed due to video being scanned in 10 min intervals. Occasionally equipment failure occurred resulting in periods of missed data collection, which may also have been a reason for the low observed occurrence of feeding and drinking behavior. There was no significant treatment effect for water consumption, both the T+ and T+W drank similar amounts of water, 0.58 ± 0.09 L and 0.42 ± 0.09 L, respectively. Overall, drinking behavior was more prevalent at 16 h of transport compared to 8 and 24 h ($p < 0.05$), but drinking behavior was similar among treatments at 32 h of transport ($p > 0.05$). The increase in water consumption at 16 h and the increase consumption of feed at 24 h may have allowed the pigs to maintain their weight over the rest of the transport period.

Table 4. Least squares means ±0.54 for treatment by time interaction for eating behavior over 32 h for pigs weaned and transported ($p < 0.05$). Treatments: pigs with feed and water (T+) or feed only (T+F). n = 12 pigs/treatment.

Time, h	Treatments		
	CON	T+	T+F
2	5.83 [a]	0.33 [b]	0.67 [b]
4	6.50 [a]	0.17 [b]	0.00 [b]
6	8.50 [a]	0.5 [b]	0.00 [b]
8	5.67 [a]	0.17 [b]	0.17 [b]
10	7.83 [a]	0.50 [b]	2.00 [c]
12	8.00 [a]	0.50 [b]	0.67 [b]
14	7.17 [a]	1.17 [b]	0.17 [b]
16	8.17 [a]	0.17 [b]	0.00 [b]
18	6.17 [a]	0.50 [b]	0.00 [b]
20	5.50 [a]	0.00 [b]	0.00 [b]
22	1.67 [a]	0.00 [b]	0.00 [b]
24	4.50 [a]	1.00 [b]	0.00 [b]
26	4.17 [a]	1.00 [b]	0.33 [b]
28	4.33 [a]	0.33 [b]	0.67 [b]
30	2.67 [a]	0.50 [b]	0.50 [b]
32	0.78 [a]	0.00 [a]	0.00 [a]

[a–c] Means with different superscripts within each time point differ at $p < 0.05$.

However, it is difficult to know if water was actually consumed. There may have been some spillage, which is normally seen even when animals are not being transported.

3.3.3. Behavior Post-Treatment

CON pigs were weaned after the 32 h study period. Upon being weaned, CON pigs spent 72% of their time lying. T+ and T− pigs both spent 73% of their time lying after transport, while T+F and T+W spent 77% and 67% of their time lying, respectively.

CON pigs spent 23% of their time standing after the 32 h study period. T+ and T+W both spent 13% of their time standing while T−, T+F, and spent 19% and 17% of their time standing after transport, respectively.

Both CON and T+ pigs spent 2% of their time eating after the 32 h study period. T+W, T− and T+F pigs spent 13%, 5%, and 4% of their time eating after transport, respectively. Drinking behaviors were not observed in CON pigs post-weaning. T+, T−, T+F, and T+W pigs spent 1%, 3%, 2%, and 5% of their time drinking, respectively.

T+W pigs appeared to spend less time lying and standing and more time eating and drinking than the other transported animals. However, the difference between these behaviors was not significant. Pigs spend more time resting, feeding, and drinking during the 3 d after transport than the non-transported control animals [8]. This is likely due to exhaustion, dehydration, and hunger. In future studies, it may advantageous to increase the sampling time of the videos to be able to detect behaviors that are exhibited at low frequencies, such as eating and drinking. It may also be beneficial to observe behavior for longer than 24-h post-transport.

3.3.4. Feed and Water Consumption after Transport

Feed and water intake was recorded at 7 and 14 d post-treatment, but no differences were found among treatment groups ($p > 0.05$). Similar results in feed and water intake post-treatment were reported by Garcia *et al.* [12]. This may indicate that weaned and transported pigs, regardless of what treatment group they were, they ate and drank similarly over time.

3.3.5. Post-Transport Body Weights

Post-transport body weights were recorded at 7 and 14 d. Body weight was similar among all treatment groups at 7 and 14 d post-transport ($p > 0.05$). However, a sex by treatment interaction for post-transport body weight was observed ($p < 0.01$). Sex differences were found within the CON and the T+ treatment groups. CON males had a lower average weight compared to females, 7.77 ± 0.28 kg and 8.57 ± 0.28 kg, respectively. T+ males had a higher average weight than females, 8.34 ± 0.40 kg and 7.16 ± 0.27 kg. Male pigs have been reported to be heavier at birth than females [30], but sex differences are typically not seen at this early of an age. There were several pigs in the study that could be considered poor-doers, piglets below weaning weights after 7 d post-transport. Poor-doers are piglets that are still not meeting their physiological requirements after 7 d post-transport and have been documented to be an indicator of poor welfare [7].

4. CONCLUSIONS

Weaned pigs are exposed to similar transport stresses as finishing pigs, in conjunction to also concurrently experiencing weaning stress, which is the most stressful event in an animal's life. However, literature pertaining to the effect of transportation on the welfare of weaned pigs is limited [6,18,19,20,21,31,32,33,34,35,36,37]. Transport is an added stressor in a pig's life that many times cannot be avoided. The use of a multidisciplinary approach (e.g., performance, physiology, behavior, morbidity, and mortality) to investigate implications of transporting and different aspects of transport (e.g., duration, space allowance) on pigs at weaning should be further studied [6].

In the current study pigs weaned and transported with or without feed and water, lost significantly more body weight than CON animals by 8 h of transport. In addition, animals not provided with water lost significantly more weight than those provided with water by 32 h of transport. This suggests that for recently weaned pigs, there is no need to provide feed during transportation, but provision of water may aid in reducing stress during transport over 24 h. Furthermore, providing water may help weaned pigs habituate to their current environment. In addition, physiological and behavioral changes were apparent after weaning and during transport. N:L increased significantly in males that were not provided with water compared to females, possibly suggesting that castrated males may experience acute stress at a higher level than females. The overall appreciation of the results points towards the usefulness of providing water during long distance transport, and this is already applied in practice in some countries.

Further research is needed to determine the long-term effects of long distance transport on weaned pigs in relation to performance, physiology, and behavior. Studies investigating the provision of water in relation to stocking density, season, and transport duration would be beneficial to the pork industry. Using other measures of welfare in addition to the ones already established may help give better insight into the physiological changes associated with long distance transport of weaned pigs, and the benefits of providing water during transport on the animals' welfare.

ACKNOWLEDGMENTS

The National Pork Board (USA) funded this work. We thank student workers for collection of field data.

REFERENCES

1. Averos, X.; Herranz, A.; Sanchez, R.; Gozalvez, L.F. Effect of duration of commercial journeys 21 between rearing farms and growing-finishing farms on the physiological stress response of weaned pigs. *Livest. Sci.* **2009**, *122*, 339–344.
2. Lambooij, W. Transport of pigs. In *Livestock Handling and Transport*; Granding, T., Ed.; CAB International: Wallingford, UK, 1993; pp. 213–231.
3. Official Journal of the European Union. Available online: http://eur-lex.europa.eu/legal-content/EN/TXT/PDF/?uri=CELEX:32005R0001&from=en (accessed on 21 March 2016).
4. National Pork Producers Council. Available online: http://www.nppc.org/issues/animal-health-safety/28-h-law/ (accessed on 7 October 2014).
5. Sutherland, M.A.; Krebs, N.; Smith, J.S.; Dailey, J.W.; Caroll, J.A.; McGlone, J.J. The effect of three space allowances on the physiology and behavior of weaned pigs during transportation. *Livest. Sci.* **2009**, *126*, 183–188.
6. Sutherland, M.A.; Backus, B.L.; McGlone, J.J. Effects of transport at weaning on the behavior, physiology and performance of pigs. *Animals* **2014**, *4*, 657–669.
7. Lewis, N. Transport of early weaned pigs. *Behav. Sci.* **2008**, *110*, 128–135.
8. Lewis, N.J.; Berry, R.J. Effects of season behavior of early-weaned piglets during and immediately following transport. *Appl. Anim. Behav. Sci.* **2006**, *100*, 182–192.
9. Le Dividch, J.; Seve, B. Energy requirement of the young piglet. The weaner pig: Nutrition and management. In Proceedings of a British Society of Animal Science Occasional Meeting, Nottingham, UK, September 2000.
10. Hay, M.; Orgeur, P.; Levy, F.; Le Dividich, J.; Concordet, D.; Nowak, R.; Schaal, B.; Mormede, P. Neuroendocrine consequences of very early weaning in swine. *Physiol. Behav.* **2001**, *72*, 263–269.
11. Kanitz, E.; Tuchcherer, M.; Tuchcherer, A.; Stabenow, B.; Manteuffel, G. Neuroendocrine and immune response to acute endotoxemia in suckling piglets. *Biol. Neonate* **2002**, *81*, 203–209.
12. Garcia, A.; Pirner, G.; Picinin, G.; May, M.; Guay, K.; Backus, B.; Sutherland, M.; McGlone, J. Effect of provision of feed and water during transport on the welfare of weaned pigs. *Animals* **2015**, *5*, 407–425.
13. Transport Quality Assurance Handbook, 4th ed. Available online: http://old.pork.org/filelibrary/tqa/manual.pdf (accessed on 7 October 2014).
14. Lambooy, E.; Garson, G.J.; Walstra, P.; Mateman, G.; Merkus, G.S. Transport of pigs by car for two days; some aspects of watering and loading density. *Livest. Prod. Sci.* **1985**, *13*, 289–299.
15. Jones, S.D.; Rompala, R.E.; Haworth, C.R. Effects of fasting and water restriction on carcass shrink in pork quality. *Can. J. Anim. Sci.* **1985**, *65*, 613–618.
16. Brumm, M.C.; Jesse, G.W.; Mayes, H.F.; Sinn, G.M.; Clemens, E.T. Effects of feed and water restriction receiving diet crude protein on feeder pig performance. *J. Anim. Sci.* **1987**, *64*, 1606–1611.

17. Lambooy, E. Road transport of pigs over a long distance: Some aspects of behavior, temperature and humidity during transport and some effects of the last two factors. *Anim. Prod.* **1988**, *46*, 257–263.
18. Sutherland, M.A.; Bryer, P.J.; Davis, B.L.; McGlone, J.J. Space requirements of weaned pigs during a 60 minutes transport in summer. *J. Anim. Sci.* **2009**, *87*, 363–370.
19. Sutherland, M.A.; McDonald, A.; McGlone, J.J. Environmental, trailer, and gender effects on the percentage of dead and non-ambulatory pigs during transport. *Vet. Rec.* **2009**, *165*, 13–18.
20. Sutherland, M.A.; Bryer, P.J.; Davis, B.L.; McGlone, J.J. A multidisciplinary approach to assess the welfare of weaned pigs during transport at three space allowances. *J. Appl. Anim. Welf. Sci.* **2010**, *13*, 237–249.
21. Sutherland, M.A.; Bryer, P.J.; Davis, B.L.; Smith, J.F.; McGlone, J. The combined effects of transport and food and water deprivation on the physiology of breeding age gilts. *Livest. Sci.* **2012**, *144*, 124–131.
22. Prunier, A.; Bonneau, M.; von Boreli, E.H.; Cinotti, S.; Gunn, M.; Fredriksen, B.; Giersing, M.; Morton, D.B.; Tuyttens, F.A.M.; Velarde, A. A review of the welfare consequences of surgical castration in piglets and the evaluation of non-surgical methods. *Anim. Welf.* **2006**, *15*, 277–289.
23. Taddio, A.; Goldback, M.; Ipp, M. Stevens, B.; Koren, G. Effect of neonatal circumcision on pain responses during vaccination in boys. *Lancet* **1995**, *345*, 291–292.
24. McGlone, J.J.; Salak-Johnson, J.L. Lumpkin, E.A.; Nicholson, R.I.; Gibson, M.; Norman, R.L. Shipping stress and social status effects on pig performance, plasma cortisol, natural killer activity, and leukocyte numbers. *J. Anim. Sci.***1993**, *71*, 888–896.
25. Dunn, J.A. Psychoneuroimmunology for the psychoneuroendocrinologist: A review of animal studies of nervous system-immune system interactions. *Psychoneuro-endocrinology* **1989**, *14*, 251–274.
26. Averos, X.; Martin, S.; Riu, M.; Serratosa, J.; Gosalvez, L.F. Stress response of extensively reared young bulls being transported to growing-finishing farms under Spanish summer commercial conditions. *Livest Sci.* **2008**, *119*, 174–182.
27. Riondato, F.; D'Angelo, A.; Miniscalco, B.; Bellino, C.; Guglielmino, R. Effects of road transportation on lymphocyte subsets in calves. *Vet. J.* **2008**, *175*, 364–368.
28. Metz, J.H.M.; Gonyou, H.W. Effect of age and housing conditions on the behavioural and haemolytic reaction of piglets to weaning. *Appl. Anim. Behav. Sci.* **1990**, *27*, 299–309.
29. Dybkjaer, L. The identification of behavioral indicators of "stress" in early weaned piglets. *Appl. Anim. Behav. Sci.* **1992**, *35*, 135–147.
30. Sutherland, M.A.; McDonald, A.; McGlone, J.J. Effects of variations in the environment, length of journey and type of trailer on the mortality and morbidity of pigs being transported to slaughter. *Vet. Rec.* **2009**, *165*, 13.
31. Pettigrew, J.E.; Cornelius, S.G.; Moser, R.L.; Heeg, T.R.; Hanke, H.E.; Miller, K.P.; Hagen, C.D. Effects of oral doses of corn oil and other factors on preweaning survival and growth of piglets. *J. Anim. Sci.* **1984**, *62*, 601–612.
32. Villé, H.; Bertels, S.; Geers, R.; Janssens, S.; Goedseels, V.; Parduyns, G.; van Bael, J.; Goossens, K.; Bosschaerts, L.; de Ley, J.; *et al.* Electrocardiogram parameters of piglets during housing, handling and transport. *Anim. Prod.* **1993**, *56*, 211–216.
33. Geers, R.; Bleus, E. van Schie, E.; Villé, H.; Gerard, H.; Janssens, S.; Nackaerts, G.; Decuypere, E.; Jourquin, J. Transport of pigs different with respect to the halothane gene: Stress assessment. *J. Anim. Sci.* **1994**, *72*, 2552–2558.
34. Hicks, T.A.; Mcglone, J.J.; Whisnant, C.S.; Kattesh, H.G.; Norman, R.L. Behavioral, endocrine, immune and performance measures for pigs exposed to acute stress. *J. Anim. Sci.* **1998**, *76*, 474–483.
35. Berry, R.J.; Lewis, N.J. The effect of duration and temperature of simulated transport on the performance of early-weaned piglets. *Can. J. Anim. Sci.* **2001**, *81*, 199–204.
36. Lewis, N.J.; Berry, R.J.; Crow, G.; Wamnes, S. Assessment of the effects of season of transport on the performance of early-weaned piglets. *Can. J. Anim. Sci.* **2005**, *85*, 449–454.

37. Bench, C.; Schaefer, A.; Faucitano, L. The welfare of pigs during transport. In *Welfare of Pigs from Birth to Slaughter*; Wageningen Academic Publishers: Wageningen, The Netherlands, 2008; pp. 161–195.

CHAPTER 7

Effects of early feeding on the host rumen transcriptome and bacterial diversity in lambs

Weimin Wang[1]Chong Li[1,2,3], Fadi Li[1] Xiaojuan Wang[1,2], Xiaoxue Zhang[1], Ting Liu[1] Fang Nian[1], Xiangpeng Yue[2] Fei Li[2], Xiangyu Pan[1], Yongfu La[1], Futao Mo[1], Fangbin Wang[1] & Baosheng Li[4]

[1]*College of Animal Science and Technology, Gansu Agricultural University, Lanzhou 730000, P. R. China.*

[2]*College of Pastoral Agriculture Science and Technology, Lanzhou University, Lanzhou 730000, P. R. China.*

[3]*Engineering Laboratory of Sheep Breeding and Reproduction Biotechnology in Gansu Province, Minqin 733300, P. R. China.*

[4]*JinchangZhongtian Sheep Industry Co. Ltd., Jinchang 737100, P. R. China*

ABSTRACT

Early consumption of starter feed promotes rumen development in lambs. We examined rumen development in lambs fed starter feed for 5 weeks using histological and biochemical analyses and by performing high-throughput sequencing in rumen tissues. Additionally, rumen contents of starter feed-fed lambs were compared to those of breast milk-fed controls. Our physiological and biochemical findings revealed that early starter consumption facilitated rumen development, changed the pattern of ruminal fermentation, and increased the amylase and carboxymethylcellulase activities of rumen micro-organisms. RNA-seq analysis revealed 225 differentially expressed genes between the rumens of breast milk- and starter feed-fed lambs. These DEGs were involved in many metabolic pathways, particularly lipid and carbohydrate metabolism, and included *HMGCL* and *HMGCS2*. Sequencing analysis of 16S rRNA genes revealed that ruminal bacterial communities were more diverse in breast milk-than in starter feed-fed lambs, and each group had a distinct microbiota. We conclude that early starter feeding is beneficial to rumen development and physiological function in lambs. The underlying mechanism may involve the stimulation of ruminal ketogenesis and butanoate metabolism via *HMGCL* and *HMGCS2* combined with changes in the fermentation type induced by ruminal microbiota. Overall, this study provides insights into the molecular mechanisms of rumen development in sheep.

1. INTRODUCTION

The rumen has several important physiological functions, including absorption, transport, metabolic activity, and host protection[1]. The rumen of a new-born ruminant is essentially non-functional, as it has a smooth, stratified, squamous epithelium with no prominent papillae, and the microbiota has not yet been established[2,3]. The initiation of solid feed intake can trigger development of the rumen. The fermentation of solid feed by ruminal microorganisms results in the production of volatile fatty acids (VFAs)[4,5]. Intraruminal VFA administration stimulates the morphological development of the rumen epithelium in milk-fed animals[3,6,7]. Feeding starter along with milk is a common method for accelerating early rumen development8. It has been reported that solid feed consumption can increase total VFA concentrations and the molar proportion of acetate and butyrate in the rumen of starter + milk replacer-fed lambs when compared with lambs fed only milk replacement during 1 to 84 days of life[9]. Indeed, administering grain-based feed and orchard grass hay to neonatal ruminants significantly increase the length and density of the rumen papillae, whereas feeding milk alone resulted in little development of these structures and poor animal growth[10]. Therefore, this feeding strategy provides biologically appropriate fuels, and is thought to enhance early rumen development. However, the molecular mechanisms involved in this process remain unclear.

Daichi *et al.* identified three genes (*HMGCS2*, *AKR1C1* and *FABP3*) associated with rumen development by screening for candidate genes by *in silico* digital differential display (DDD)[11]. *TGFβ1* (transforming growth factor-beta 1) and *ESRRα* (oestrogen-related receptor alpha) can contribute to the development of the rumen epithelium and energy metabolism, respectively, as transcriptional regulators. These factors also trigger activation of downstream targets that can mediate rumen development and function in growing calves[12]. Changes of monocarboxylate transporter 1 (*MCT1*) expression occur in an age-dependent manner in the rumen epithelium of neonatal calves, which suggests that the expression of *MCT1* may be affected by the provision of liquid feed[13]. During the studies described above, many changes in gene expression related to rumen development were identified in calves.

Recently, interest in the diversity and function of ruminal bacteria has grown. The ruminal bacteria digest both complex and simple carbohydrates in the rumen and produce nutrients, such as volatile fatty acids (VFA), microbial proteins, and vitamins, for the host. Diet is a major factor that influences the structure and function of microbial communities that are present in the rumen[14]. The effect of diet on the structure of rumen microbial communities has been widely investigated using both culture-based and standard molecular methods, such as PCR, PCR-DGGE, and real-time PCR[15-25]. Currently, next-generation sequencing is widely used to study rumen microbial ecology. The application of such an approach would allow for much greater coverage of the microbial diversity and physiology of a complex environment, such as the rumen in calves[26], cows[27,28], and goats[29].

In this present study, next-generation sequencing (RNA-seq and 16S rRNA gene sequencing) was used to assess the effects of early feeding on host rumen gene expression and microbial community structure in lambs. We identify 225 differentially expressed genes between starter feed- and breast milk-fed lambs. These differentially expressed genes (DEGs) are involved in many metabolic pathways, such as carbohydrate and lipid metabolism. We also elucidate diet-dependent rumen microbial taxonomical composition changes at the phylum and genus levels. Our findings offer insights into the complexity of rumen development in sheep.

2. RESULTS

2.1 Growth performance, rumen morphology, rumen fermentation parameters, and enzymatic activity

Growth performance, rumen morphology, rumen fermentation parameters, and the enzymatic activity of rumen microorganisms from the beast milk- and starter feed-fed groups are shown in Table 1. The body weight (BW) and average daily gain (ADG) were higher in the starter feed- than in the breast milk-fed group. For rumen morphology, starter feed did not affect the papilla width (PW) or muscular thickness (MT) of the rumen. The weight of the reticulo-rumen (WRR), the volume of the reticulo-rumen (VRR), and the papilla height (PH) of the rumen were higher in the starter feed-fed group than in the breast milk-fed group. Regarding the rumen fermentation parameters, the starter feed-fed group had higher concentrations of total VFA, propionate, butyrate, and valerate, and lower concentrations of acetate, isobutyrate, and A:P compared with the breast milk-fed group. Concentrations of nitrogen, ammonium nitrogen, urea nitrogen, and protein nitrogen were higher in the starter feed-fed group than in the breast milk-fed group. For the enzymatic activity potentials of rumen contents, proteinase activity in the starter feed-fed group was lower than in the breast milk-fed group. The amylase and carboxymethylcellulase activities in the starter feed-fed group were higher than in breast milk-fed group.

2.2 RNA sequencing—data mapping and annotation

A total of 6 cDNA libraries were sequenced from the rumen of starter feed- and breast milk-fed groups ($n = 3$ per group). Libraries were sequenced using a HiSeq 2500 sequencing platform at Personalbio-Shanghai, China, and 6 sets of reads were obtained. After removing adaptors and filtering, we obtained a total of 149.9 M and 123.3 M high-quality clean reads for the breast milk- and starter feed-fed groups, respectively. RNA-seq yielded 24.6 M to 68.7 M reads for all 6 samples, and more than 90% of reads met our quality control criteria (Supplementary Table S1).

After mapping clean reads to the ovine genome, 68.81–84.24% of reads were successfully aligned, and 94.38–95.68% of mapped reads had unique genomic locations. Moreover, 64.95–78.05% of reads mapped to a gene, 87.29–93.10% of reads mapped to an exon, and 21.95–35.05% of reads mapped to introns or intergenic regions (Supplementary Table S2 and Figure S1).

Table 1: Animal performance of Hu sheep used for RNA and metagenomics sequencing.

Item			Breast milk-fed	Starter feed-fed	*P*-value
Growth performance	BW (kg)		8.49 ± 1.94	12.05 ± 2.72	0.026
	ADG (g/d)		0.13 ± 0.01	0.18 ± 0.01	0.010
Rumen morphology	WRR (g)		98.32 ± 27.59	196.72 ± 35.62	<0.001
	VRR (mL)		894.6 ± 320.86	1500.7 ± 291.92	0.007
	PH (μm)		659.40 ± 19.97	1003.33 ± 24.95	<0.001
	PW (μm)		624.74 ± 40.02	474.90 ± 18.79	0.053
	MT (μm)		807.45 ± 35.26	849.62 ± 39.59	0.429
Rumen fermentation parameters	pH		5.91 ± 0.10	5.30 ± 0.04	0.130
	Total VFA (mmol/L)		33.05 ± 5.52	113.24 ± 7.99	<0.001
	VFA proportion (mol/100 mol)	Acetate (A)	58.45 ± 9.24	50.67 ± 5.64	0.015
		Propionate (P)	26.98 ± 5.13	32.05 ± 6.63	0.025
		Butyrate	8.65 ± 1.50	11.78 ± 2.66	<0.001
		Isobutyrate	0.97 ± 0.81	0.31 ± 0.23	0.011
		Valerate	2.63 ± 1.21	3.84 ± 0.96	0.007
		Isovalerate	2.32 ± 2.07	1.34 ± 1.22	0.152
		A:P	2.32 ± 0.88	1.67 ± 0.49	0.028
	ruminal metabolic parameters (mg/100 mL)	Nitrogen	929.44 ± 290.70	4267.08 ± 558.99	<0.001
		Ammonium nitrogen	13.48 ± 3.55	22.66 ± 1.70	<0.001
		Urea nitrogen	2.68 ± 0.04	8.97 ± 1.44	<0.001
		Protein nitrogen	913.29 ± 236.48	4235.44 ± 423.10	<0.001
Enzymes activity of rumen microorganisms	Proteinase (U/mg)		0.017 ± 0.001	0.007 ± 0.000	<0.001
	Amylase (U/mg)		0.294 ± 0.021	0.585 ± 0.034	<0.001
	carboxymethylcellulase (U/mg)		0.019 ± 0.002	0.034 ± 0.001	<0.001

2.3 Differentially expressed genes between starter feed- and breast milk-fed lambs

In this present RNA-seq study, 18,716 genes were detected in the rumen of all 6 individuals (Supplementary Table S3). To characterize the relationship of genome-wide expression profiles between starter feed- and breast milk-fed lambs, all annotated unigenes from 6 rumen samples were clustered using the software Cluster 3.0 (Fig. 1). The genes detected in the different feeding treatment groups were clearly distinct. A total of 225 genes were differentially expressed, with a criteria of at least a 2-fold difference and a P-value less than 0.05 ($|log2FC| \geq 1$, $p < 0.05$, $FDR < 0.05$), among which 188 and 37 genes were up- and down-regulated in the starter feed-fed group (Supplementary Table S4). Table 2 shows the top 20 differentially expressed (DE) genes, including the top 10 genes with either higher or lower expression in the starter feed-fed compared to the breast milk-fed lambs.

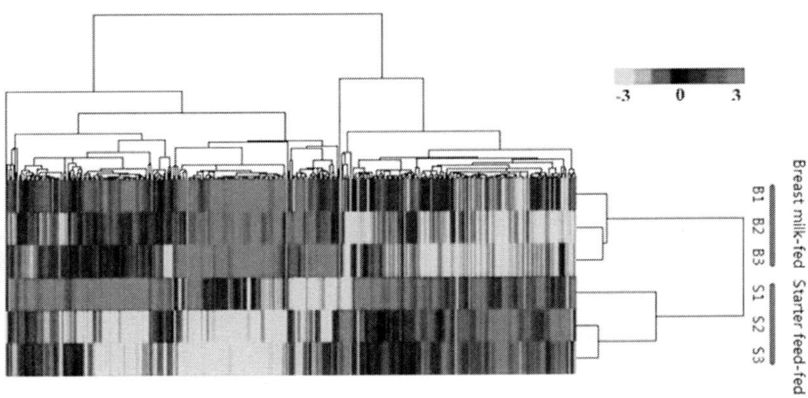

Figure 1: Average linkage hierarchical clustering analysis of the log_2 transformed change ratio of 18,716 unigenes generated with Cluster 3.0 software.

Data are shown in a tree analysis generated using the software package Java Treeview. Each row represents a differentially expressed gene and each column represents a sample. Green and red colour gradients indicate a reduction or increase in transcript abundance, respectively.

Table 2: List of 20 rumen DEGs between Hu sheep that received starter feed or breast milk.

Gene	Starter feed-fed	Breast milk-fed	FC (S/B)	*P*-value	Full name
GPNMB	35.30	1998.39	−56.61	4.76E-02	Glycoprotein (transmenbrane) nmb
KRT4	92.20	4519.14	−49.01	3.64E-04	Keratin 4, type II
Unknown1	4.04	110.80	−27.42	1.06E-02	—
TRDC	1.23	31.40	−25.57	4.34E-02	T cell receptor delta constant
Unknown2	1.55	36.35	−23.43	4.14E-02	—
SRGN	2.85	58.12	−20.41	1.91E-02	Serglycin
RGS4	7.71	148.34	−19.23	1.41E-02	Regulator of G-protein signaling 4
FGR	2.20	40.02	−18.19	4.80E-02	FGR proto-oncogene, Src family tyrosine kinase
NXPE2	2.80	50.11	−17.88	9.10E-03	Neurexophilin and PC-esterase domain family, member 2
Unknown3	332.15	4734.76	−14.25	2.02E-06	—
COX8C	1.86	229.54	123.39	3.03E-07	Cytochrome c oxidase subunit VIIIC
Unknown4	0.23	32.03	137.98	1.20E-02	—
MYL10	0.31	53.46	172.41	1.58E-02	Myosin, light chain 10, regulatory
Unknown5	182.28	31458.78	172.59	6.24E-03	—
Unknown6	242.21	56941.99	235.09	6.97E-03	—
Unknown7	5.05	1532.46	303.38	1.25E-02	Ribosomal protein
Unknown8	1.20	372.79	311.68	2.13E-02	—
Unknown9	0.65	239.04	366.04	3.50E-02	—
Unknown10	1.23	625.32	509.61	5.94E-03	—
Unknown11	0.31	295.88	954.32	4.82E-02	—

To validate the differential expression of genes, we selected seven genes for qRT-PCR analysis. Compared with the breast milk-fed group, expression of *ECM2*, *Serpinb1*, and *KRT4* mRNA transcripts were lower in rumens of starter feed-fed lambs, whereas the expression of *Cox8c*, *SLC14A1*, *HMGCL*, and *HMGCS2* mRNA transcripts were higher in the rumens of starter feed-fed lambs (Fig. 2). Expression of 7 selected genes (*ECM2*, *Serpinb1*, *KRT4*, *Cox8c*, *SLC14A1*, *HMGCL*, and *HMGCS2*) showed significant differences between the breast milk- and starter feed-fed groups. Accordingly, the qRT-PCR analyses largely confirmed the RNA-seq findings, with the correlation coefficient of the fold-change (FC) values from the two methods being 0.99 and the R^2 value for the linear regression also being 0.99, indicating the robustness of the RNA-seq data.

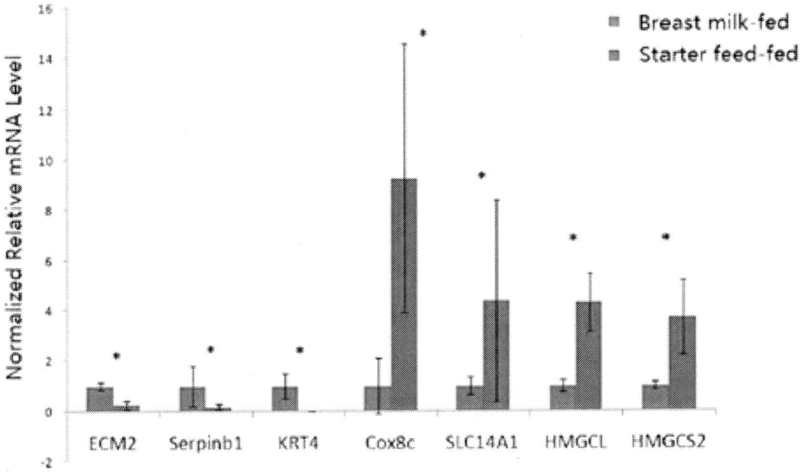

Figure 2: Validation of differentially expressed genes in the rumen from breast milk-and starter feed-fed lambs.The qPCR measurements of the expression of *Cox8c*, *SLC14A1*, *Serpinb1*, *ECM2*, *KRT4*, *HMGCL*, and *HMGCS2* mRNA transcripts were analysed using the ΔΔCt method; *significant difference between the groups of lambs.

2.4 Pathway analysis of differentially expressed genes

To explore the biological functions of the 225 DEGs, GO (Gene Ontology) enrichment analysis was performed. These DEGs were classified into three main categories: biological process, cellular component, and molecular function (Fig. 3 and Supplementary Table S5). Among all 86 GO terms, "Oxidoreductase activity" ($P = 3.12E\text{-}13$) was the most enriched cluster, followed by "Lipid metabolic process" ($P = 6.05E\text{-}07$), "Lyase activity" ($P = 8.00E\text{-}05$), "Cofactor metabolic process" ($P = 2.28E\text{-}03$), "Sulfur compound metabolic process" ($P = 3.37E\text{-}03$), "Transferase activity, transferring acyl groups" ($P = 3.40E\text{-}03$), "Biological process" ($P = 7.53E\text{-}03$), "Isomerase activity" ($P = 9.37E\text{-}03$), "Small molecule metabolic process" ($P = 2.29E\text{-}02$), "extracellular region" ($P = 1.21E\text{-}04$), "organelle" ($P = 4.16E\text{-}02$), "extracellular space" ($P = 4.29E\text{-}02$), and "molecular function" ($P = 1.15E\text{-}02$).

The red line indicates $P = 0.05$. Details of the GO enrichment analysis are presented in Supplementary Table S5.

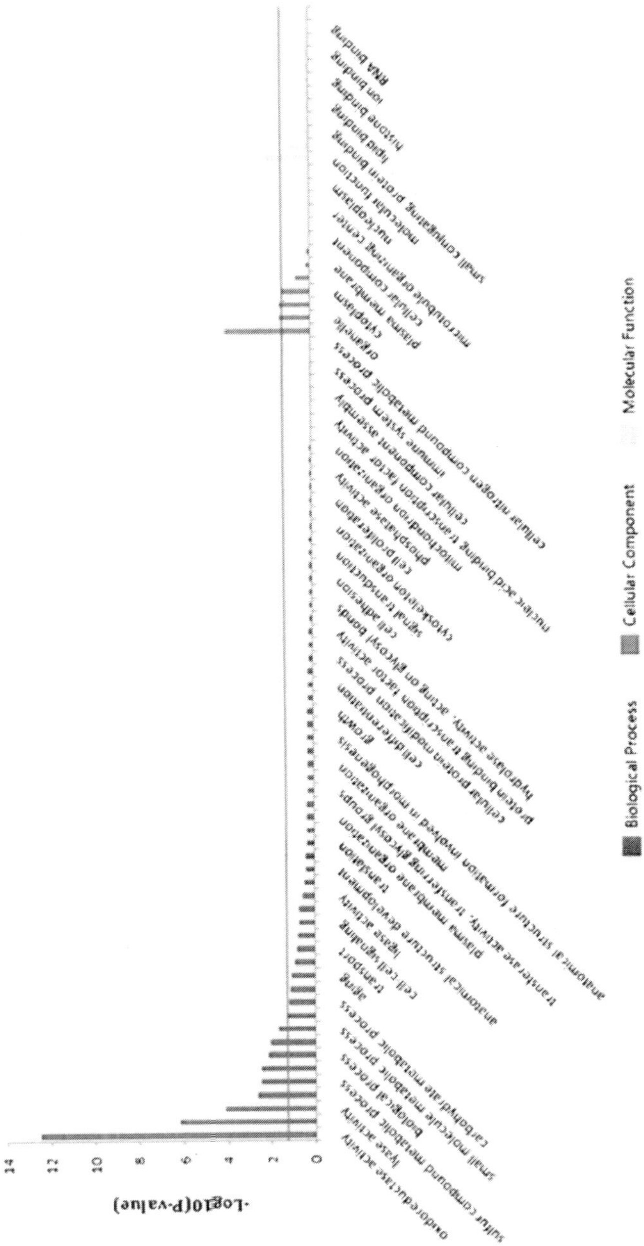

Figure 3: Gene ontology (GO) classification of differentially expressed genes.

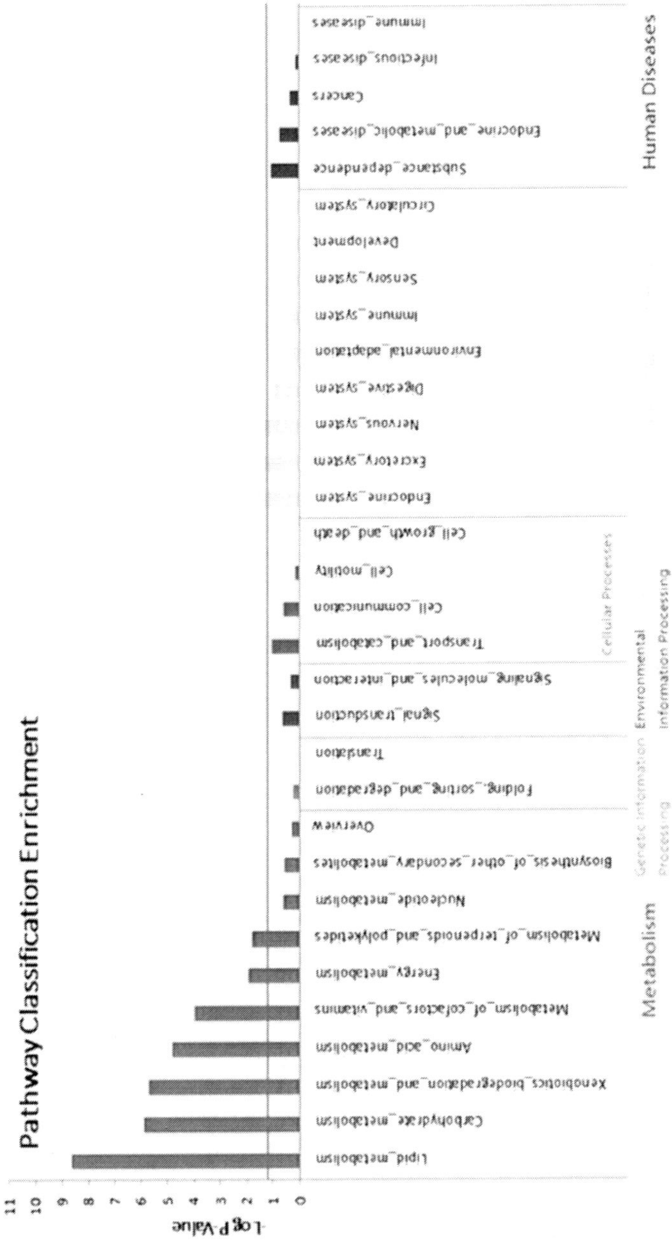

Figure 4: KEGG enrichment analysis of differentially expressed genes.The red line indicates $P = 0.05$. Details of the KEGG enrichment analysis can be found in Supplementary Table S6.

KEGG (Kyoto Encyclopaedia of Genes and Genomes) enrichment analysis was used to align all DE genes to two specific pathways—Metabolism and Organismal Systems (Fig. 4 and Supplementary Table S6). The most enriched category was lipid metabolism ($P = 2.42E\text{-}09$, 331 genes, including 19 DEGs), followed by carbohydrate metabolism ($P = 1.27E\text{-}06$, 337 genes, including 16 DEGs), xenobiotic biodegradation and metabolism ($P = 2.08E\text{-}06$, 102 genes, including 9 DEGs), amino acid metabolism ($P = 1.47E\text{-}05$, 237 genes, including 12 DEGs), metabolism of cofactors and vitamins ($P = 1.05E\text{-}04$, 130 genes, including 8 DEGs), energy metabolism ($P = 1.14E\text{-}02$, 121 genes, including 5 DEGs), metabolism of terpenoids and polyketides ($P = 1.68E\text{-}02$, 18 genes, including 2 DEGs), the endocrine system ($P = 2.32E\text{-}02$, 760 genes, including 15 DEGs), the excretory system ($P = 2.89E\text{-}02$, 104 genes, including 4 DEGs), and the nervous system ($P = 4.68E\text{-}02$, 555 genes, including 11 DEGs).

2.5 Rumen microbial 16S rRNA sequencing data and alpha diversity

After data filtering, quality control, assembling of pair-end reads, and the removal of primers, chimeras, and low confidence singletons, a total of 189,357 V1-V3 16S rRNA sequence reads from 12 samples, with an average of 15,780 sequence reads for each sample (the minimum and maximum values for one sample were 11,153 and 25,008 sequence reads, respectively), were used in this study. The average length of sequence reads after primer removal was 395 bp. The overall number of OUTs detected by our analysis was reached 6798 based on 97% nucleotide sequence identity between reads.

The rarefaction curve was constructed by random sampling for all sequences, which (Supplementary Figure S2) revealed that most of our sampling efforts yielded sufficient OUT coverage to accurately describe the bacterial composition of each group of lambs. Alpha-diversity measures (Table 3) indicated that the Chao, ACE, and Shannon index of the starter feed-fed lambs were higher than those of the breast milk-fed lambs ($P < 0.01$), while the Simpson index of the starter feed-fed lambs was lower than that of breast milk-fed lambs ($P < 0.05$).

Table 3: Alpha diversity measures of bacterial communities in the ruminal contents between the starter feed- and breast milk-fed groups.

Alpha diversity index	Breast milk-fed	Starter feed-fed	*P*-value
OTU	1601.7±108.5	798.0±65.6	<0.001
Chao	1940.8±110.8	972.7±65.1	<0.001
ACE	2045.8±117.1	1019.3±67.6	<0.001
Simpson	0.024±0.004	0.104±0.031	0.028
Shannon	5.560±0.172	3.878±0.283	<0.001
Coverage	0.961±0.002	0.986±0.002	<0.001

2.6 Taxonomic composition of ruminal bacterial communities between starter feed- and breast milk-fed lambs

In total, 21 phyla were identified within the ruminal microbiota. The abundance of 11 phyla in all of the samples was <0.5%, which included *Armatimonadetes, Chloroflexi, Cyanobacteria, Elusimicrobia, Fusobacteria, GN02, LD1, Planctomycetes, SR1, TM7,* and *WPS2*. Among the 21 phyla, *Bacteroidetes, Firmicutes, Proteobacteria,* and *Spirochaetes*were detected as the dominant phyla among all of the samples (Table 4). The phylum *Bacteroidetes* was the most abundant in both groups (66.01% and 58.47% in the starter feed- and breast milk-fed groups, respectively), although there was no significant difference between these two groups. Within 11 phylum with an observed >0.5% relative abundance, the *Fibrobacteres* ($P = 0.036$), *Lentisphaerae* ($P = 0.002$), and *Verrucomicrobia* ($P = 0.024$) abundances were significantly lower in the starter feed-fed than in the breast milk-fed lambs; however, the abundance of other phyla was not significantly different between the two groups ($P > 0.05$).

Among the 17 genera observed to have >0.5% relative abundance, 5 changed significantly between the starter feed- and breast milk-fed groups (Table 5), while the others remained relatively stable. Moreover, many of the sequences (27.89% and 63.48% in the starter feed-and breast milk-fed groups, respectively) remained unclassified at the genus level. Within the *Bacteroidetes* phylum, the abundance of genus *CF231*was significantly lower in the starter feed-fed group than that in the breast milk-fed group ($P = 0.019$). Within the *Fibrobacteres* phylum, the abundance of genus *Fibrobacter* was significantly lower in the starter feed-fed group than that in the breast milk-fed group ($P = 0.039$). Within the *Firmicutes* phylum, genus *Acidaminococcus* ($P = 0.004$) and *Succiniclasticum* ($P = 0.013$) were significantly higher in the starter feed-fed group than in the breast milk-fed group. Within the *Proteobacteria*phylum, the abundance of genus *Dechloromonas* was significantly lower in the starter feed-fed group than in the breast milk-fed group ($P = 0.050$).

Table 4: Phylum-level taxonomic composition of bacterial communities in the ruminal contents between the starter feed- and breast milk-fed groups.

Phylum	% of sequences in starter feed-fed group	% of sequences in breast milk-fed group	*P*-value
Actinobacteria	0.132 ± 0.072	0.335 ± 0.226	0.413
Bacteroidetes	66.007 ± 4.187	58.471 ± 1.385	0.118
Fibrobacteres	0.084 ± 0.050	0.769 ± 0.279	**0.036**
Firmicutes	19.367 ± 3.551	14.99 ± 1.484	0.283
Lentisphaerae	0.119 ± 0.088	3.265 ± 0.752	**0.002**
Proteobacteria	2.593 ± 1.231	6.016 ± 1.441	0.101
Spirochaetes	4.690 ± 1.202	9.583 ± 4.780	0.344
Synergistetes	1.200 ± 0.450	1.070 ± 0.529	0.862
Tenericutes	0.677 ± 0.004	1.186 ± 0.335	0.363
Verrucomicrobia	0.149 ± 0.146	7.481 ± 2.762	**0.024**
Unclassified	0	0.003 ± 0.002	0.154
Others (<0.5%)	0.088 ± 0.067	1.715 ± 0.290	<0.001

2.7 OUT diversity and similarity analyses

In this study, beta diversity analysis was performed. PCoA analysis using the Bray–Curtis similarity metric revealed that the samples clustered according to group (Fig. 5). PC1, PC2, and PC3 accounted for 43%, 15%, and 8.7% of variation, respectively.

Table 5: Genus-level taxonomic composition of the bacterial communities in the ruminal contents between the starter feed- and breast milk-fed groups.

Phylum	Genus	% of sequences in starter feed-fed group	% of sequences in breast milk-fed group	*P*-value
Bacteroidetes	*Prevotella*	40.370 ± 9.576	17.196 ± 5.925	0.067
	CF231	0.064 ± 0.000	1.613 ± 0.553	**0.019**
Fibrobacteres	*Fibrobacter*	0.084 ± 0.050	0.750 ± 0.277	**0.039**
Firmicutes	*Butyrivibrio*	0.821 ± 0.241	0.328 ± 0.075	0.079
	Ruminococcus	3.240 ± 2.658	0.441 ± 0.123	0.318
	Acidaminococcus	1.580 ± 0.426	0.001 ± 0.001	**0.004**
	Dialister	3.721 ± 1.777	0.001 ± 0.001	0.063
	Succiniclasticum	5.043 ± 1.399	0.765 ± 0.223	**0.013**
	Bulleidia	2.512 ± 1.875	0.011 ± 0.006	0.210
	RFN20	0.006 ± 0.006	0.923 ± 0.473	0.081
Proteobacteria	*Dechloromonas*	0.014 ± 0.007	1.420 ± 0.632	**0.050**
	Ruminobacter	0.001 ± 0.001	0.864 ± 0.549	0.147
	Succinivibrio	1.981 ± 1.178	1.014 ± 0.988	0.544
Spirochaetes	*Sphaerochaeta*	6.617 ± 4.016	1.364 ± 0.537	0.224
	Treponema	2.964 ± 2.951	3.187 ± 1.423	0.947
Synergistetes	*Pyramidobacter*	1.048 ± 0.490	0.213 ± 0.060	0.122
Tenericutes	*Anaeroplasma*	0	0.567 ± 0.325	0.112
Unclassified (Genus)		27.919 ± 8.595	53.156 ± 6.044	**0.037**
Others (<0.5%)		0.019 ± 0.004	0.157 ± 0.027	**0.000**

2.8 Quantification of total ruminal bacteria and abundant bacterial genera

As indicated by qPCR, the total ruminal bacterial copy numbers were higher in the starter feed-fed group than in the breast milk-fed group ($P = 0.000$). The abundance of phylum *Firmicutes* ($P = 0.039$) and *Bacteroidetes* ($P = 0.000$), and the genera *Prevotella* ($P = 0.036$), *Bacteroides* ($P = 0.041$), *and Selenomonas ruminantium-Mitsuokella multiacida* ($P = 0.004$) were significantly higher in the starter feed-fed group than in the breast milk-fed group (Table 6).

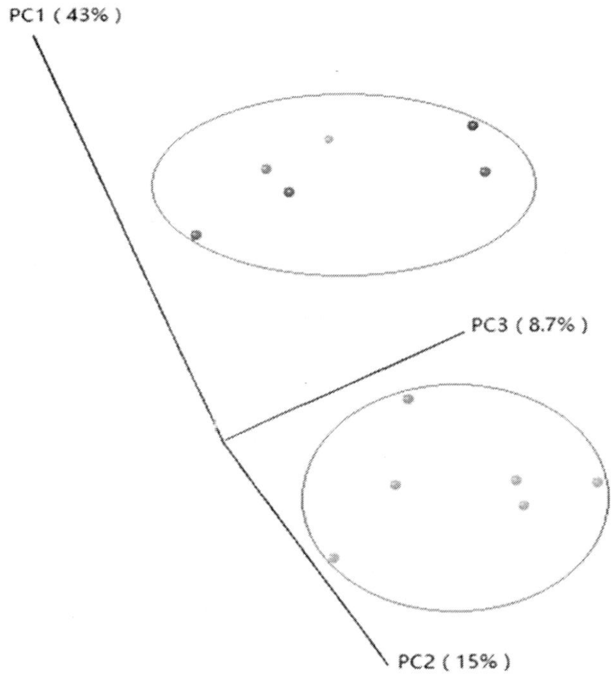

Figure 5: Principal coordinate analysis (PCoA) of ruminal bacterial OTUs between the starter feed- and breast milk-fed groups.

Table 6: qPCR results for total bacteria, two phyla, and four genera of ruminal bacterial communities in the starter feed- and breast milk-fed groups.

Taxon	Abundance of bacteria[a]		P-value
	Starter feed-fed group	Breast milk-fed group	
Bacteria	30.76 ± 2.77	5.23 ± 1.25	**<0.001**
Bacteroidetes	16.38 ± 1.13	3.12 ± 0.77	**<0.001**
Firmicutes	3.90 ± 1.24	0.87 ± 0.30	**0.039**
Prevotella	11.53 ± 4.46	0.66 ± 0.43	**0.036**
Clostridium Butyricum	0.22 ± 0.20	0.0015 ± 0.00075	0.285
Bacteroides spp.	0.02 ± 0.004	0.009 ± 0.003	**0.041**
Selenomonas ruminantium-Mitsuokella multiacida	0.195 ± 0.052	0.0017 ± 0.00083	**0.004**

2.9 Relationship between bacterial communities and functional variables

The abundance of bacterial communities at the genus level and functional variables were considered to be correlated with each other if the correlation coefficients were above 0.55. The relative abundance of the genera *Fibrobacter*, *RNF20*, and *Dechloromonas* were negatively correlated with the TVFA concentration, where as *Succiniclasticum*, *Prevotella*, *Bulleidia*, *Dialister*, and *Acidaminococcus* were positively correlated with TVFA (Fig. 6). The acetate and propionate concentrations were positively correlated with the relative abundance of *Ruminobacter*, *CF231*, *Prevotella*, and *Dialister*, respectively. Butyrate concentrations were positively correlated with the relative abundances of *Ruminococcus*, *Pyramidobacter*, *Butyrivibrio*, and *Bulleidia*. The proteinase activity, isovalerate and isobutyrate concentration, and pH were all positively correlated with those of *Fibrobacter*, *RFN20*, and *Dechloromonas*. The nitrogen concentration was negatively correlated with the relative abundance of *Fibrobacter*, *RNF20*, and *Dechloromonas*, but was positively correlated with *Pyramidobacter*, *Butyrivibrio*, *Bulleidia*, and *Acidaminococcus*. Amylase activity was positively correlated with the relative abundance of *Prevotella*, *Dialister*, and *Acidaminococcus*. By contrast, the relative abundance of *Ruminobacter*was negatively correlated with carboxymethylcellulase activity, while the relative abundances of *Dialister* and *Acidaminococcus* were positively correlated with carboxymethylcellulase activity.

3. DISCUSSION

Development of the rumen is an important physiological event for young ruminants. It entails the growth and cellular differentiation of the rumen, which results in a major shift in the pattern of nutrients that are delivered to the intestines and liver, and then to the peripheral tissues of the animal[30]. Many studies have shown that the rumen development process includes anatomic development (increase in rumen mass and growth of the rumen papillae)[31,32], functional achievement (fermentation capacity and enzyme activity)[33,34], and microbial colonization (bacteria, fungi, archaea and protozoa)[35,36]. Together, these observations suggest that anatomic, functional, and microbial development in the rumen represents an integrated system, and they should be studied together to better understand the process of rumen development.

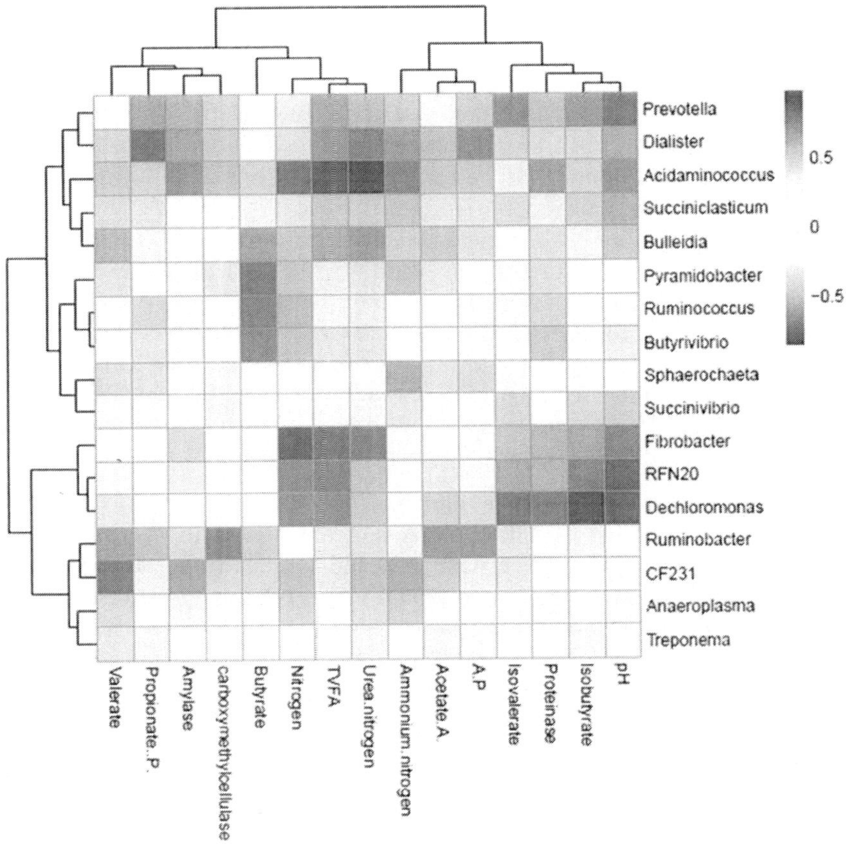

Figure 6: Coefficients of correlation between the relative abundances of ruminal bacterial genera and functional variables.

A previous study established that an early feeding strategy benefits rumen development in lambs[9]. In this present study, rumen fluid in the starter feed-fed group had greater concentrations of TVFA, and increased molar proportions of acetate, propionate, butyrate, and valerate than in the breast milk-fed group. This finding indicated that the consumption of starter from 7 days of life resulted in an earlier initiation of rumen fermentation31. Increased amounts of TVFA, acetate, propionate, butyrate, and valerate in the rumen of starter feed-fed lambs likely accounted for the greater WRR, VRR, and PH documented in this present study. The ruminal fluid pH value represents an important index of rumen health. Rumen pH values below 5.0 to 5.5 are considered to be abnormal and suggestive of SARA (subacute ruminal acidosis), whereas rumen pH values of 5.6 to 5.8 are considered to be marginal[37,38]. The ruminal fluid pH value in the starter feed-fed group was 5.3, which might be at risk for SARA. However, Li *et al.* demonstrated that SARA is not solely rumen pH-dependent, and it should be combined with clinical symptoms for diagnosis[39]. Because the starter feed-fed lambs in this study were without clinical symptoms of SARA and the development of rumen morphology in the starter feed-fed lambs was better than that in breast milk-fed lambs, we propose that the starter feed-fed lambs were not in SARA. The reason for the lower pH value in the starter feed-fed group may be the sampling time. Moreover, greater starter consumption can lead to higher amylase and carboxymethylcellulase activity potential[40]. Therefore, it is not surprising that we found that the amylase and carboxymethylcellulase activity potentials were greater in the starter feed-fed group. This finding implied that optimal fibre-degrading capacity occurred when the starter was administered[33]. In summary, the physical form of forage could accelerate rumen development in lambs.

Our present study along with previous findings have established that solid feed consumption increases total VFA concentrations in the rumen of starter feed- compared with breast milk-fed lambs. Some studies have reported that VFAs in the rumen, especially butyrate, enhance the growth of rumen papillae[3,6,7]; however, the underlying molecular mechanisms involved remain poorly characterized. In this present study, we used deep RNA sequencing to analyse the rumen transcriptomes of starter feed- and breast milk-fed lambs. We identified 225 DEGs in rumen tissues between starter feed- and breast milk-fed lambs. To investigate the biological functions of the DEGs, we performed GO annotation and KEGG pathway analysis. This study clearly revealed that most DGEs enriched many metabolism pathways, among which the most enriched categories were lipid metabolism ($P = 2.42E\text{-}09$, 331 genes, including 19 DEGs) and carbohydrate metabolism ($P = 1.27E\text{-}06$, 337 genes, including 16 DEGs). Lipid and carbohydrate metabolism may be two key pathways responsible for starter feed-dependent acceleration of rumen development in lambs.

We found that genes associated with the carbohydrate and lipid metabolism pathways were all up-regulated in starter feed-fed lambs, including *HMGCL*, *HMGCS2*, *PCK2*, and *MCEE* suggesting that rumen carbohydrate and lipid metabolism were more active in starter feed-fed lambs than in breast milk-fed lambs. *HMGCL* and *HMGCS2* were enriched in the butanoate metabolism pathway. The mature ruminal epithelium captures most of its energy from the oxidation of VFA. More than 90% of butyrate produced during microbial fermentation is oxidized and used for ketogenesis41. *HMGCS2* plays a central role in coordinating ruminal ketogenic flux, similar to its role in the liver[41]. *HMGCL* contributes to ketogenesis by converting HMG-CoA to acetoacetate. Ketogenesis is a hallmark of the metabolic development of ruminal epithelium tissue[42].

Based on rumen morphology and fermentation parameters, these up-regulated genes that are related to ketogenesis may be regulated by increased VFA, especially butyrate, in the rumen of starter feed-fed lambs.

The rumen readily metabolizes pyruvate, which contributes carbon for the production of oxaloacetate, lactate, or ketone bodies[43]. Zhang *et al.* reported the expression of *PCK2* mRNA transcripts was induced by propionate in a concentration-dependent manner, suggesting that propionate can directly regulate its own metabolism in young ruminant hepatocytes via the upregulation of *PCK2* mRNA transcript expression[44]. Additionally, MCEE (methylmalonyl-CoA epimerase enzyme) is found throughout nature and functions as a catabolic enzyme in higher animals[45]. Additionally, the micro-organisms that colonize animals may use this enzyme for various other functions, such as propionate fermentation and glyoxylate regeneration[46]. In this present study, expression of *PCK2* and *MCEE* mRNA transcripts and the proportion of propionate in starter feed-fed lambs were both higher than in breast milk-fed lambs. Thus, starter feed consumption can induce more propionate metabolism through the upregulation of *PCK2*and *MCEE* mRNA transcripts.

Associations between gastrointestinal microbial communities and their hosts have been recently shown to play an important role in host health and physiological function. The ruminal bacteria can digest complex and simple carbohydrates in the rumen and produce nutrients, such as volatile fatty acids (VFA), microbial proteins, and vitamins, for the host. Diet is a major factor that influences the structure and function of microbial communities in the rumen14. In this present study, we elucidated changes in the rumen microbial taxonomical composition in starter feed- and breast milk-fed lambs at the phylum and genus levels. Using the Roche-454 Titanium platform, we obtained an average of 15,780 reads for each sample with good coverage (>97.3%). Furthermore, our findings indicate that each group of lambs has a distinct ruminal microbiota, as is reflected by the clustering of samples between breast milk- and starter feed-fed lambs using PCoA. Moreover, most alpha-diversity indices (except the Simpson index) were higher in the breast milk-fed group than in the starter feed-fed group, suggesting that the ruminal microbiota in the breast milk-fed group is more diverse than in the starter feed-fed group.

It is commonly thought that a more diverse rumen microflora promotes greater stability in the rumen environment. However, the establishment of ruminal bacterial communities in the first days after birth might involve a more complex process. Li *et al.* reported that the rumen microbiota of younger calves (14 days) exhibited a heterogeneous microbial composition and contained more numerous, yet transient bacterial species and genera than older calves (42 days). Moreover, a significantly higher percentage of input 16S sequence reads from microbiota from younger calves could be assigned to any genus, indicating that more unknown bacteria may exist in the microbiome of younger calves[47]. This possibility is similar to that of the breast milk-fed lambs in this study, indicating that early starter feeding promotes the establishment of predominant microflora and the depletion of transient bacterial species and genera. In this present study, only 3 of the bacterial genus that disappeared in the starter feed-fed group but not in the breast milk-fed group have a fermentation function; most of the 'disappeared' bacterial genus are aerobic bacteria or bacteria without fermentation function. Jami *et al.* reported that the most significant change occurred between the first and third days of life in the rumen was the reduction in taxa associated with aerobic or facultative anaerobic function and an increase in those taxa associated with obligatory anaerobic function. Furthermore, almost all genera that showed a sharp reduction on day 3 were either aerobic or facultative anaerobic, indicating the emergence of a new anaerobic environmental niche as early as 3 days after birth[48]. However, in this present study, anaerobic environment of the rumen is not yet fully established in breast milk-fed lambs by 42 days of age, as many aerobic or facultative anaerobic bacterial genus still remain. These observations indicate that early starter feeding promotes the establishment of the ruminal anaerobic environment and stability of the rumen microflora.

In both groups, *Bacteroidetes, Firmicutes, and Proteobacteria* were the dominant phyla in ruminal microbiota, in accord with previous studies[49,50,51]. As previously reported, *Verrucomicrobia* could be found at a higher proportion in calves exclusively fed milk and were represented exclusively by the genus *Akkermansia*26,48; we also found that the relative abundance of *Verrucomicrobia* was significantly higher in breast milk-fed lambs. This observation may reflect an opportunistic capacity for this phylum in young lambs. The genus *Prevotella* was prominent in starter feed-fed lambs, reaching up to 40.37% of total reads, whereas in breast milk-fed lambs the relative abundance was only 17.20%. As previously reported, *Prevotella* is the most abundant genus in the adult rumen[52] and is thought to account for a large portion of the rumen microbial genetic and metabolic diversity[53]. Jami *et al.* reported that the *Bacteroidetes* phylum is less abundant when high-calorie diets are consumed, and its composition changes to being predominantly composed of genus *Prevotella* when high-fibre diets are introduced[48]. As described above, differences in the abundance of prominent genus between starter feed- and breast milk-fed groups of lambs could be a consequence of differences in fermentable substrates in the rumen.

Based on the data on rumen function variables, we explored the relationship between ruminal microbiota and rumen functions. The abundances of the genera *Prevotella, Bulleidia, Dialister,* and *Acidaminococcus* were each found to be positively correlated with TVFA and urea nitrogen concentrations, while the abundances of the genera *Fibrobacter, RFN20,* and *Dechloromonas* were each negatively correlated with TVFA and nitrogen concentrations, suggesting that they might be involved in nitrogen and volatile fatty acid metabolism. Similarly, the genera *Dialister* and *Acidaminococcus* might also participate in fibrolytic enzyme secretion and starch degradation.

Although ruminal microbiota in breast milk-fed lambs was more diverse than in starter feed-fed lambs, the absolute abundance of total bacteria in starter feed-fed lambs was markedly higher than in breast milk-fed lambs, indicating that the amount of fermentable substrate is critical for bacterial proliferation. Early starter feeding provides ample organic matter for micro-organism fermentation, thereby promoting the reproduction of bacteria and establishment of predominant microflora, while increasing the abundance of fermented products, such as VFA, NH_3, and other small molecules. As revealed by the results of RNA-seq in this present study, fermented products in the rumen enhance the growth of rumen papillae by affecting many metabolic pathways, including lipid and carbohydrate metabolism.

Thus, based on the overall mRNA expression profiles and abundance of bacteria, we conclude that improvements in rumen morphology and function in starter feed-fed lambs are a consequence of stimulation of ruminal ketogenesis and butanoate metabolism via the *HMGCL* and *HMGCS2* genes combined with changes in the fermentation type because of the ruminal microbiota. Confirmation of a causative relationship between gene expression and rumen development stimulated by starter feeding in sheep may require further studies that use knock-in or knock-out animals to up- or down-regulate the expression of related genes.

4. CONCLUSION

Overall, we identified 225 DEGs in the rumen between starter feed- and breast milk-fed lambs. Functionally, these genes were related to metabolism, particularly of lipids and carbohydrates. The genes *HMGCL*and *HMGCS2* were enriched in the butanoate metabolism pathway and play critical roles in ketogenesis. Additionally, the rumen microbiota of pre-ruminant lambs was responsive to dietary modifications, as well as structural and physiological changes in the host. The ruminal bacterial communities were more diverse in the breast milk-fed group than in the starter feed-fed group, as each group had a distinct microbiota. After starter feed consumption, many aerobic bacteria or non-fermenting bacteria disappeared. We propose that early administration of starter feed is beneficial to the development of rumen morphology and function in lambs, and the underlying mechanism may involve the stimulation of ruminal ketogenesis and butanoate metabolism via the *HMGCL* and *HMGCS2* genes combined with changes in the fermentation type mediated by the ruminal microbiota. This study enhances our understanding of the molecular mechanism that regulates rumen development in sheep.

5. MATERIALS AND METHODS

5.1 Ethics Statement

All experiments in this study were carried out in accordance with the approved guidelines from the Regulation of the Standing Committee of Gansu People's Congress. All experimental protocols and the collection of samples were approved by the Ethics Committee of Gansu Agricultural University.

5.2 Animals and sample collection

A total of 12 purebred male Hu lambs were used in this study that were obtained from a commercial sheep farm (Jinchang Zhongtian Sheep Industry Co., Gansu, China). Lambs weighed 3.51 ± 0.57 kg at birth. The beginning date of experimentation was adjusted for each lamb to account for different birth dates. To ensure that environmental conditions were similar throughout the experiment, all lambs were housed in a well-ventilated room with controlled temperature and humidity. The 12 lambs were divided into 2 groups (breast milk- and starter feed-fed groups) under the homogeneity principle (6 lambs per group). Starter feed-fed lambs received the starter diet (Table S7) from day 7 of life. Ewes were fed the diet three times per day at 06:30–08:30, 12:30–14:30, and 18:00–20:00. When feeding ewes, the lambs and ewes were separated and the lambs were fed a starter diet, but they did not touch the ewes' feed. After ewe feeding, the feeders were removed. Lambs were released from the feeding fence, and were fed the starter diet *ad libitum* with free access to water. Breast milk-fed lambs only suckled milk without receiving the starter diet. Body weights were measured before the morning feeding. All lambs were slaughtered at 42 days according to a standard procedure that was approved by the Biological Studies Animal Care and Use Committee, Gansu Province, P.R. of China.

For RNA sequencing, the rumen tissues of three starter feed-fed lambs and three breast milk-fed lambs were selected by referring to the index of rumen morphology, rumen fermentation parameters, and enzymatic activity of rumen micro-organisms. For rumen microbial 16S rRNA sequencing, 6 starter feed-fed lambs and 6 breast milk-fed lambs were selected. Immediately after slaughter, a section of rumen tissue from the ventral blind sac (2 × 2 cm in size) was quickly excised. Tissue samples were snap-frozen in liquid nitrogen and stored at −80 °C for subsequent total RNA analysis. Rumen fluid was collected and strained through four layers of cheesecloth, transferred into plastic bottles, and the stored at −80 °C for subsequent DNA analysis.

5.3 Rumen morphology

After slaughter, the rumen was separated and ligatured using cotton thread. The displacement method was used to measure the rumen volume. The rumen weight was measured after cleaning and eliminating the rumen contents. Rumen tissue specimens (~1 × 1 cm) were obtained from the cranial dorsal sac and fixed in 40% formaldehyde. After fixation, tissue specimens were trimmed and processed according to standard histological procedures, then were stained with haematoxylin and eosin. For each tissue specimen, a total of three sections with 5+ papillae length (PL), papillae width (PW), and muscular layer thickness (ML) were examined. PL and PW were measured in well-oriented papillae for each rumen sac, while the ML was measured at random locations in each rumen sac. PL was defined as the distance from the tip to the base of the papillae, while the PW was defined as the average width of the papillae base, middle, and tip. Morphometric analyses were performed at a magnification of 4 × 10 times (Olympus BX-51; Olympus Corporation, Tokyo, Japan) using Image Pro plus 6.0 (Media Cybernetics, Silver Spring, MD, USA).

5.4 Volatile fatty acid assay

The ruminal fluid pH was measured immediately after collection using a digital pH meter (PB21, Sartorius, Goettingen, Germany). Individual and total VFA in aliquots of strained ruminal fluid were quantified by gas chromatography[54]. Samples were injected using an auto-sampler (AI 3000, Thermo Scientific, Waltham, MA, USA) into an AE-FFAP capillary column (30 m × 0.25 mm × 0.33 μm, ATEO, LanZhou, China) on a Varian GC (TRACE 1300, Thermo Scientific, MA, USA). Samples were run at a split-ratio of 20:1 with a column temperature of 45 °C to 150 °C with an increase of 10 °C/min followed by a 5-min hold. The injector and detector temperatures were 200 °C and 250 °C, respectively. Peak integration was performed using Chromeleon® Software. All ruminal fluid samples were assayed in duplicate.

5.5 Enzymatic activity assay

Samples for enzyme activity potential measurements were prepared as suggested by Rey *et al*.34. Enzyme activity potentials of carboxymethylcellulase and amylase in rumen contents were determined by measuring the release of reducing sugars from substrates (carboxymethylcellulose and starch, respectively)34; reaction times were 30, 15, and 15 min, respectively. One enzyme activity unit (U) was defined as the amount of enzyme required to release 1 μmol reducing sugars (xylose or glucose equivalents)/min per g of wet rumen content[55]. Protease activity was assayed using azocasein as a substrate, according to the method of Eun and Beauchemin[56]. In this assay, the hydrolysis of azocasein released an azo group, which induced a colour change that could be measured by spectrophotometry at 420 nm.

5.6 RNA preparation and sequencing

Total RNA was extracted using TransZol reagent (TransGen Biotech, Beijing, China) according to the manufacturer's instructions. Paired-end libraries were prepared for each RNA-seq sample using a "Truseq®Stranded Total RNA Sample Preparation Kit (Illumina®)"; all of these procedures were performed according to the manufacturer's instructions. After quality control, sequencing of all the libraries was performed using an Illumina HiSeq 2500 instrument. RNA library construction and sequencing was performed by Shanghai Personal Biotechnology Co., Ltd.

5.7 Analyses of RNA-Seq data

The RNA-Seq results were transferred from an Illumina fastq format to a standard Sanger fastq format with fq_all2std.pl; data were processed using the Tophat–Cufflinks pipeline[57]. The *Ovis aries* reference genome and gtf annotation file were downloaded from Ensembl (*O. aries* 3.1) and the build index was implemented with bowtie version 2.1.0. TOPHAT (version 2.0.9) was used for transcriptome assembly, and the Cuffdiff script from Cufflinks was used for gene expression analysis with the option-classic-fpkm. Expression levels of each gene were represented by the FPKM value, which indicates the fragments per kilobase of exon per million fragments mapped, and was calculated by the following formula[58]:

$$\text{FPKM} = \frac{\textit{reads count} \times 1{,}000{,}000}{\textit{total reads map to genome} \times \textit{gene length (Kb)}}$$

Finally, genes were classified as differentially expressed if they exhibited two-fold or greater changes between the starter feed- and breast milk-fed groups, and if they showed statistical significance at $p \leq 0.05$ based on the Audic–Claverie method[59,60,61] with a false discovery rate (FDR) < 0.0562.

5.8 Cluster analysis

Hierarchical clustering analysis was carried out for all annotated transcripts from the breast milk- and starter feed-fed groups. The RPKM counts for each transcript were clustered using the Cluster 3.0 software package, and JAVA Treeview was used to view the cluster images. Results were visualized using JAVA Treeview[63].

5.9 Gene ontology and KEGG enrichment analysis of differentially expressed genes

The functional categories of DE genes were established with the GO (Gene Ontology) and KEGG (Kyoto Encyclopedia of Genes and Genomes). GO and KEGG enrichment analyses were performed via http://www.geneontology.org and http://www.genome.jp/kegg/. Analyses were conducted as described by Ashburner *et al.*[64] and Kanehisa *et al.*[65].

5.10 Q-RT-PCR validation of DE genes

Relative expression levels of DE genes in the rumen were quantified by real-time PCR. *GAPDH* was selected as an internal control for qRT-PCR validation because of the stable expression of its mRNA transcript in rumen tissues. The primer sequences and PCR conditions for analysed genes are listed in Supplementary Table S8. The $2^{-\Delta\Delta Ct}$ method was used for relative gene expression level analysis[66]. For each gene, the average ΔCt value of the breast feed-fed group was used as reference to calculate the −ΔΔCt value, and Student's *t*-test was used to analyze expression differences between two groups. Correlations were calculated by Pearson's correlation (SPSS 16.0, SPSS Inc., Chicago, IL, USA), using data from RNA-seq and qRT-PCR.

5.11 DNA extraction, PCR amplification, and sequencing

Total genomic DNA was isolated from rumen contents using the Omega E.Z.N.A™ Stoll DNA Kit according to the manufacturer's instructions (Omega Bio-Tek, USA). Concentrations of extracted DNA were determined using a Nano-Drop 2000 spectrophotometer (Thermo Scientific, Wilmington, DE, USA). The V1-V3 hypervariable region of the bacterial 16S ribosomal RNA gene was amplified by PCR from microbial genomic DNA that had been harvested from rumen fluid samples using barcoded fusion primers (forward: 5′-AGAGTTTGATCCTGGCTCAG-3′; reverse: 5′-TTACCGCGGCTGCTGGCAC-3′)[67]. PCR thermo cycling conditions were 94 °C for 5 min; 94 °C for 30 sec, 55 °C for 60 sec and 72 °C for the extension time, repeated for 30 cycles; 72 °C 7 min. PCR products were excised from a 1.5% agarose gel and purified using a QIAquick Gel Extraction Kit (QIAGEN, cat #28706). Barcoded V1-V3 amplicons were sequenced using the Roche GS FLX + platform (454 Life Sciences-A Roche Company, Branford, CT, USA).

5.12 Analysis of 16S rRNA sequencing data

Raw sequences were filtered through a quality control pipeline using the Quantitative Insight into Microbial Ecology (QIIME) tool kit[68] and mother[69] and bases with quality scores above 30 were retained for further analyses. High-quality reads were assigned to operational taxonomic units (OTUs) at a 97% identity threshold using the QIIME Uclust algorithm68, and taxonomical groups were assigned using the mother-based implementation of the RDP Bayesian classifier (http://rdp.cme.msu.edu/) with a 0.80 confidence threshold[70]. Chimeric DNA sequences were detected using UCHIME and removed[71]. Alpha-diversity values were obtained using various diversity indices (observed species, the Chao estimate, abundance-based coverage estimator [ACE], the Shannon and Simpson diversity indices). Principal coordinate analysis (PCoA) of microbial communities was performed using the Bray–Curtis distance[72].

5.13 Quantification of total bacteria and selected bacterial genera

Absolute quantitative real-time PCR (qPCR) was performed to measure copy numbers of the 16S rRNA genes of total bacteria and six selected bacterial genera (*Prevotella spp., Clostridium butyricum, Bacteroides spp., Firmicutes, Bacteroidetes,* and *Selenomonas ruminantium-Mitsuokella multiacida*). The primers, which were previously validated, are listed in

Table S9 in the Supplemental material. All qPCR assays were performed using SYBR Premix Ex Taq (Perfect Real Time; TaKaRa, Japan) on an ABI 7900HT Fast real-time PCR system (Applied Biosystems, Foster City, CA, USA). Standard curves for total bacteria and for each bacterial genus were prepared using plasmid DNA that contained each unique 16S rRNA insert. The copy number of the 16S rRNA gene per gram of fresh tissue was calculated as described by Zhou *et al.*[73]. The relative abundances of bacterial genera were calculated by dividing the copy number of the 16S rRNA gene for each genus by the copy number of the 16S rRNA genes for total bacteria.

5.14 Correlations between ruminal bacterial communities and anatomic and functions variables

Spearman's rank correlations between ruminal bacterial communities (Roch 454 relative abundance) and anatomic and functions variables were analysed using the PROC CORR procedure of SAS. Only those bacterial groups that represented >2% of the total community in at least one sample and that were detected in >50% of the rumen tissue samples were included in the analysis[29].

5.15 Statistical analysis

Differences in growth performance, rumen morphology, rumen fermentation parameters, and the enzymatic activity of rumen micro-organisms between the beast milk- and starter feed-fed groups were analysed using SPSS Version 16.0 for Windows (SPSS, Chicago, IL, USA). A one-way analysis of variance and the least significant difference (LSD) *t*-test were used for multiple comparisons. A P value of <0.05 was considered to indicate a statistically significant difference. Data are presented as means ± standard deviation.

5.16 Nucleotide sequence accession numbers

All RNA sequence reads have been submitted to the NCBI BioProject under accession code PRJNA317746 and PRJNA317540. All 16S rRNA sequence reads have been submitted to the NCBI BioProject under accession code PRJNA317793.

ACKNOWLEDGEMENTS

We thank Jinchang Zhongtian Sheep Industry Co. Ltd. (Jinchang) for providing facilities and equipment for animal feeding and slaughter. This study was supported financially by the National Natural Science Foundation of China (31260564), the Specialized Research Fund for the Doctoral Program of Higher Education (20130211110033), and the earmarked fund for China Agriculture Research System (CARS-39).

REFERENCES

1. Roh, S. G. et al. *Identification of differentially expressed transcripts in bovine rumen and abomasum using a differential display method.* Journal of animal science 85, 395–403 (2007).
2. Bryant, M. P., Nola, S., Cecelia, B. & Isadore, R. *Studies on the composition of the ruminal flora and fauna of young calves.* Journal of dairy science 41, 21 (1958).
3. Tamate, H., McGilliard, A. D., Jacoson, N. L. & Getty, R. *Effect of various dietaries on the anatomical development of the stomach in the calf.* Journal of dairy science 45, 13 (1962).
4. van Houtert, M. F. J. *The production and metabolism of volatilefatty acids by ruminants fed roughages: A review.* Anim Feed Sci Tech 43, 37 (1993).
5. Flatt, W. P., Warner, R. G. & Loosli, J. K. *The influence of purified materials on the development of the ruminant stomach.* Journal of dairy science 41, 8 (1958).
6. Lane, M. A. & Jesse, B. W. *Effect of volatile fatty acid infusion on development of the rumen epithelium in neonatal sheep.* Journal of dairy science 80, 740–746 (1997).
7. Sander, E. G., Warner, R. G., Harrison, H. N. & Loosli, J. K. *The stimulatory effect of sodium butyrate and sodium propionateon the development of rumen mucosa in the young calf.* Journal of dairy science 42, 6 (1959).
8. Drackley, J. K. *Calf nutrition from birth to breeding. The Veterinary clinics of North America.* Food animal practice 24, 55–86 (2008).
9. Lane, M. A., Baldwin, R. L. t. & Jesse, B. W. *Sheep rumen metabolic development in response to age and dietary treatments.* Journal of animal science 78, 1990–1996 (2000).
10. Connor, E. E., Baldwin, R. L. t., Li, C. J., Li, R. W. & Chung, H. *Gene expression in bovine rumen epithelium during weaning identifies molecular regulators of rumen development and growth.* Functional & integrative genomics 13, 133–142 (2013).
11. Kato, D. et al. *Utilization of digital differential display to identify differentially expressed genes related to rumen development.* Animal science journal=Nihon chikusan Gakkaiho (2015).
12. Connor, E. E. et al. *Transcriptional regulators transforming growth factor-beta1 and estrogen-related receptor-alpha identified as putative mediators of calf rumen epithelial tissue development and function during weaning.* Journal of dairy science 97, 4193–4207 (2014).
13. Flaga, J., Gorka, P., Zabielski, R. & Kowalski, Z. M. *Differences in monocarboxylic acid transporter type 1 expression in rumen epithelium of newborn calves due to age and milk or milk replacer feeding.* Journal of animal physiology and animal nutrition 99, 521–530 (2015).
14. Kocherginskaya, S. A., Aminov, R. I. & White, B. A. *Analysis of the Rumen Bacterial Diversity under two Different Diet Conditions using Denaturing Gradient Gel Electrophoresis, Random Sequencing, and Statistical Ecology Approaches.* Anaerobe 7, 119–134 (2001).

15. McAllister, T. A., Cheng, K. J., Okine, E. K. & Mathison, G. W.*Dietary, environmental and microbiological aspects of methane production in ruminants.* Can J Anim Sci 76, 13 (1996).
16. Tajima, K. et al. *Rumen Bacterial Community Transition During Adaptation to High-grain Diet.* Anaerobe 6, 12 (2000).
17. Lee, C. W. et al. *Temporal patterns of gene expression after acute hindlimb ischemia in mice: insights into the genomic program for collateral vessel development.* J Am Coll Cardiol 43, 474–482 (2004).
18. McSweeney, C. S. & Denman, S E. *Effect of sulfur supplements on cellulolytic rumen micro-organisms and microbial protein synthesis in cattle fed a high fibre diet.* Journal of applied microbiology 103, 1757–1765 (2007).
19. Edwards, J. E., R., M. N., J., T. A. & Wallace, R. J. *16S rDNA library-based analysis of ruminal bacterial diversity.* Antonie van Leeuwenhoek 86, 263–281 (2004).
20. Wright, A. D., Auckland, C. H. & Lynn, D. H. *Molecular diversity of methanogens in feedlot cattle from Ontario and Prince Edward Island, Canada.* Applied and environmental microbiology 73, 4206–4210 (2007).
21. Ellis, J. L. et al. *Aspects of rumen microbiology central to mechanistic modelling of methane production in cattle.* J Agr Sci146, 213–233 (2008).
22. Belanche, A., Ramos-Morales, E. & Newbold, C. J. *In vitroscreening of natural feed additives from crustaceans, diatoms, seaweeds and plant extracts to manipulate rumen fermentation.* Journal of the science of food and agriculture (2015).
23. Hart, K. J., Yanez-Ruiz, D. R., Duval, S. M., McEwan, N. R. & Newbold, C. J. *Plant extracts to manipulate rumen fermentation.* Anim Feed Sci Tech 147, 8–35 (2008).
24. Patra, A. K. & Saxena, J. *Dietary phytochemicals as rumen modifiers: a review of the effects on microbial populations.* Antonie van Leeuwenhoek 96, 363–375 (2009).
25. Wanapat, M. & Cherdthong, A. *Use of real-time PCR technique in studying rumen cellulolytic bacteria population as affected by level of roughage in swamp buffalo.* Current microbiology 58, 294–299 (2009).
26. Li, R. W., Connor, E. E., Li, C., Baldwin Vi, R. L. & Sparks, M. E.*Characterization of the rumen microbiota of pre-ruminant calves using metagenomic tools.* Environmental microbiology 14, 129–139 (2012).
27. Hristov, A. N., Callaway, T. R., Lee, C. & Dowd, S. E. *Rumen bacterial, archaeal, and fungal diversity of dairy cows in response to ingestion of lauric or myristic acid.* Journal of animal science90, 4449–4457 (2012).
28. Chen, Y., Penner, G. B., Li, M., Oba, M. & Guan, L. L. *Changes in bacterial diversity associated with epithelial tissue in the beef cow rumen during the transition to a high-grain diet.* Applied and environmental microbiology 77, 5770–5781 (2011).
29. Jiao, J., Huang, J., Zhou, C. & Tan, Z. *Taxonomic Identification of Ruminal Epithelial Bacterial Diversity during Rumen Development in Goats.* Applied and environmental microbiology 81, 3502–3509 (2015).
30. Baldwin Vi, R. L., McLeod, K. R., Klotz, J. L. & Heitmann, R. N.*Rumen development, intestinal growth and hepatic metabolism in the pre-and postweaning ruminant.* Journal of dairy science 87, 55–65 (2004).
31. Wang, Y. H. et al. *Effect of dietary starch on rumen and small intestine morphology and digesta pH in goats.* Livest Sci 122, 48–52 (2009).
32. Reynolds, C. K., Durst, B., Lupoli, B., Humphries, D. J. & Beever, D. E. *Visceral tissue mass and rumen volume in dairy cows during the transition from late gestation to early*
33. Faubladier, C., Julliand, V., Danel, J. & Philippeau, C. *Bacterial carbohydrate-degrading capacity in foal faeces: changes from birth to pre-weaning and the impact of maternal supplementation with fermented feed products.* The British journal of nutrition 110, 1040–1052 (2013).
34. Rey, M., Enjalbert, F. & Monteils, V. *Establishment of ruminal enzyme activities and fermentation capacity in dairy calves from birth through weaning.* Journal of dairy science 95, 1500–1512 (2012).
35. Fouts, D. E. et al. *Next generation sequencing to define prokaryotic and fungal diversity in the bovine rumen.* PloS one 7, e48289 (2012).

36. Fonty, G., Gouet, P., Jouany, J. P. & Senaud, J. *Establishment of the microflora and anaerobic fungi in the rumen of lambs*. Microbioloy 133, 1835–1843 (1987).
37. Duffield, T. et al. *Comparison of techniques for measurement of rumen pH in lactating dairy cows*. Journal of dairy science 87, 59–66 (2004).
38. Garrett, E. F. et al. *Diagnostic methods for the detection of subacute ruminal acidosis in dairy cows*. Journal of dairy science82, 1170–1178 (1999).
39. Li, S. et al. *Effects of subacute ruminal acidosis challenges on fermentation and endotoxins in the rumen and hindgut of dairy cows*. Journal of dairy science 95, 294–303 (2012).
40. Suarez, B. J. et al. *Effects of supplementing concentrates differing in carbohydrate composition in veal calf diets: I. Animal performance and rumen fermentation characteristics*. Journal of dairy science 89, 4365–4375 (2006).
41. Naeem, A., Drackley, J. K., Stamey, J. & Loor, J. J. *Role of metabolic and cellular proliferation genes in ruminal development in response to enhanced plane of nutrition in neonatal Holstein calves*. Journal of dairy science 95, 1807–1820 (2012).
42. Lane, M. A., Baldwin, R. L. t. & Jesse, B. W. *Developmental changes in ketogenic enzyme gene expression during sheep rumen development*. Journal of animal science 80, 1538–1544 (2002).
43. Pennington, R. J. & Sutherland, T. M. *The production of succinate from propionate by carbon dioxide fixation in sheep rumen epithelial tissue*. The Biochemical journal 60, xxxvii (1955).
44. Zhang, Q., Koser, S. L. & Donkin, S. S. *Propionate induces mRNA expression of gluconeogenic genes in bovine calf hepatocytes*. Journal of dairy science (2016).
45. Bobik, T. A. & Rasche, M. E. *Identification of the human methylmalonyl-CoA racemase gene based on the analysis of prokaryotic gene arrangements. Implications for decoding the human genome*. The Journal of biological chemistry 276, 37194–37198 (2001).
46. Bikker, H. et al. *A homozygous nonsense mutation in the methylmalonyl-CoA epimerase gene (MCEE) results in mild methylmalonic aciduria*. Human mutation 27, 640–643 (2006).
47. Liu, H., Vaddella, V. & Zhou, D. *Effects of chestnut tannins and coconut oil on growth performance, methane emission, ruminal fermentation, and microbial populations in sheep*. Journal of dairy science 94, 6069–6077 (2011).
48. Jami, E., Israel, A., Kotser, A. & Mizrahi, I. *Exploring the bovine rumen bacterial community from birth to adulthood*. The ISME journal 7, 1069–1079 (2013).
49. Petri, R. M. et al. *Characterization of the core rumen microbiome in cattle during transition from forage to concentrate as well as during and after an acidotic challenge*. PloS one 8, e83424 (2013).
50. Sadet, S., Martin, C., Meunier, B. & Morgavi, D. P. *PCR-DGGE analysis reveals a distinct diversity in the bacterial population attached to the rumen epithelium*. Animal: an international journal of animal bioscience 1, 939–944 (2007).
51. Rey, M. et al. *Establishment of ruminal bacterial community in dairy calves from birth to weaning is sequential*. Journal of applied microbiology 116, 245–257 (2014).
52. Stevenson, D. M. & Weimer, P. J. *Dominance of Prevotella and low abundance of classical ruminal bacterial species in the bovine rumen revealed by relative quantification real-time PCR*. Appl Microbiol Biotechnol 75, 165–174 (2007).
53. Purushe, J. et al. *Comparative genome analysis of Prevotella ruminicola and Prevotella bryantii: insights into their environmental niche*. Microbial ecology 60, 721–729 (2010).
54. Jiao, J. et al. *In vitro evaluation on neutral detergent fiber and cellulose digestion by post-ruminal microorganisms in goats*. Journal of the science of food and agriculture 94, 1745–1752 (2014).
55. Jiao, J. et al. *Rumen development process in goats as affected by supplemental feeding v. grazing: age-related anatomic development, functional achievement and microbial colonisation*. The British journal of nutrition 113, 888–900 (2015).
56. Eun, J. S. & Beauchemin, K. A. *Effects of a proteolytic feed enzyme on intake, digestion, ruminal fermentation, and milk production*. Journal of dairy science 88, 2140–2153 (2005).

57. Trapnell, C. et al. *Differential gene and transcript expression analysis of RNA-seq experiments with TopHat and Cufflinks*. Nature protocols 7, 562–578 (2012).
58. Trapnell, C. et al. *Transcript assembly and quantification by RNA-Seq reveals unannotated transcripts and isoform switching during cell differentiation*. Nature biotechnology 28, 511–515 (2010).
59. Audic, S. & Claverie, J. M. *The significance of digital gene expression profiles*. Genome research 7, 986–995 (1997).
60. Li, H. et al. *RNA-Seq derived identification of differential transcription in the chrysanthemum leaf following inoculation with Alternaria tenuissima*. BMC genomics 15, 9 (2014).
61. Young, N. D. et al. *Whole-genome sequence of Schistosoma haematobium*. Nature genetics 44, 221–225 (2012).
62. Buschmann, T., Zhang, R., Brash, D. E. & Bystrykh, L. V. *Enhancing the detection of barcoded reads in high throughput DNA sequencing data by controlling the false discovery rate*. BMC bioinformatics 15, 264 (2014).
63. Saldanha, A. J. *Java Treeview–extensible visualization of microarray data*. Bioinformatics 20, 3246–3248 (2004).
64. Ashburner, M. et al. *Gene ontology: tool for the unification of biology. The Gene Ontology Consortium*. Nature genetics 25, 25–29 (2000).
65. Kanehisa, M., Goto, S., Kawashima, S., Okuno, Y. & Hattori, M. *The KEGG resource for deciphering the genome*. Nucleic acids research 32, D277–D280 (2004).
66. Livak, K. J. & Schmittgen, T. D. *Analysis of relative gene expression data using real-time quantitative PCR and the 2(-Delta Delta C(T)) Method*. Methods 25, 402–408 (2001).
67. Zhao, L. et al. *Quantitative genetic background of the host influences gut microbiomes in chickens*. Scientific reports 3, 1163 (2013).
68. Caporaso, J. G. et al. *QIIME allows analysis of high-throughput community sequencing data*. Nature methods 7, 335–336 (2010).
69. Schloss, P. D. et al. *Introducing mothur: open-source, platform-independent, community-supported software for describing and comparing microbial communities*. Applied and environmental microbiology 75, 7537–7541 (2009).
70. Cole, J. R. et al. *The Ribosomal Database Project: improved alignments and new tools for rRNA analysis*. Nucleic acids research 37, D141–D145 (2009).
71. Edgar, R. C., Haas, B. J., Clemente, J. C., Quince, C. & Knight, R.*UCHIME improves sensitivity and speed of chimera detection*. Bioinformatics 27, 2194–2200 (2011).
72. Lozupone, C. & Knight, R. *UniFrac: a new phylogenetic method for comparing microbial communities*. Applied and environmental microbiology 71, 8228–8235 (2005).
73. Zhou, M., Hernandez-Sanabria, E. & Guan, L. L. *Assessment of the microbial ecology of ruminal methanogens in cattle with different feed efficiencies*. Applied and environmental microbiology 75, 6524–6533 (2009).

CHAPTER 8

House Dust Mite Allergy Models—Reliability for Research of Airway Defensive Mechanisms

Tomas Buday, Jana Plevkova

Department of Pathophysiology, Jessenius Faculty of Medicine, Comenius University in Bratislava, Martin, Slovakia

ABSTRACT

Extensive research of respiratory physiology and diseases of the airways and lungs provides better understanding of mechanisms of diseases and it is a source of new data with potential clinical application. The majority of the data in the field of airway defensive reflexes were obtained from guinea pig model, as the most convenient one. Neurophysiology and neuropharmacology of the cough reflex—which is exclusively mediated by the vagus nerve, share close similarities between guinea pigs and humans. The models used to study pathological processes and their influence on airway-defensive reflexes use sensitization with ovalbumin—the protein from chicken egg, which does not mimic allergies within human respiratory system. House dust mites (HDM) (Dermatophagoides pteronyssius and Dermatophagoides farinae) represent frequent human aeroallergens, however the HDM models are used considerably less than ovalbumin models. The primary objective of this review is to focus on already validated models of HDM-induced airway diseases to see, whether some of them are suitable to study mechanisms of peripheral cough plasticity in a condition of HDM-induced pathological processes. The main purpose of future use and validation of HDM model is to produce higher ability to translate the results obtained in animal models to human cough research.

Keywords: House Dust Mite, Allergy, Asthma, Animal Model, Ovalbumin, Cough

1. INTRODUCTION

Extensive research of respiratory physiology and diseases of the airways and lungs provides better understanding of mechanisms of diseases and it is a source of new data with potential clinical application. The key purpose of the basic research is to produce the results with the ability to translate and reproduce the data obtained in HEK (human embryonic kidney) cells, expression systems or animal models to man. The rationale for using animal models in biomedical research is scientific and animal models are likely to remain necessary until science develops alternative models and systems that are equally sound and robust [1] . The data obtained from animal models should be compared and translated to the conditions of human diseases.

Ultimate condition for successful translation is a similarity of the physiology of used animal models and human physiology. However, the substance of studied disease or pathological condition is supposed to be as similar as possible [2] .

Cough is the most common symptom of many inflammatory diseases such as asthma and chronic obstructive pulmonary disease (COPD) and as yet, there is no effective treatment (regardless of the cause) to target this symptom. It is also essential to profile, and to establish the efficacy of novel antitussive therapies being developed by the pharmaceutical industry in animal models prior to testing in man. Most pre-clinical studies of neural pathways in the cough reflex and its pharmacological regulation have been conducted in guinea pigs, rats, rabbits, cats and dogs. Although some investigators in the past used the rat and murine models for cough, recent understanding of the airways sensory nerves, regulation of cough limits their use as not relevant models as they lack certain types of nerve endings mediating and modulating cough in humans [3] .

One of the most useful and convenient animal models that have been widely used in the last few years and where much information has been gathered regarding the pharmacological modulation of the cough reflex, is the conscious guinea pigs. Extensive work in this field clearly shows that the neurophysiology and neuropharmacology of the guinea pig vagus nerve is very similar to man. A lot of data about the modulation of cough as a defensive mechanisms and nerves involved in their regulatory circuits were obtained from animal models using ovalbumin-induced airway hyperresponsiveness [4] -[7] .

Animal models using ovalbumin-induced allergic inflammation and subsequent hyper-sensitivity hyperresponsiveness may have limitations as to translation of the results. Airway hypersensitivity/hyperresponsiveness in humans is mediated by aeroallergens such as pollens or house dust mite (HDM) particles. These allergens probably better mimic the natural airway exposure and therefore may be more suitable models to study airway defense mechanisms on the background of airway inflammation.

The basic concept of this review is to focus on validated models of the house dust mite (HDM)-induced airway hypersensitivity/hyperresponsiveness in guinea pigs to assess, whether they can be successfully used to study the nature and modulation of airway defensive reflexes.

2. OVALBUMIN VERSUS HDM MODELS

Nowadays, ovalbumin (OVA) is standard model allergen used in respiratory research in animal models—it induces allergic reaction in animals and even tough OVA-induced allergy reproduces many features of asthmatic lung (e.g. specific IgE levels, Th2-associated eosinophilic inflammation, early (EAR) and late (LAR) asthmatic responses, tissue remodeling of the airway with associated hyperresponsiveness), the clinical significance of ovalbumin as model antigen is dubious.

Ovalbumin models of acute and chronic airway inflammation have been extensively reviewed by Kumar [8] with the list of advantages and disadvantages regarding the clinical application of the data, all together with the description of the limitation of OVA use. OVA, however, is seldom implicated in human asthma, and other groups have used alternative allergens that may have greater clinical relevance, for example house dust mite (HDM) and cockroach extracts [9] .

Actually, systemic delivery of OVA (with or without adjuvant—aluminum hydroxide) does not simulate the manner of sensitization to aeroallergens (e.g. pollen, molds or HDM particles) in everyday life. Another problem is that human subjects are basically not allergic to chicken egg protein (ovalbumin) unless it is a food allergy.

One of the alternative models is model of topically applied HDM allergens—a route that, incidentally, is reported to induce tolerance when using OVA.

Vast majority of asthmatic patients have increased levels of HDMspecific IgE antibodies and after contact with HDM allergens they manifest with symptoms of asthma (e.g. EAR, LAR and increase in inflammation of airway).

For reasons mentioned, the choice of HDM as model allergen used in sensitization of animal models seems logical—both choice of model antigen and manner of sensitization simulate situation from real life much better than OVA model of airway hypersensitivity. In general, currently used HDM models rely on topical administration of HDM into airways (usually intranasally) over course of multiple weeks, which results in airway inflammation accompanied with increased eosinophilia, but there is not enough evidence to show allergic nature of inflammation (HDM-specific IgE antibodies, Band T-cells). Therefore it is not clear if the inflammation is truly allergic or just a consequence of repeated nasal insults.

In these models, which are based on repeated exposure of model animal to HDM, it is not possible to clearly differentiate between sensitization phase and challenge phase. There would be more reports of occurrence of asthma phenotypes (HDM-specific IgE antibodies, EAR and LAR) if these models have strong allergic component, the same which is observed in OVA-$Al(OH)_3$, which shows specific IgE antibodies and OVA-induced degranulation of mast cells (hyperresponsivity type I). Majority of studies done on HDM models have either assessed specific IgE antibody levels or have reported only weak specific IgE response. This fact is in contrast with clinical experience, where in patients with HDM-induced asthma the levels of specific IgE antibodies is around 100 times higher than in non-allergic individuals. In addition, if specific IgG1 is used as a marker of HDM-specific clonal expansion of Band T-cells, it seems like this occurs after observation of large airway inflammation (eosinophilia, production of IL-5 and IL-13).

As of now, there is only little knowledge of the role of different cell populations in these models (such as mast cells, dendritic cells, B-cells and Th2-cells) and studying these effectors should be able to elucidate the mechanisms behind repeated HDM sensitization model [10] .

3. HOUSE DUST MITES AND THEIR ANTIGENIC POTENTIAL

House dust mite allergens are most important indoor allergens for humans, which induce asthma, rhinitis, dermatitis and other allergic diseases [11] . Mites are part of subclass Acari, where more than 50,000 species are identified to date. For purposes of research, relationship to diseases in humans and as a main perennial indoor allergen, the mites are of particular interest. Most important species, which are the main source of HDM allergens are Dermatophagoides pteronyssius and Dermatophagoides farinae. In addition to these two species, in regions with moderate climate Euroglyphus maynei and in tropical and subtropical areas Blomia tropicalis can be occasionally important. Lepidoglyphus destructor is the source of occupational allergens, which can in some locations contribute to house-dust allergy [12] . The potential to stimulate the immune system to hyperresponsive state of HDM is in mites themselves and in addition, in their faecal pellets as well [13] .

HDM induce allergic reaction via allergen specific CD4+ Th2 lymphocytes, however, the role of innate immunity is also critically important. This could be also the reason, why some of the Th2 directed therapeutic approaches have not been successful in clinical trials, because, some other mechanism (outside of the Th2 polarisation) may play an important role.

Immunoblotting studies of HDM allergens in extracts of mites have differed in detail, but in whole there is complexity and the existence of some dominant specificities. Western blotting using sera selected on the basis of anti-mite IgE has revealed 32 different IgE binding sites. Most sera recognized a unique combination of 6 to 10 bands. Out of 32 bands, nine of them were recognized at high frequency. Subsequent study revealed that most sera reacted to 1-7 bands, although some of them did have a broader reactivity spectrum. This pattern is consistent with several similar studies using either D. pteronyssius or D. farinae. However, the limitations of blotting studies should be recognized—the concentration of allergens vary from extract to extract, some allergens (e.g. Der p2) have their binding site enhanced by denaturation, while others (e.g. Der p1) lose their antigenicity potential and some of the allergens (e.g. Der p 7) show multiple bands [12] .

Two groups of HDM allergens have been studied extensively—25 kDa cysteine protease (group 1) and 14 kDa epididymal protein (group 2). More than 80% of humans with HDM allergy mount an IgE response to group 1 and more than 90% to group 2. However, about 20% of patients do not have an IgE antibody to either of two group allergens. There are many other HDM allergens containing IgE activity binding sites, although these are present in low and variable concentrations in mite extracts, usually at less than 1% of the group 1 and 2 allergens—they are therefore called minor allergens [11] . Antigenic potential of HDM is summarized in Figure 1, list of identified allergens from Dermatophagoides pteronyssius and Dermatophagoides farinae is present in Table1

4. HDM VERSUS AIRWAY—AN INTERACTION

The airway mucosa responds to the presence of HDM particles and the response basically starts with the recognition of the immunogenic epitopes by tissue immune cells, mainly dendritic cells.

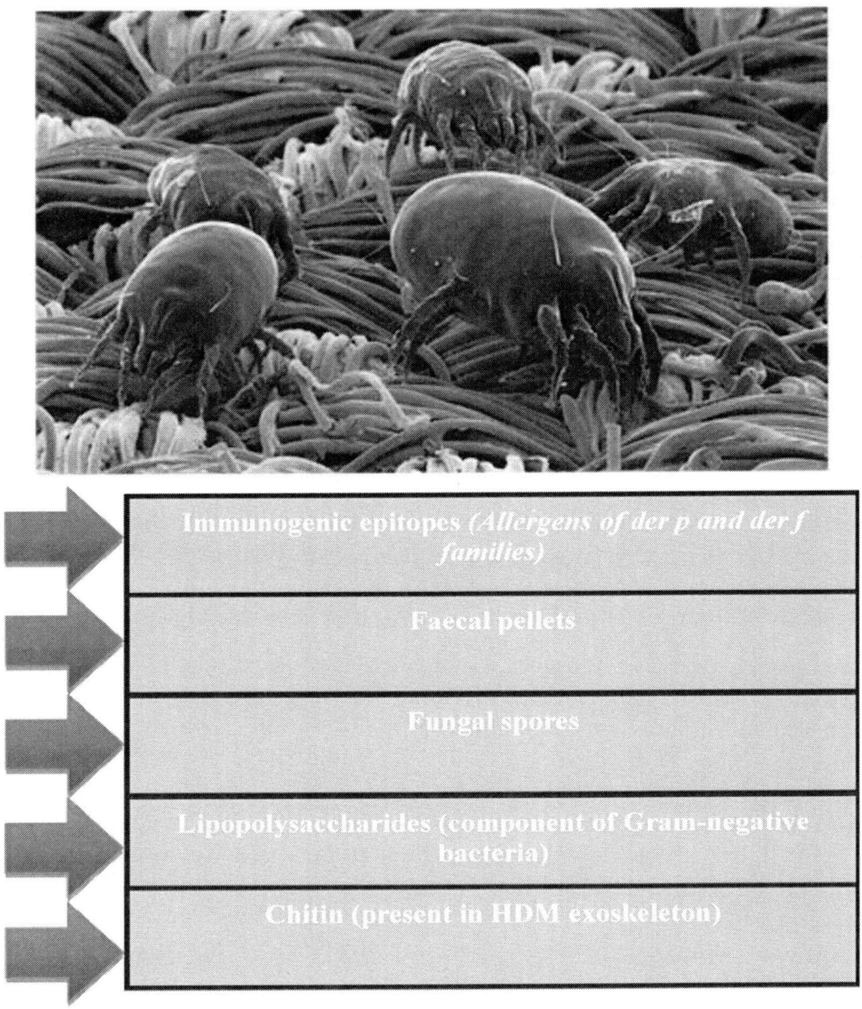

Figure 1. House dust mite allergenicity. Various components of HDM, and their associated faecal pellets and dust, which activate the immune system and initiate an inflammatory response. Modified according to source [13] . Image credit: http://www.blogas.lt/uploads/a/agiesha/204184.jpg

Epithelial cells of bronchi send signals to dendritic cells to induce Th2 immunity to inhaled HDM through release of innate pro-Th2 cytokines (e.g. GM-CSF, TSLP, IL-25 and IL-33) [14] -[16] . Recruitment of eosinophils and IL-4 competent basophils to the lung, which together with activated dendritic cells can traffic to mediastinal lymph nodes is the result of HDM inhalation. Only recently it was shown that FcεRI+ dendritic cells are necessity to initiation of Th2 immunity in a model of mucosal sensitization to inhaled HDM allergens and also the role of basophils is limited to amplification of response in asthmatic lung [17] .

Protease activity is very common amongst many human allergens, such as fungi, pollen, animal dander or bee venom. HDM and their faecal pellets contain proteolytic enzymes as well—group 1 (Der p 1 and Der f 1) are cysteine proteases and groups 3, 6 and 9 (Der p 3, Der p 6, Der p 9, Der f 3, Der f 6 and Der f 9) are serine proteases [18] . Together, they are responsible for 79% of proteolytic activity of HDM [19] . Der p 1 (Der f 1, respectively) disrupts intercellular tight junctions (proteins occluding and claudin-1 [20]), which allows delivery of antigen to submucosal antigen-presenting cells, whereas the serine proteases activate intracellular degradation of occludin and ZO-1 [21] . This impairment of barrier function of epithelium has not been replicated in murine in vivo model after single dose of HDM [22] as in poorly controlled asthmatic patients [23] , which suggests, that this impairment is a result of chronic exposure to HDM, rather than an acute insult.

Moreover, proteases originating from inhaled HDM are able to react with cell surface protease-activated receptors (PARs) in airways, in order to create infiltration of leukocytes and amplification of response to allergens [24] . Activation of these receptors on eosinophils and mast cells leads to their degranulation.

Extracts prepared from mites contain endotoxin. Gram-negative Bartonella sp. are thought to be the source of endotoxin [25] . It can activate both TLR2 and TLR4 directly—these receptors are expressed on epithelium of bronchi and on dendritic cells as well [26] [27] . Moreover, the expression of TLR4 is increased after HDM airway challenge on bronchial epithelium [28] . HDM-induced epithelial TLR4 signaling activates NF-κB and induces GM-CSF, which in turn results in alveolar macrophage maturation and up-regulation of the co-stimulatory molecules CD40 and CD86. There is a tight correlation between epithelial TLR4 expression and expression of these co-stimulatory molecules on macrophages [28] .

TLR activation on epithelial cells also results in trans-activation of epidermal growth factor receptor (EGFR) [30] which can result in cell migration and production of proteins involved in innate immune response including cytokines and mucins.

Table 1. Identified allergens of D. pteronyssius and D. farinae [29] .

Dermatophagoides pteronyssius		*Dermatophagoides farinae*	
Allergen	Biochemical Name	Allergen	Biochemical Name
Der p 1	Cysteine protease	Der f 1	Cysteine protease
Der p 2	NPC2 family	Der f 2	NPC2 family
Der p 3	Trypsin	Der f 3	Trypsin
Der p 4	Alpha amylase		
Der p 5			
Der p 6	Chymotrypsin	Der f 6	Chymotrypsin
Der p 7		Der f 7	
Der p 8	Glutathione s-transferase		
Der p 9	Collagenolytic serine protease		
Der p 10	Tropomyosin	Der f 10	Tropomyosin
Der p 11	Paramyosin	Der f 11	Paramyosin
		Der f 13	Fatty acid binding protein
Der p 14	Apolipophorin	Der f 14	Apolipophorin
Der p 15	Chitinase-like protein	Der f 15	Chitinase
		Der f 16	Gelsolin/villin
		Der f 17	Calcium binding protein
Der p 18	Chitin binding protein	Der f 18	Chitin binding protein
Der p 20	Arginine kinase		
Der p 21			
		Der f 22	
Der p 23	Peritrophin-like protein domain		
		Der f 24	Ubiquinol-cytochrome c reductase binding protein homologue

Mites carry fungal spores on their surface and also, the fungi are inhabitants of their gut [31] . C-type lectin receptors (dectin-1 and -2) are present of surface of myeloid cells. Their role is to function as classic PRRs by binding to β-glucans, which are present in fungal and bacterial cell walls, thus creating a link between innate and adaptive responses. Exposure of airway epithelium in vitro to HDM leads to β-glucan-dependent production of CCL20 [32] , which is a chemokine for immature dendritic cells. Extract prepared from dust mites is a specific stimulus for this chemokine and its levels are not affected by cell treatment with either purified Der p 1, chitin, endotoxin or other allergen extracts.

Chitin is the main component of fungal cell wall. It also provides rigidity of exoskeleton of many species (e.g. crustaceans, helminthes, insects and mites). Chitin is fairly potent adjuvant, which regulates both innate and adaptive (Th1, Th2 and Th17) immune responses [33] . It also acts as an element of recognition of tissue infiltration by cells responsible for innate immunity and also induces accumulation of basophils, which express IL-4 in murine model [34] . Exogenous chitin can stimulate macrophages by interacting with different cell surface receptors including TLR2 and dectin-1, thus providing another pathway by which HDM can elicit activation of immune system Allergens Der p 15 and Der p 18 are chitinases. These antigens display high frequency of binding to IgE antibodies in human sera—they are detected in up to 70% of HDM-allergic patients, which identifies them as HDM allergens of potential high importance [35] .

In addition to mite-derived allergens, innate pattern recognition receptors (PRR) derived from microbial compounds, including lipopolysaccharide and β-glucans are also routinely found in HDM extracts. Moreover, chitin, a glucosamine-based polymer that forms the mite exoskeleton as well as fungal cell walls also stimulates the immune system [13] .

Mite faecal pellets consist of three to five balls that contain food, debris and proteolytic enzymes. They are bound together by mucus and are covered in a chitinous peritropic membrane and egested. In a lifetime of 10 weeks, a house dust mite will produce approximately 2000 faecal particles (of average diameter 10 to 40 μm) and even larger number of partially digested enzyme-covered dust particles, which can be inhaled and deposited in the conducting airways [13] .

5. HDM RHINITIS MODEL

Studies using mice, especially transgenics and knockout strains, have been important for understanding the role of cytokines, adhesion molecules and cell receptors in allergic inflammatory responses [36] .

Preclinical research on allergic airway disease is focused predominately on lower airways and asthma—animal models of allergic rhinitis are relatively underdeveloped and understudied by comparison. Until recently, allergic rhinitis models have relied on short-term protocols and therefore present weaknesses of focusing mainly on acute inflammation and less attention is drawn on airway remodelling. However, more effort is being put into developing chronic models to address nasal obstruction, rhinorrhoea and remodelling which define allergic rhinitis in humans [36] .

Primary laboratory animals used to describe nasal responses to allergic stimuli are Dunkin-Hartley guinea pigs and, to lesser extent, the Brown Norway rats and BALB/c mice [36] .

Nasal obstruction as response to allergic stimulus is characterized by early and late phases of inflammation. An immediate and transient episode of itching and sneezing begins within seconds of exposure and lasts for 5 up to 30 minutes. Late phase is characterized by rhinorrhoea and airway obstruction that can last for hours [36] .

The initial irritation and sneeze reflex is promoted by mediators released from mast cells and basophils (histamine, tryptase, cysteinyl leukotrienes, platelet activating factor). Mucus hypersecretion with airway obstruction in secondary phase is accompanied by progression in mucosal swelling, tissue infiltration of eosinophils and neutrophils and the synthesis and release of prostaglandins, interleukins and reactive oxygen species [36] .

In allergic guinea pigs, the enumeration of nasal rubbings is subjective but useful measure, especially for testing early phase mechanisms and therapies involving histamineand leukotriene-dependent pathways. However, the histopathology in both early and late phases in allergic rhinitis hasn't been extensively studied. This contrasts with extensive descriptions of airway remodelling and pathology that drives analogous responses in lower airways (early and late bronchoconstriction) that are well-studied in mice [36] .

6. HDM ASTHMA MODELS

Various models have been developed by increasing the number of allergen exposures over many weeks-socalled chronic allergen challenge models. OVA as well as other allergens such as HDM extract and/or grass pollen have been used. In many of these studies long-term challenge leads to the development of tolerance and progressive decrease in airway hyperresponsiveness and airway eosinophilia. The induction of tolerance may be dependent on the strain of mouse used and the nature and route of allergen administration [37] .

Chronic models have been shown to reproduce some of the hallmarks of asthma such as goblet cell metaplasia, epithelial hypertrophy, subepithelial fibrosis and limited smooth muscle hyperplasia. Certain atypical features are observed in these chronic allergen challenge models that are not observed in humans, for example inflammation is not restricted to the conducting airways in the mouse, whereas it is restricted in humans. In addition, there are few mast cells in the airway walls or epithelium of mice [37] .

Despite these concerns, development of a standardized chronic model in mice is important as many of the clinical studies were initiated based on outcomes in primary allergen challenge models [37] .

7. APPLICATION TO THE COUGH RESEARCH

Cough is an exclusively vagal reflex mediated by the airway sensory neurons with the bodies located in the vagal nodose and jugular ganglia. These fibers are branching in the lower airways mucosa and their complex sensitivity to either mechanical stimulation or acid (nodose Aδ fibers) and different chemicals (jugular C fibres) make them perfect sensors for potentially dangerous events (acid aspiration, foreign body aspiration, accumulation of mucus, or inflammatory mediators). All of these stimuli are relevant to provoke coughing [38] .

Our team has studied cough reflex and its modulation under many physiological and pathological conditions and in the last ten years there has been a growing pool of evidence about such modulation named cough plasticity. The plasticity of the cough neural pathways may occur at the level of airway sensory nerves mediating the cough reflex, at the level of ganglia of the vagus nerve, and also at the level of brainstem circuits regulating neurogenesis of cough [39] . The result of such a modulation is that certain characteristics of cough reflex are modulated either in quantitative or qualitative aspect. The most important and probably clinically most relevant one is up-regulation of the cough reflex, which is usually a mechanism leading to intractable coughing with poor response to cough medication seriously reducing the patients' quality of life (syndrome of chronic cough, chronic hypersensitive cough, and idiopathic chronic cough concepts) [40] [41] .

Neuroplasticity changes at different levels are consequences of ongoing inflammation of various aetiologies, including viral, bacterial allergic and neurogenic inflammation. Inflammatory mediators, cytokines, other signal molecules and nerve growth factors are responsible for the changes in sensory channel protein expressions, changes of the nerve fibre phenotypes, branching and growing of existing nerves—thereby increasing the sensitivity of the airways to the nonspecific stimuli [42] [43] .

This hypersensitivity has been repeatedly documented in laboratory conditions and clinical settings in animal models and humans suffering from chronic cough. Since these models are in majority ovalbumin-based, there is a circumstantial suggestion that it may not be relevant for human respiratory tract disease. We do believe that validation of the HDM models for future cough research will be more relevant for translational research in this field.

ACKNOWLEDGEMENTS

Tomas Buday, Jana Plevkova Supported by VEGA 1/0107/2014 and Comenius University Grant No. UK/120/2014.

REFERENCES

1. Chow, P.K.H., Ng, R.T.H. and Ogden, B.E. (2008) Using Animal Models in Biomedical Research: A Primer for the Investigator. World Scientific, Hackensack.
2. Howells, D.W., Sena, E.S., Porritt, M.J., Rewell, S., et al. (2008) Can Animal Models of Disease Reliably Inform Human Studies? PLoS Medecine, 7, e1000245
3. Belvisi, M.G. and Bolser, D.C. (2002) Summary: Animal Models for Cough. Pulmonary Pharmacology & Therapeutics, 15, 249-250. http://dx.doi.org/10.1006/pupt.2002.0349
4. Mokry, J., Joskova, M., Mokra, D., Christensen, I. and Nosalova, G. (2013) Effects of Selective Inhibition of PDE4 and PDE7 on Airway Reactivity and Cough in Healthy and Ovalbumin-Sensitized Guinea Pigs. Advances in Experimental Medicine and Biology, 756, 57-64. http://dx.doi.org/10.1007/978-94-007-4549-0_8
5. Sutovska, M., Adamkov, M., Kocmalova, M., Mesarosova, L., Oravec, M. and Franova, S. (2013) CRAC Ion Channels and Airway Defense Reflexes in Experimental Allergic Inflammation. Advances in Experimental Medicine and Biology, 756, 39-48. http://dx.doi.org/10.1007/978-94-007-4549-0_6
6. Hori, A., Fujimura, M., Ohkura, N. and Tokuda, A. (2011) Involvement of Nitric Oxide (NO) in Cough Reflex Sensitivity between Non-Sensitized and OVA-Sensitized Guinea Pigs. Cough, 7, 5. http://dx.doi.org/10.1186/1745-9974-7-5
7. Brozmanova, M., Plevkova, J., Tatar, M. and Kollarik, M. (2008) Cough Reflex Sensitivity Is Increased in the Guinea Pig Model of Allergic Rhinitis. Journal of Physiology and Pharmacology, 59, 153-161.
8. Kumar, R.K., Herbert, C. and Foster, P.S. (2008) The "Classical" Ovalbumin Challenge Model of Asthma in Mice. Current Drug Targets, 9, 485-494. http://dx.doi.org/10.2174/138945008784533561
9. Nials, A.T. and Uddin, S. (2008) Mouse Models of Allergic Asthma: Acute and Chronic Allergen Challenge. Disease Models Mechanisms, 1, 213-220 http://dx.doi.org/10.1242/dmm.000323
10. Birrell, M.A., Van Oosterhout, A.J.M. and Belvisi, M.G. (2010) Do the Current House Dust Mite-Driven Models Really Mimic Allergic Asthma? European Respiratory Journal, 36, 1220-1223 http://dx.doi.org/10.1183/09031936.00069110
11. An, S., Chen, L., Long, C., Liu, X., Xu, X., Lu, X., Rong, M., Liu, Z. and Lai, R. (2013) Dermatophagoides farinae Allergens Diversity Identification by Proteomics. Molecular & Cellular Proteomics, 12, 1818-1828. http://dx.doi.org/10.1074/mcp.M112.027136
12. Thomas, W.R. and Smith, W. (1998) House-Dust-Mite Allergens. Allergy, 53, 821-832. http://dx.doi.org/10.1111/j.1398-9995.1998.tb03987.x
13. Gregory, L.G. and Lloyd, C.M. (2011) Orchestrating House Dust Mite-Associated Allergy in the Lung. Trends in Immunology, 32, 402-411. http://dx.doi.org/10.1016/j.it.2011.06.006
14. Barrett, N.A., et al. (2009) Dectin-2 Recognition of House Dust Mite Triggers Cysteinyl Leukotriene Generation by Dendritic Cells. The Journal of Immunology, 182, 1119-1128.http://dx.doi.org/10.4049/jimmunol.182.2.1119
15. Hammad, H. and Lambrecht, B.N. (2008) Dendritic Cells and Epithelial Cells: Linking Innate and Adaptive Immunity in Asthma. Nature Reviews Immunology, 8, 193-204. http://dx.doi.org/10.1038/nri2275

16. Lambrecht, B.N. and Hammad, H. (2009) Biology of Lung Dendritic Cells at the Origin of Asthma. Immunity, 31, 412- 424. http://dx.doi.org/10.1016/j.immuni.2009.08.008
17. Hammad, H., et al. (2010) Inflammatory Dendritic Cells—Not Basophils—Are Necessary and Sufficient for Induction of Th2 Immunity to Inhaled House Dust Mite Allergen. The Journal of Experimental Medicine, 207, 2097-2111. http://dx.doi.org/10.1084/jem.20101563
18. Chapman, M.D., et al. (2007) Proteases as Th2 Adjuvants. Current Allergy and Asthma Reports, 7, 363-367. http://dx.doi.org/10.1007/s11882-007-0055-6
19. Stewart, G.A., et al. (1994) Immunobiology of the Serine Protease Allergens from House Dust Mites. American Journal of Industrial Medicine, 25, 105-107. http://dx.doi.org/10.1002/ajim.4700250128
20. Wan, H., et al. (1999) Der p 1 Facilitates Transepithelial Allergen Delivery by Disruption of Tight Junctions. Journal of Clinical Investigation, 104, 123-133. http://dx.doi.org/10.1172/JCI5844
21. Wan, H., et al. (2001) The Transmembrane Protein Occludin of Epithelial Tight Junctions Is a Functional Target for Serine Peptidases from Faecal Pellets of Dermatophagoides Pteronyssinus. Clinical & Experimental Allergy, 31, 279- 294. http://dx.doi.org/10.1046/j.1365-2222.2001.00970.x
22. Turi, G.J., et al. (2011) The Effects of Inhaled House Dust Mite on Airway Barrier Function and Sensitivity to Inhaled Methacholine in Mice. American Journal of Physiology-Lung Cellular and Molecular Physiology, 300, L185-L190. http://dx.doi.org/10.1152/ajplung.00271.2010
23. Bhure, U.N., et al. (2009) Lung Epithelial Permeability and Inhaled Furosemide: Added Dimensions in Asthmatics. Annals of Nuclear Medicine, 23, 549-557. http://dx.doi.org/10.1007/s12149-009-0275-z
24. Reed, C.E. and Kita, H. (2004) The Role of Protease Activation of Inflammation in Allergic Respiratory Diseases. Journal of Allergy and Clinical Immunology, 114, 997-1008. http://dx.doi.org/10.1016/j.jaci.2004.07.060
25. Valerio, C.R., et al. (2005) Bacterial 16S Ribosomal DNA in House Dust Mite Cultures. Journal of Allergy and Clinical Immunology, 116, 1296-1300. http://dx.doi.org/10.1016/j.jaci.2005.09.046
26. Hammad, H., et al. (2009) House Dust Mite Allergen Induces Asthma via Toll-Like Receptor 4 Triggering of Airway Structural Cells. Nature Medicine, 15, 410-416. http://dx.doi.org/10.1038/nm.1946
27. Phipps, S., et al. (2009) Toll/IL-1 Signaling Is Critical for House Dust Mite-Specific Helper T Cell Type 2 and Type 17 [Corrected] Responses. American Journal of Respiratory and Critical Care Medicine, 179, 883-893. http://dx.doi.org/10.1164/rccm.200806-974OC
28. Hongjia, L., et al. (2010) House Dust Mite Regulate the Lung Inflammation of Asthmatic Mice through TLR4 Pathway in Airway Epithelial Cells. Cell Biochemistry and Function, 28, 597-603. http://dx.doi.org/10.1002/cbf.1697
29. Allergen Nomenclature: IUIS Allergen Nomenclature Sub-Committee. http://www.allergen.org
30. Koff, J.L., et al. (2008) Multiple TLRs Activate EGFR via a Signaling Cascade to Produce Innate Immune Responses in Airway Epithelium. American Journal of Physiology-Lung Cellular and Molecular Physiology, 294, L1068-L1075.http://dx.doi.org/ 10.1152/ajplung.00025.2008
31. Hay, D.B., et al. (1992) How Relevant Are House Dust Mite-Fungal Interactions in Laboratory Culture to the Natural Dust System? Experimental and Applied Acarology, 16, 37-47. http://dx.doi.org/10.1007/BF01201491
32. Nathan, A.T., et al. (2009) Innate Immune Responses of Airway Epithelium to House Dust Mite Are Mediated through Beta-Glucan-Dependent Pathways. Journal of Allergy and Clinical Immunology, 123, 612-618. http://dx.doi.org/10.1016/j.jaci.2008.12.006
33. Lee, C.G., et al. (2011) Role of Chitin and Chitinase/Chitinase-Like Proteins in Inflammation, Tissue Remodeling, and Injury. Annual Review of Physiology, 73, 479-501. http://dx.doi.org/10.1146/annurev-physiol-012110-142250

34. Reese, T.A., et al. (2007) Chitin Induces Accumulation in Tissue of Innate Immune Cells Associated with Allergy. Nature, 447, 92-96. http://dx.doi.org/10.1038/nature05746
35. O'Neil, S.E., et al. (2006) The Chitinase Allergens Der p 15 and Der p 18 from Dermatophagoides pteronyssinus. Clinical & Experimental Allergy, 36, 831-839. http://dx.doi.org/10.1111/j.1365-2222.2006.02497.x
36. Wagner, J.G. and Harkema, J.R. (2012) From Mouse to Man: Translational Value of Animal Models of Allergic Rhinitis. In: Kowalski, M., Ed., Allergic Rhinitis, InTech, 7-11. http://dx.doi.org/10.5772/25352
37. Shin, Y.S., Takeda, K. and Gelfand, E.W. (2009) Understanding Asthma Using Animal Models. Allergy, Asthma Immunology Research, 1, 10-18. http://dx.doi.org/10.4168/aair.2009.1.1.10
38. Canning, B.J. and Chou, Y.L. (2009) Cough Sensors. I. Physiological and Pharmacological Properties of the Afferent Nerves Regulating Cough. Handbook of Experimental Pharmacology, 187, 23-47. http://dx.doi.org/10.1007/978-3-540-79842-2_2
39. Mazzone, S.B. and Canning, B.J. (2002) Synergistic Interactions between Airway Afferent Nerve Subtypes Mediating Reflex Bronchospasm in Guinea Pigs. American Journal of Physiology-Regulatory, Integrative and Comparative Physiology, 283, R86-R98.
40. Morice, A.H. (2013) Chronic Cough Hypersensitivity Syndrome. Cough, 9, 14. http://dx.doi.org/10.1186/1745-9974-9-14
41. McGarvey, L.P. (2008) Does Idiopathic Cough Exist? Lung, 186, S78-S81. http://dx.doi.org/10.1007/s00408-007-9048-4
42. Lieu, T. and Undem, B.J. (2011) Neuroplasticity in Vagal Afferent Neurons Involved in Cough. Pulmonary Pharmacology & Therapeutics, 24, 276-279. http://dx.doi.org/10.1016/j.pupt.2011.02.003
43. Chen, C.Y., Joad, J.P., Bric, J. and Bonham, A.C. (2009) Central Mechanisms I: Plasticity of Central Pathways. Handbook of Experimental Pharmacology, 187, 187-201. http://dx.doi.org/10.1007/978-3-540-79842-2_9

CHAPTER 9

Functional characterization and differential nutritional regulation of putative Elovl5 and Elovl4 elongases in large yellow croaker (*Larimichthys crocea*)

Songlin Li[1], Óscar Monroig[2], Tianjiao Wang[1], Yuhui Yuan[1], Juan Carlos Navarro[3], Francisco Hontoria[3], Kai Liao[1], Douglas R. Tocher[2], Kangsen Mai[1], Wei Xu[1] & Qinghui Ai[1,4]

[1]*Key Laboratory of Aquaculture Nutrition and Feed (Ministry of Agriculture) and Key Laboratory of Mariculture (Ministry of Education), Ocean University of China, Qingdao, 266003, People's Republic of China*

[2]*Institute of Aquaculture, School of Natural Sciences, University of Stirling, Stirling, FK9 4LA, Scotland, UK*

[3]*Instituto de Acuicultura Torre de la Sal (IATS-CSIC), Ribera de Cabanes, 12595, Castellón, Spain*

[4]*Laboratory for Marine Fisheries and Aquaculture, Qingdao National Laboratory for Marine Science and Technology, Qingdao, 266003, People's Republic of China*

ABSTRACT

In the present study, two elongases, Elovl4 and Elovl5, were functionally characterized and their transcriptional regulation in response to n-3 LC-PUFA administration were investigated *in vivo* and *in vitro*. We previously described the molecular characterization of croaker *elovl5*. Here, we report the full-length cDNA sequence of croaker *elovl4*, which contained 1794 bp (excluding the polyA tail), including 909 bp of coding region that encoded a polypeptide of 302 amino acids possessing all the characteristic features of Elovl proteins. Functional studies showed that croaker Elovl5, displayed high elongation activity towards C_{18} and C_{20}PUFA, with only low activity towards C_{22} PUFA. In contrast, croaker Elovl4 could effectively convert both C_{20} and C_{22} PUFA to longer polyenoic products up to C_{34}. n-3 LC-PUFA suppressed transcription of the two elongase genes, as well as *srebp-1* and *lxrα*, major regulators of hepatic lipid metabolism. The results of dual-luciferase reporter assays and *in vitro* studies both indicated that the transcriptions of *elovl5* and *elovl4* elongases could be regulated by Lxrα. Moreover, Lxrα could mediate the transcription of *elovl4* directly or indirectly through regulating the transcription of *srebp-1*. The above findings contribute further insight and understanding of the mechanisms regulating LC-PUFA biosynthesis in marine fish species.

1. INTRODUCTION

Long chain (C_{20-24}) polyunsaturated fatty acids (LC-PUFA), especially the omega-3 (n-3) fatty acids eicosapentaenoic acid (20:5n-3, EPA) and docosahexaenoic acid (22:6n-3, DHA), play pivotal roles in promoting cognitive development, enhancing immune function and minimizing the risk of cardiovascular disease[1,2,3]. The limited biosynthetic capacity for LC-PUFA in humans makes consumption of fish, especially oily fish, a key component of the diet to guarantee adequate supply of these physiologically essential nutrients4. Farmed fish that now contribute significantly to seafood intake in the human diet have been traditionally produced with diets containing high levels of marine fishmeal (FM) and fish oil (FO), ingredients that ensured good growth rates and high levels of n-3 LC-PUFA in the flesh. However, the ever-increasing use of vegetable oils (VO) in aquafeeds has led to reduced deposition of n-3 LC-PUFA in farmed fish and, consequently, to a decrease in their nutritional value for human consumers[4,5,6]. This has prompted interest in elucidating the mechanisms underlying the endogenous LC-PUFA biosynthetic pathways in farmed finfish species[7, 8].

Elongases of very long-chain fatty acids (Elovl) catalyze the rate-limiting condensation step in the elongation of fatty acids including LC-PUFA biosynthesis[9, 10]. The crucial role of Elovl enzymes has also been confirmed in fish species[7, 11], and the overexpression of fish elongases elevated the endogenous production of LC-PUFA in transgenic zebrafish and nibe croaker[12, 13]. Three fatty acid elongases, Elovl2, Elovl4 and Elovl5, have been identified and functionally characterized as crucial enzymes involved in the biosynthetic pathway of LC-PUFA[11]. While Elovl2 appears to be lost in Acanthopterygii, a phylogenic group that includes the vast majority of the most important farmed marine fish species, Elovl4 and Elovl5 are present in virtually all fish species[11]. Importantly, functional studies have shown that Elovl5 can effectively elongate both C_{18} and C_{20} PUFA, whereas Elovl4 is mainly involved in the elongation of C_{20-22} LC-PUFA producing polyenes up to 36 carbons[7,11].

Dietary fatty acids can regulate the transcription of *elovl5* and *elovl4* in fish and, generally, expression of these elongase genes is suppressed by dietary n-3 LC-PUFA[14,15,16,17,18,19,20]. Additionally, PUFA and their metabolites regulate lipid metabolism, including inhibition of lipogenesis and activation of fatty acid oxidation, through several transcription factors such as the nuclear receptors liver X receptor α (Lxrα) and sterol regulatory element-binding proteins (Srebp)[21]. Srebp-1c can stimulate expression of target genes directly through binding to sterol response elements (SREs) in the promoter region of target genes[22]. In addition to the direct regulation mechanism, Lxrα can also indirectly mediate the expression of target genes through regulating the transcription of *srebp-1* and certain other transcription factors[23]. However, the role of these transcription mediators on the expression of elongases Elovl4 and Elovl5 remains poorly understood in fish species. It has been demonstrated that grouper *elovl5* promoter activity was elevated by over-expression of *lxrα*, but not when *srebp-1* was over-expressed, indicating the direct regulation role of Lxrα on *elovl5*expression[16]. However, mouse *elovl5* was activated by Srebp-1c directly, and also by Lxrα indirectly elevating the transcription of *elovl5* via Srebp-1c[24]. Elovl4 is the member of Elovl family that has been investigated most recently in fish and no information is available on the underlying regulation mechanism of *elovl4*[25].

Large yellow croaker (*Larimichthys crocea*) is an important carnivorous marine fish species widely cultured in southeast China. Recently, low retention of n-3 LC-PUFA has been observed after a large proportion of dietary FO was replaced by VO, which seriously affected the nutritional quality of the fillet of farmed croaker[26]. Recently an *elovl5*-like gene (cDNA) was cloned and its mRNA expression was investigated in response to dietary fatty acids, although no functional data was reported[20]. The aims of the present study were to characterize the function of the previously cloned *elovl5*, and the molecular cloning and functional characterization of an *elovl4* cDNA from *L. crocea*. In addition, we also investigated the underlying mechanisms regulating the expression of these two elongases. The results may contribute to increased insight into the potential regulatory mechanisms involved, which will be key to understanding and applying strategies for enhancing the biosynthesis of LC-PUFA in large yellow croaker.

2. RESULTS

2.1 Molecular cloning and phylogenetics of the *L. crocea* Elovl4

The full-length cDNA of croaker *elovl4* encompassed 1794 bp including a 262 bp 5′ untranslated terminal region (UTR), a 909 bp coding region that encoded 302 aa and a 623 bp 3′ UTR (GenBank Accession No. KP681700). BLAST analysis revealed that the croaker *elovl4* shared high sequence identity with *elovl4* sequences from other teleosts including the orange-spotted grouper (*Epinephelus coioides*, 94%), rabbitfish (*Siganus canaliculatus*, 94%), cobia (*Rachycentron canadum*, 92%) and Atlantic salmon (*Salmo salar*, 86%). Additionally, the croaker *elovl4*cDNA was 40% identical to croaker *elovl5*.

Characteristically, the croaker *elovl4* deduced protein contained one histidine box (HXXHH), five putative membrane-spanning domains, and single lysine and arginine residues (RXKXX) at the carboxyl terminus (Fig. 1). Phylogenetic tree analysis showed that the croaker Elovl4 clustered with several other Elovl4-like sequences from fish and mammals, and more distantly with Elovl2 and Elovl5 from these vertebrate lineages (Fig. 2). The results strongly suggested that the newly cloned cDNA encoded an Elovl4 elongase.

2.2 Functional characterization of the Elovl4 and Elovl5 elongases from *L. crocea*

The functions of croaker Elovl4 and Elovl5 were characterized by determining the FA profiles of *S. cerevisiae* transformed with pYES2-Elo4 or pYES2-Elo5, respectively, and grown in the presence of potential FA substrates. For Elovl4, its role in the biosynthesis of saturated VLC-FA was studied by comparing the saturated FA ($>C_{24}$) profiles of yeast transformed with pYES2-Elo4 and pYES2 (control) (Table 1). No significant differences were found in the content of 24:0, 26:0, 28:0, 30:0 or 32:0, indicating that the croaker Elovl4 does not have a role in the biosynthesis of saturated VLC-FA. In order to investigate the role of the *L. crocea* Elovl4 in the biosynthesis of VLC-PUFA, pYES2-Elo4 transformed yeast were grown in the presence of C_{18} (18:4n-3 and 18:3n-6), C_{20} (20:5n-3 and 20:4n-6) and C_{22} (22:5n-3, 22:4n-6 and 22:6n-3) substrates.

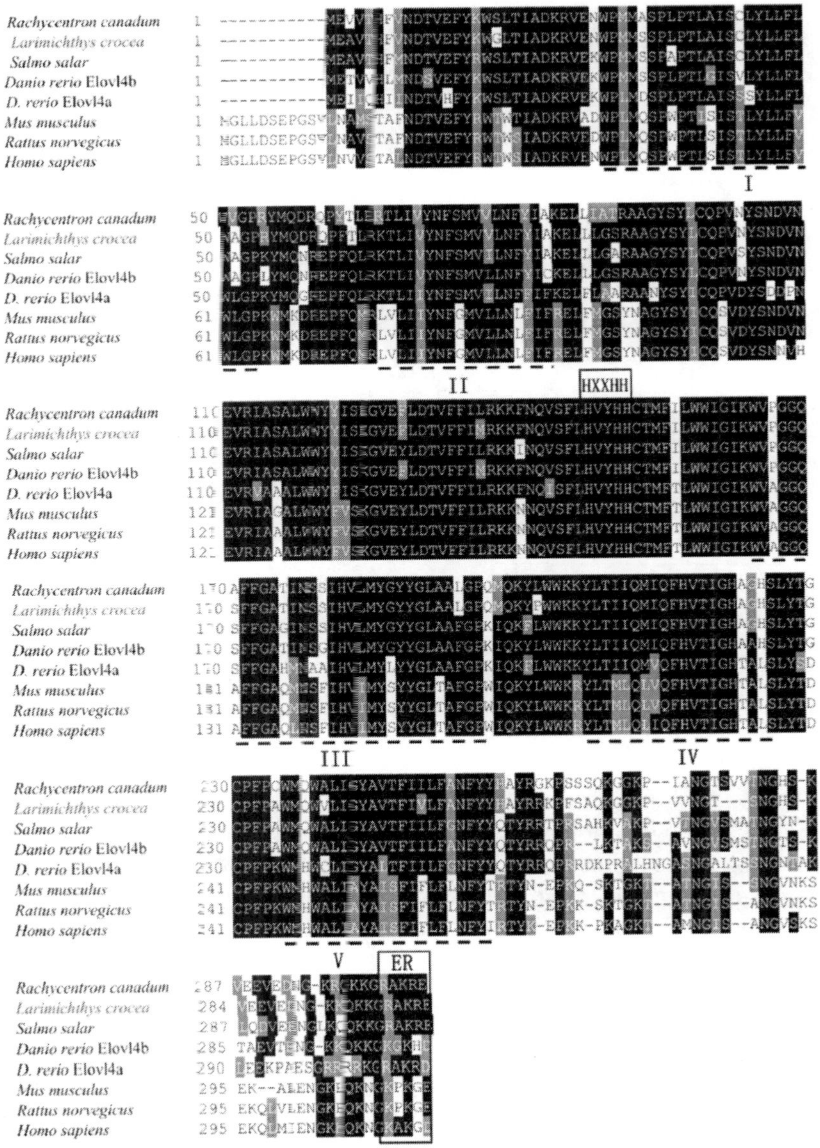

Figure 1 Comparison of the deduced amino acid (AA) sequences of Elovl4 from large yellow croaker, other fish, mouse and human. The AA sequences were aligned using ClustalX, and identity/similarity shading was based on a 75% identity threshold. Identical residues are shaded black and similar residues are shaded grey. Indicated are the conserved HXXHH histidine box motif, five (I–V) putative membrane-spanning domains and the ER retrieval signal.

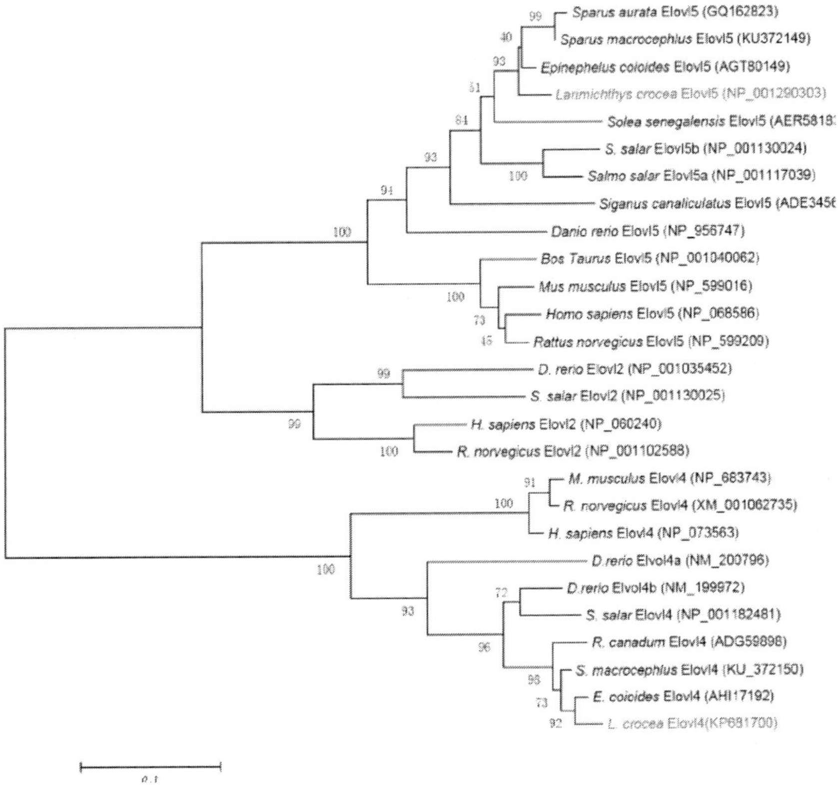

Figure 2 Phylogenetic tree comparing the large yellow croaker Elovl4 with elongase proteins from other organisms. The tree was constructed using the Neighbor Joining method [45] using MEGA4. The numbers represent the frequencies (%) with which the tree topology presented was replicated after 1000 iterations.

The results confirmed that croaker Elovl4 was able to elongate all PUFA substrates producing in most cases polyenes of up to 32 or 34 carbons, except for DHA that was only elongated to 24:6n-3 (Table 2). Interestingly, the elongation towards C_{28} and C_{30} substrates showed the highest conversion values (%) regardless of the initial PUFA substrate. However, despite limited activity towards DHA, the croaker Elovl4 could elongate both 20:5n-3 and 22:5n-3 to 24:5n-3, the C_{24}substrate for DHA synthesis.

Table 1: Functional characterization of the large yellow croaker Elovl4: Role in biosynthesis of very long-chain saturated fatty acids (FA).

FA	Control	Elovl4
24:0	8.4 ± 2.2	10.8 ± 2.0
26:0	79.4 ± 7.0	78.3 ± 4.9
28:0	7.8 ± 1.7	6.8 ± 3.0
30:0	3.7 ± 2.4	3.4 ± 0.5
32:0	0.6 ± 0.6	0.7 ± 0.7
34:0	nd	nd

Table 2: Functional characterization of the large yellow croaker Elovl4: conversions of polyunsaturated fatty acid (FA) substrates.

FA substrate	Product	% Conversion
18:4n-3	20:4n-3	8.5
	22:4n-3	32.6
	24:4n-3	71.5
	26:4n-3	100.0
	28:4n-3	100.0
	30:4n-3	94.5
	32:4n-3	72.2
18:3n-6	20:3n-6	6.1
	22:3n-6	36.4
	24:3n-6	67.1
	26:3n-6	90.7
	28:3n-6	93.2
	30:3n-6	89.7
	32:3n-6	21.9
20:5n-3	22:5n-3	14.7
	24:5n-3	47.0
	26:5n-3	69.2
	28:5n-3	92.7
	30:5n-3	98.2
	32:5n-3	81.9
	34:5n-3	11.1

20:4n-6	22:4n-6	21.1
	24:4n-6	50.7
	26:4n-6	65.7
	28:4n-6	85.0
	30:4n-6	91.4
	32:4n-6	41.1
22:5n-3	24:5n-3	23.5
	26:5n-3	59.8
	28:5n-3	90.4
	30:5n-3	97.6
	32:5n-3	75.5
22:4n-6	24:4n-6	35.8
	26:4n-6	71.0
	28:4n-6	89.7
	30:4n-6	94.2
	32:4n-6	49.4
22:6n-3	24:6n-3	1.2

The function of *L. crocea* Elovl5 was characterized by growing the transgenic yeast expressing its coding region in the presence of C_{18}(18:3n-3, 18:2n-6, 18:4n-3 and 18:3n-6), C_{20} (20:5n-3 and 20:4n-6) and C_{22} (22:5n-3 and 22:4n-6) PUFA substrates. The results from the yeast assays revealed that the *L. crocea* Elovl5 exhibited high conversion towards C_{18} PUFA, particularly 18:4n-3 (86.7%) and 18:3n-6 (77.9%), followed by C_{20} substrates including 20:5n-3 (74.6%) and 20:4n-6 (70.0%). The *L. crocea* Elovl5 showed relatively low activity towards C_{22}substrates such as 22:5n-3 (7.6%) and 22:4n-6 (2.3%) (Table 3). Fatty acid composition of yeast transformed with the empty pYES2 vector and grown in the presence of PUFA substrates used in both Elovl4 and Elovl5 functional assays primarily consisted of the four major endogenous FA, namely 16:0, 16:1n-7, 18:0 and 18:1 (including 18:1n-9 and 18:1n-7 isomers), together with any exogenously added FA (data not shown).

2.3 Tissue distribution of *elovl4*

The distribution of *elovl4* mRNA in *L. crocea* was detected in several tissues, including eye, brain, testis, heart, liver, kidney, spleen, stomach, intestine and muscle. Gene expression analysis revealed that the highest levels of *elovl4* transcript were found in eye, follow by brain and testis (Fig. 3).

Table 3: Functional characterization of the large yellow croaker Elovl5: conversions of polyunsaturated fatty acid (FA) substrates.

FA substrate	FA product	% Conversion
18:3n-3	20:3n-3	48.3
	22:3n-3	6.3
	24:3n-3	7.4
18:2n-6	20:2n-6	24.1
	22:2n-6	6.9
18:4n-3	20:4n-3	86.7
	22:4n-3	28.0
	24:4n-3	3.1
18:3n-6	20:3n-6	77.9
	22:3n-6	21.8
	24:3n-6	4.8
20:5n-3	22:5n-3	74.6
	24:5n-3	11.1
	26:5n-3	6.2
20:4n-6	22:4n-6	70.0
	24:4n-6	4.6
	26:4n-6	4.8
22:5n-3	24:5n-3	7.6
	26:5n-3	6.3
22:4n-6	24:4n-6	2.3

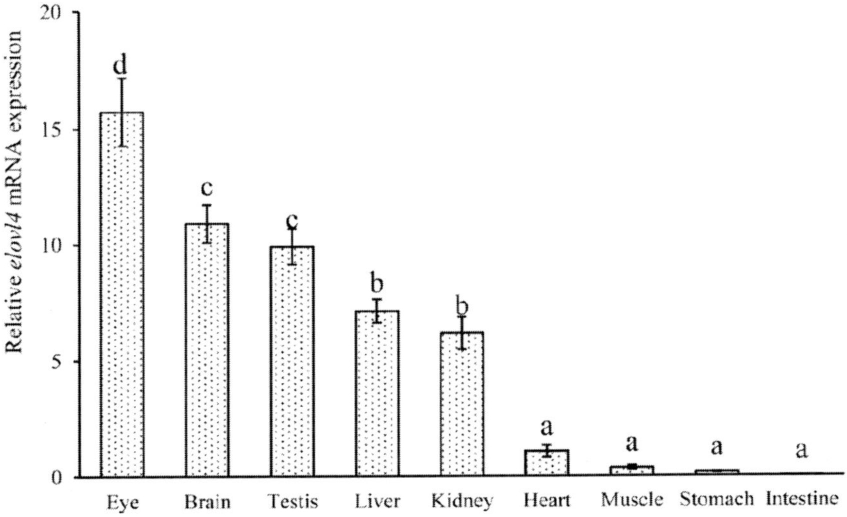

Figure 3 Tissue expression of elovl4 in large yellow croaker. Values (means ± standard error of the mean, SEM) in bars that have the same letter are not significantly different ($P > 0.05$; Tukey's test) among treatments (n = 3).

2.4 Nutritional regulation of the *L. crocea elovl4* and *elovl5* and transcription factors: *in vivo* and *in vitro* trials

The expression of *elovl4* and *elovl5* from fish fed the experimental diets containing varying levels of LC-PUFA is shown in Fig. 4A. The results showed that the expression of both genes was down-regulated in fish fed diets containing both high and moderate n-3 LC-PUFA levels in comparison to diets with low n-3 LC-PUFA ($P < 0.05$). *In vitro* assays performed with hepatocyte primary cultures prepared from fish fed a standard diet indicated that the expression of *elovl4*, but not *elovl5*, was down-regulated by DHA supplementation ($P < 0.05$) (Fig. 4B). Additionally, supplementation of EPA to hepatocytes suppressed the expression of both elongases ($P < 0.05$) (Fig. 4C).

With regards to the regulation of transcription factors, dietary n-3 LC-PUFA significantly decreased the expression of *srebp-1* and *lxrα* ($P < 0.05$) (Fig. 5A). The *in vitro* studies clearly showed that both DHA and EPA significantly suppressed the transcription of *srebp-1* and *lxrα* ($P < 0.05$) (Fig. 5B and C).

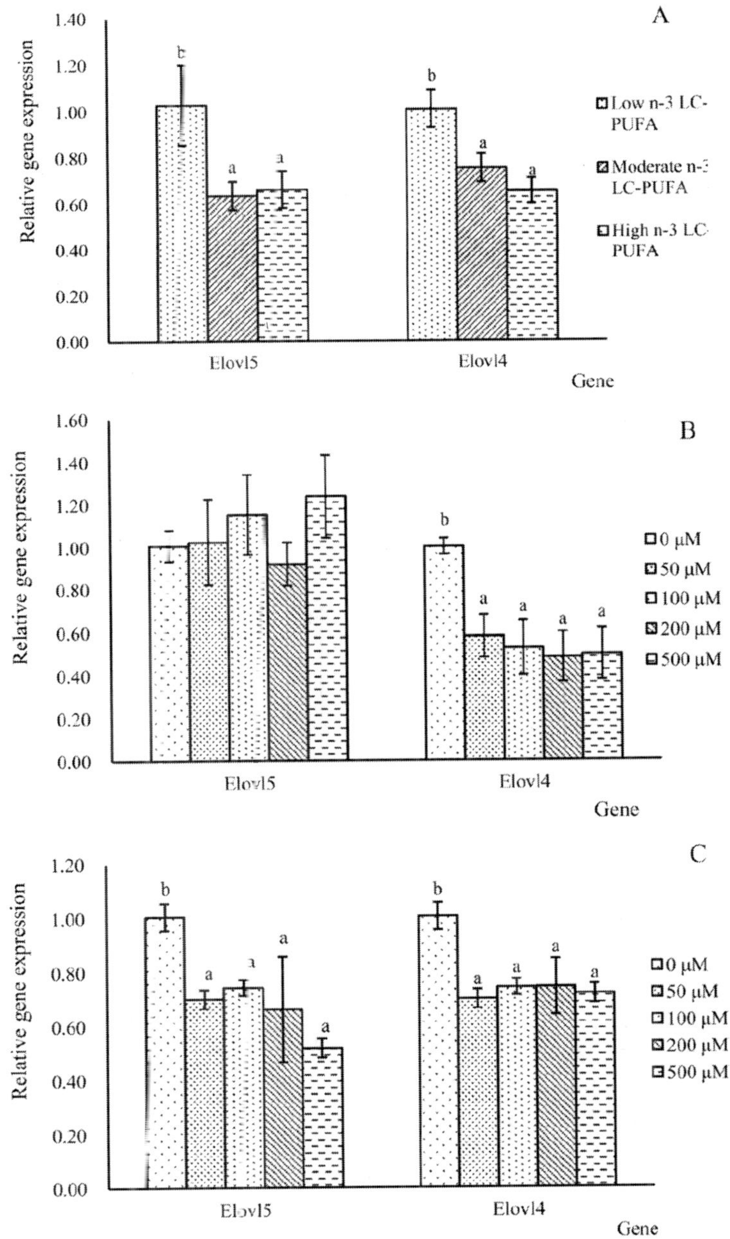

Figure 4 Effects of dietary n-3 LC-PUFA (**A**), DHA (**B**) and EPA (**C**) contents on the expression of *elovl4* and *elovl5*. Values (means ± SEM) in bars that have the same letter are not significantly different ($P > 0.05$; Tukey's test) among treatments (n = 3).

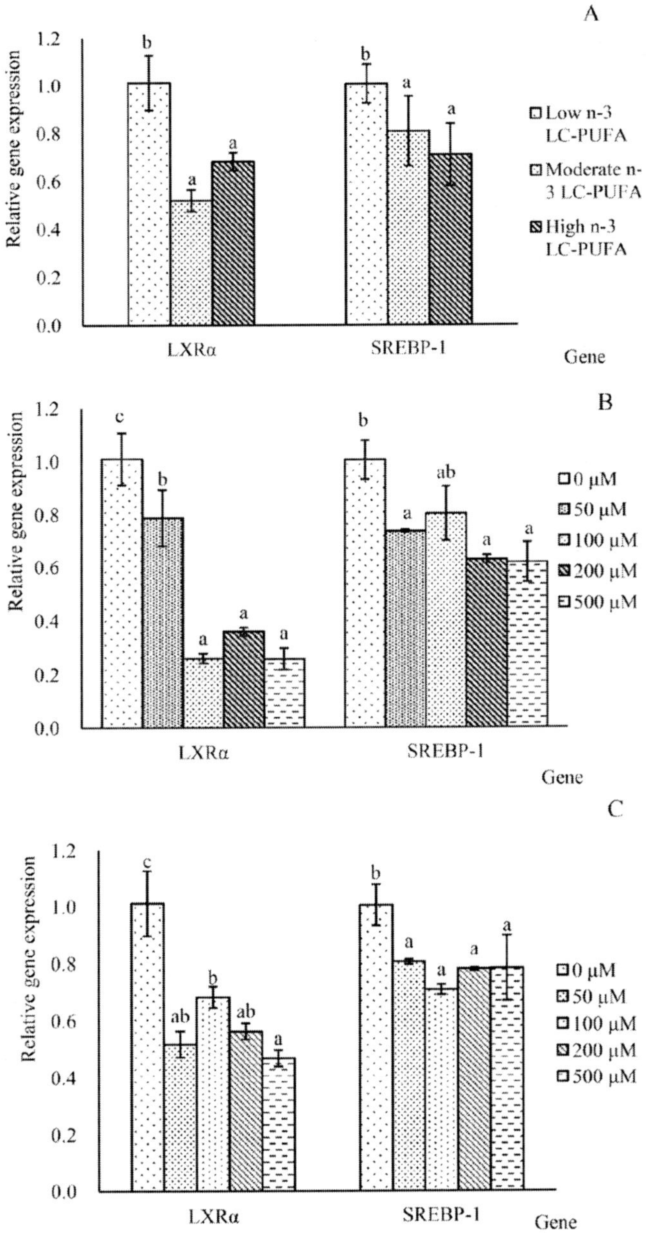

Figure 5 Effects of dietary n-3 LC-PUFA (**A**), DHA (**B**) and EPA (**C**) contents on the expression of the transcription factors *lxrα* and *srebp-1*. Values (means ± standard error of the mean, SEM) in bars that have the same letter are not significantly different ($P > 0.05$; Tukey's test) among treatments (n = 3).

2.5 Transcriptional regulation of the *L. crocea* elovl4 and elovl5 by Lxrα and Srebp-1

elovl4 ($P < 0.05$), but the expression of *elovl4* was significantly suppressed through the inhibition of Srebp-1 ($P < 0.05$) in yellow croaker hepatocytes (Fig. 6A). These results indicated that both *lxrα* and *srebp*up-regulated *elovl4*.

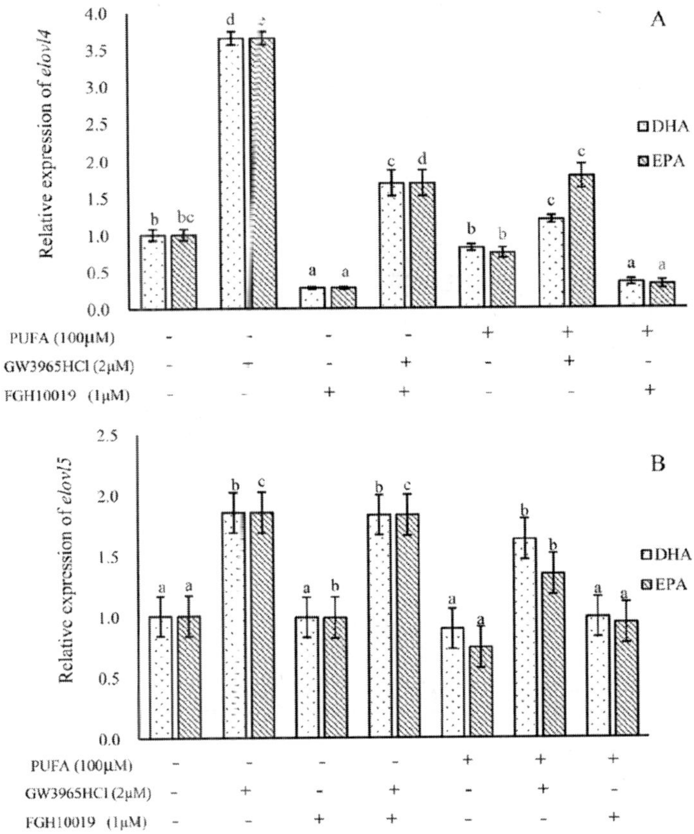

Figure 6 Transcriptional regulation of *L. crocea elovl4* (**A**) and *elovl5* (**B**) by Lxrα and Srebp-1. The Lxrα activator, GW3965HCl, and Srebp-1 inhibitor, FGH10019, were used to investigate effects on the expression of *L. crocea elovl4* and *elovl5*. "PUFA" indicates that hepatocytes were incubated with either docosahexaenoic acid (DHA) or eicosapentaenoic acid (EPA). Values (means ± standard error of the mean, SEM) in bars that have the same letter are not significantly different ($P > 0.05$; Tukey's test) among treatments (n = 3).

The addition of DHA or EPA significantly reduced the abovementioned up-regulation effects of *lxrα* and *srebp* ($P < 0.05$) (Fig. 6A). In addition, the inhibition of Srebp-1 significantly decreased the regulatory role of LXRα on the transcription of *elovl4* ($P < 0.05$) (Fig. 6A).

Regarding the transcription of *elovl5*, over-expression of *lxrα*significantly increased the transcription of *elovl5* ($P < 0.05$), while the inhibition of Srebp-1 had no significant effect on *elovl5* expression ($P > 0.05$) (Fig. 6B). Moreover, the addition of DHA or EPA suppressed the up-regulation of *elovl5* by Lxrα ($P < 0.05$) (Fig. 6B). Nevertheless, inhibition of Srebp-1 did not affect the regulation of *elovl5* by Lxrα ($P < 0.05$) (Fig. 6B).

2.6 Dual-luciferase reporter assays

In order to investigate the molecular mechanisms involved in the regulation of *L. crocea elovl4* and *elovl5*, regions of 1921 (GenBank Accession No. KY863452) and 2291 bp (GenBank Accession No. KY863451), respectively, upstream of the initiation codon of the genes were cloned to perform dual-luciferase reporter assays. For *elovl4*, the reporter activity was 2.41-fold than the control (Fig. 7A). The croaker *elovl4* reporter activity was significantly elevated by over-expression of *lxrα* and *srebp-1* (Fig. 7A). For *elovl5*, the reporter activity was 1.66-fold than the control (Fig. 7B). Similarly, the *L. crocea elovl5* reporter activity was significantly elevated by over-expression of *lxrα* (Fig. 7B). However, the over-expression of *srebp-1* had no significant effect on the activity of the *elovl5* promoter (Fig. 7B).

3. DISCUSSION

The large yellow croaker is widely cultured in China due to its great commercial value. Recently, a study investigating the effects of FO replacement in diets for *L. crocea* showed that low retention of LC-PUFA in muscle occurred, thus reducing the nutritional quality of the product[26]. This was possibly due to the low LC-PUFA biosynthetic ability of large yellow croaker as generally reported for most marine species[27]. Elongases play key roles in the biosynthesis of LC-PUFA in fish[7, 11] and thus a better understanding of the underlying mechanisms regulating the transcription of elongases would contribute to strategies designed to enhance the endogenous LC-PUFA synthetic ability of the large yellow croaker that may lead to increased nutritional value of farmed products for human consumers.

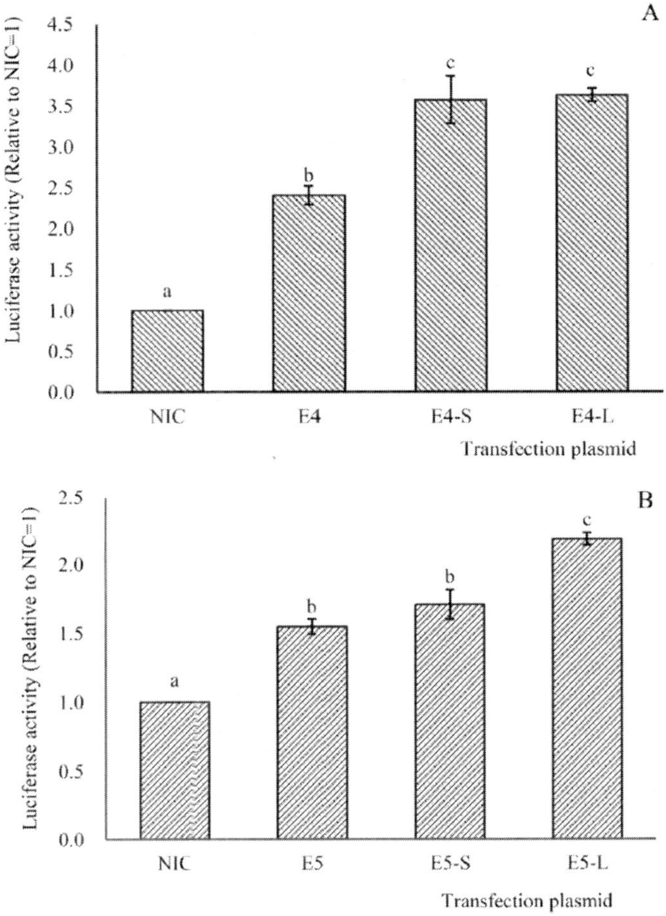

Figure 7 The dual-luciferase reporter assays of *elovl4* (**A**) and *elovl5* (**B**). The type of transfection plasmid applied to each group was as follows: NIC: PGL3 – basic + PCS2 + PRL – CMV; E: PGL-Elo promoter + PCS2 + PRL – CMV; E – S: PGL-Elo promoter + PCS2 – SREBP-1 + PRL – CMV; E – L: PGL-Elo promoter + PCS2 – LXRα + PRL – CMV. Values (means ± standard error of the mean, SEM) in bars that have the same letter are not significantly different ($P > 0.05$; Tukey's test) among treatments (n = 3).

Seven members of the Elovl protein family have been described in vertebrates[9, 28]. Typically, Elovl2 and Elovl5 play roles in the biosynthesis of LC-PUFA in vertebrates, having somewhat overlapping functions possibly related to their common evolutionary origin[29]. Whereas Elovl5 is present in virtually all teleosts, Elovl2 appears to be lost during evolution and is absent in Acanthopterygii that encompass the majority of farmed marine fish[11]. Fish generally possess two Elovl4-like elongases[11], but only the so-called "isoform b" has PUFA as preferred substrates[25]. In the present study, we conducted the molecular cloning of an *elovl4* and functionally characterized the functions of the herein reported Elovl4 and the previously cloned *elovl5*[20]. In addition, we investigated regulatory mechanisms of transcription factors in both elongases and the influence of "dietary" LC-PUFA in both *in vivo* and *in vitro* systems. The results demonstrated that the expression of the two elongases was regulated by n-3 LC-PUFA via specific transcriptional factors that exert different actions on each elongase.

The isolated croaker Elovl4 possessed all the features of the elongase family including transmembrane domains, a histidine box (HXXHH), and an arginine residue and single lysine residue (RXKXX) in the canonical C-terminal, indicating its role in LC-PUFA biosynthesis[30]. Zebrafish possessed two Elovl4 enzymes, Elovl4a and Elovl4b, with different functions[25]. Both the Elovl4 in zebrafish had roles in the biosynthesis of saturated VLC-FA, while Elovl4b was also able to operate on PUFA substrates producing polyenes of up to 36 carbons and thus participated in the biosynthesis of VLC-PUFA[25]. Phylogenetic analysis of the newly cloned *L. crocea elovl4* showed that it was an orthologue of zebrafish Elovl4b and not Elovl4a. Consistent with these genes being orthologues of the zebrafish Elovl4b, previously cloned *elovl4*-like elongases isolated from cobia[26], Atlantic salmon[31], rabbitfish[32] and orange-spotted grouper[15], had similar functions compared to zebrafish Elovl4b and they were all capable of elongating both saturated and polyunsaturated FA. In contrast, the *L. crocea* Elovl4-b like elongase did not show any activity towards endogenous saturated FA in yeast and thus does not appear to have a role in the biosynthesis of saturated VLC-FA as its previously characterized counterparts. The reasons why Elovl4 in *L. crocea* does not play a role in the biosynthesis of saturated VLC-FA remain uncertain, but a recent study reporting the functional characterization of Elovl4 in nibe croaker *Nibea mitsukurii* did not report any activity towards saturated FA using a similar expression system as that used herein. While it is possible that the activity of the nibe croaker Elovl4 towards endogenous saturated FA in yeast was simply not measured and/or reported, the close phylogenetic relationship between *L. crocea* and *N. mitsukurii*, both from the Sciaenidae family, possibly indicated that some Elovl4 from certain species or families within teleosts have subfunctionalized during evolution, as has occurred in the fatty acyl desaturase Fads2[33], a protein family that also plays key roles in LC-PUFA biosynthesis in vertebrates[11,28].

In agreement with the functional data obtained with fish Elovl4[15, 26, 31,32], the *L. crocea* Elovl4 showed activity towards PUFA substrates and thus confirmed its role in the biosynthesis of VLC-PUFA. Thus, the *L. crocea*

Elovl4 exhibited high elongation efficiency towards a range of PUFA substrates, namely C_{20} (20:5n-3 and 20:4n-6) and C_{22} (22:5n-3 and 22:4n-6) PUFA, with C_{18} substrates (18:4n-3 and 18:3n-3) being elongated to a lesser extent. The functions of VLC-PUFA are not fully understood, but studies in mammals have suggested that these compounds play pivotal roles in vision[34] and reproduction[35]. Importantly, the large yellow croaker Elovl4 could produce 24:5n-3 from directly supplemented 22:5n-3 or by step-wise elongation of 20:5n-3. Such an ability of Elovl4 has been noted as an important feature by which this enzyme can contribute to DHA biosynthesis through the Sprecher pathway in species that, like most marine farmed species including the large yellow croaker, might have lost Elovl2. Thus, the elongation product of Elovl4, namely 24:5n-3, can be desaturated to 24:6n-3 and subsequently chain shortened to 22:6n-3 by partial β-oxidation[36]Similar activities were also demonstrated for Elovl4 from zebrafish (isoform b). Atlantic salmon, cobia, rabbitfish, nibe croaker and orange-spotted grouper[15, 25, 31, 32, 37, 38].

The large yellow croaker Elovl5 could efficiently elongate C_{18} (18:4n-3 and 18:3n-6) and C_{20} (20:5n-3 and 20:4n-6) substrates, although C_{22}PUFA including 22:5n-3 and 22:4n-6 were only elongated to a small degree. Such substrate preference as observed here with the *L. crocea*Elovl5 is highly conserved among teleost Elovl5[11]. Furthermore, the *L. crocea* Elovl5 was also able to elongate 18:3n-3 and 18:2n-6, FA substrates not only for Δ6 desaturases to initiate LC-PUFA biosynthesis through the classical "Δ6 pathway" but also for elongases like Elovl5 to initiate biosynthesis through the alternative "Δ8 pathway"[7]. The ability of the *L. crocea* Elovl5 to actively elongate 18:3n-3 and 18:2n-6 has also been reported in Elovl5 elongases characterized from the southern bluefin tuna (*Thunnus maccoyii*), Japanese eel (*Anguilla japonica*), striped snakehead (*Channa striata*) and orange-spotted grouper (*E. coioides*)[14, 16, 39, 40]. While further studies are required to elucidate the desaturase gene and function repertoire, the results from the present study clearly indicated that *L. crocea* expressed genes encoding elongases with all the activities required for the biosynthesis of C_{20-22}LC-PUFA from C_{18} precursors supplied in the diet.

Aquafeeds are currently formulated with decreasing levels of the marine ingredients FM and FO due to increased demand and unpredictable availability, leading to high price volatility. Nutritional regulation of genes encoding desaturases and elongases has been proposed as a possible strategy to enhance the endogenous production on n-3 LC-PUFA and optimise their retention in farmed fish[8, 41, 42]. In general, n-3 LC-PUFA can suppress the transcription of the Elovl4 and Elovl5 encoding genes in fish[14,15,16,17,18,19,20]. Additionally, previous studies found that LC-PUFA and their metabolites can regulate transcription of lipid metabolism related genes through modulation of transcription factors including, among others, Lxrα and Srebp-1[24]. However, the regulatory mechanisms of these transcription factors on the two herein studied elongases has been barely investigated. It has been demonstrated that n-3 LC-PUFA can suppress the transcription of the *elovl4* and *elovl5* elongases both *in vivo* and *in vitro*, indicating the existence of negative feedback regulation in the LC-PUFA biosynthetic pathway. Moreover, the expression of *srebp-1* and *lxrα* were also suppressed by n-3 LC-PUFA both *in vivo* and *in vitro* in the present study, and this may largely account for the feedback suppression of expression of both *elovl4* and *elovl5*.

In mammals, Srebp-1 positively regulates transcription via interacting with sterol response elements (SRE) in the promoter region of target genes[22]. Furthermore, Lxrα can activate transcription directly and/or indirectly via regulating Srebp-1 or other transcription factors[23, 43]. Dual-luciferase reporter assays were conducted in the present study to clarify the regulatory mechanisms by which Lxrα and Srebp-1 modulate the expression of *elovl4* and *elovl5*. The results showed that the large yellow croaker Elovl5 reporter activity was induced by over-expression of *lxrα*, but not by over-expression of *srebp-1*. This may indicate that the transcription of *elovl5* was up-regulated by Lxrα directly, but not dependent on Srebp-1. This observation is consistent with the results of studies carried out in Atlantic salmon and orange-spotted grouper[16, 44]. The results of the *in vitro* experiment conducted in the present study to confirm the above findings showed that the expression of *elovl5* in *L. crocea* hepatocytes was induced by over-expression of *lxrα*, while inhibition of the expression of *srebp-1* had no effect on the expression of *elovl5* and could not repress the activation effect of Lxrα on *elovl5*. This confirmed the direct stimulatory role of Lxrα on *elovl5*, and suggested that such regulatory mechanisms operated differently compared to mammals[24]. Promoter studies showed that the large yellow croaker Elovl4 reporter activity was induced by over-expression of both *lxrα* and *srebp-1*. This observation indicated that the expression of the *L. crocea elovl4* can be up-regulated through Lxrα and Srebp-1. Additionally, the *in vitro* results showed that the transcription of *elovl4*in hepatocyte was significantly induced by over-expression of *lxrα*, but significantly repressed by inhibiting the expression of *srebp-1*. The inhibition of the expression of *srebp-1* would significantly repress the activation effect of Lxrα on *elovl4*. These results suggested that Lxrα can promote the expression of croaker *elovl4* via a Srebp-dependent pathway. The above observations indicated differential regulation of Lxrα and Srebp-1 on Elovl4 and Elovl5 in the large yellow croaker. Knowledge of the underlying regulation mechanism of these two elongases can contribute to the development of novel practical approaches to enhance endogenous LC-PUFA production via both nutritional strategies and genetic modification.

In summary, the large yellow croaker Elovl4 exhibited all the structural features of Elovl proteins. Functional studies showed that Elovl4 and Elovl5 of *L. crocea* had complementary functions since the latter could effectively elongate both C_{18} and C_{20} PUFA substrates, while Elovl4 was more effective in the elongation of C_{20} and C_{22} FA substrates. Moreover, the functional characterization of the *L. crocea* Elovl4 demonstrated its role in the biosynthesis of VLC-PUFA. In addition, n-3 LC-PUFA can regulate the transcription of *elovl5* and *elovl4* elongases through Lxrα. Moreover, Lxrα could mediate the transcription of *elovl4* directly or indirectly via regulating the transcription of *srebp-1*.

4. MATERIALS AND METHODS

4.1 Cloning and phylogenetic analysis of the *L. crocea* full-length *elovl4*cDNA

Total RNA was extracted from isolated croaker liver using Trizol Reagent (Takara, Japan). Thereafter, first strand cDNA was synthesized using Prime Script™ RT reagent Kit (Takara). Two degenerate primers (Elo4-F and Elo4-R) (Supplementary Table) were designed based on highly conserved regions of *elovl4* cDNA sequences of other fish and used for PCR (Eppendorf Mastercycler Gradient, Eppendorf, Hamburg) using PrimeSTAR® HS DNA Polymerase (Takara, Japan). The amplified products were purified (SK8131, Sangon Biotech, Shanghai), ligated into the pEASY-T1 vector (TransGen Biotech, China), transformed into competent Trans1-T1 cells (TransGen Biotech, China) and sequenced (Biosune Biotech, Shanghai, China), followed by blasting on GenBank to confirm its high similarity with other Elovl4 proteins. Gene specific primers (Supplementary Table) were designed to obtain the full-length *elovl4* cDNA sequence by 5′-and 3′-rapid amplification of cDNA ends (RACE) PCR according to the manufacturer's instructions (SMARTer™ RACE cDNA Amplification Kit, Clontech, CA, USA). RACE PCR products were purified, cloned into pEASY-T1 vector, and sequenced as described above.

The deduced amino acid (aa) sequence of the newly cloned croaker *elovl4* cDNA was aligned with other orthologues including human (*Homo sapiens*, NP_073563), mouse (*Mus musculus*, NP_683743), rat (*Rattus norvegicus*, NP_001178725), zebrafish (*Danio rerio*, NP_957090 and NP_956266), cobia (*Rachycentron canadum*, ADG59898), and Atlantic salmon (*Salmo salar*, NP_001182481). Multiple sequence alignment was performed with Mega 6.0. A phylogenetic tree was constructed on the basis of aa sequences using the neighbor-joining method and including the newly cloned croaker Elovl4 sequence and those from other vertebrate Elovl proteins[45]. Confidence in the resulting phylogenetic tree branch topology was measured by bootstrapping the data through 1000 iterations.

4.2 Functional characterization of the *L. crocea* Elovl5 and Elovl4 elongases

PCR fragments corresponding to the open reading frames (ORF) of the Elovl5 and Elovl4 elongases were amplified from croaker liver cDNA, using High Fidelity PrimeScript® RT-PCR Kit with primers containing restriction sites for *Hin*dIII and *Xho*I (Supplementary Table). The DNA fragments containing the croaker *elovl5* or *elovl4* ORF were digested with corresponding restriction endonucleases (Takara, Japan) and then ligated into a similarly restricted pYES2 vector (Invitrogen, UK) to yield the resulting plasmid constructs pYES2-Elo5 and pYES2-Elo4. *Saccharomyces cerevisiae* competent cells were transformed (S.c. EasyComp Transformation Kit, Invitrogen, USA) with the purified recombinant plasmid constructs pYES2-Elo5 or pYES2-Elo4. Transformation and selection of yeast with recombinant plasmids, and yeast culture were conducted as described in detail previously[46, 47]. A single colony of transgenic yeast was grown in *S. cerevisiae* minimal medium$^{-uracil}$ supplemented with potential fatty acid (FA) substrates. For Elovl5, each culture was supplemented with one of the following substrates: 18:3n-3, 18:2n-6, 18:4n-3, 18:3n-6, 20:5n-3, 20:4n-6, 22:5n-3 and 22:4n-6. For Elovl4, each culture was supplemented with one of the following substrates: 18:4n-3, 18:3n-6, 20:5n-3, 20:4n-6, 22:5n-3, 22:4n-6 and 22:6n-3. In order to compensate for differences in FA update efficiency by yeast, final concentrations of FA substrates varied according to their fatty acyl chain lengths, 0.5 mM (C_{18}), 0.75 mM (C_{20}) and 1.0 mM (C_{22})[25]. Elovl4 plays a role in the biosynthesis of very long-chain (>C_{24}) saturated FA (VLC-FA)[25] and thus pYES2-Elo4 transformed yeast were also grown in the absence of exogenously added fatty acids to enable comparison with saturated FA profiles of yeast transformed with the empty pYES2. For both Elovl5 and Elovl4 assays, yeast were grown for 2 days and then harvested, washed twice with 5 mL ice-cold HBSS (Invitrogen, UK) and freeze dried for 24 h for further analyses. Yeast transformed with empty pYES2 were additionally grown in the presence of all PUFA substrates tested in both assays for Elovl5 and Elovl4 as further control treatments.

4.3 Fatty acid analysis

Total lipids were extracted from yeast samples and fatty acids derivatized to methyl esters (FAME) as described in detail previously[48]. FAME were identified and quantified after splitless injection and run with temperature programming in an Agilent 6850 Gas Chromatograph system equipped with a Sapiens-5MS (30 m × 0.25 μm × 0.25 μm) capillary column (Teknokroma, Barcelona, Spain) coupled to a 5975 series MSD (Agilent Technologies, Santa Clara, CA, USA) as described by Monroig *et al.*[27]. The elongation of exogenously added PUFA substrates (18:3n-3, 18:2n-6, 18:4n-3, 18:3n-6, 18:4n-3, 20:5n-3, 20:4n-6, 22:5n-3 and 22:6n-3) was calculated by the step-wise proportion of substrate FA converted to elongated product as [areas of first product and longer chain products/(areas of all products with longer chain than substrate + substrate area)] × 100[15].

4.4 Animal experiments

The present study was carried out in strict accordance with the recommendations in the Guide for the Use of Experimental Animals of Ocean University of China. The protocols for animal care and handling, methods and experimental procedures used in this study were approved by the Institutional Animal Care and Use Committee of Ocean University of China. A 10-week feeding trial was conducted to investigate the effect of dietary n-3 LC-PUFA on expression of *elovl4*and *elovl5*, and the regulation of certain transcription factors on the two enzymes. DHA-enriched oil (DHA content: 40.6% of total fatty acid in the form of DHA-methyl ester, Jiangsu Tiankai Biotechnology Company Limited, China) and EPA-enriched oil (EPA and DHA at 45.9% and 23.8%, respectively, of total fatty acids both in the form of TAG; Hebei Haiyuan Health Biological Science and Technology Company Limited, China) were utilized in different ratios to form three isoproteic (43% crude protein) and isolipidic (18% crude lipid) diets with low (0.46%), moderate (1.05%) and high (2.44%) n-3 LC-PUFA content (the ratio of DHA to EPA was approximately 2.0). Large yellow croaker were obtained from a local farm located in Xiangshan Bay, Ningbo, China. At the beginning of the experiment, the fish were fasted for 24 h and weighed after being anesthetized with eugenol (1:10,000) (Shanghai Reagent, China). Triplicate groups of croaker (individual weight of 10.0 ± 0.6 g) were distributed into each sea cage (1 m × 1 m × 1.5 m) at a stocking density of 60 individuals, and fed twice daily (05:00 and 17:00 h) to apparent satiation for 70 days. At the end of the experiment, croaker were fasted for 24 h and anesthetized and culled by an overdose of MS222 (ethyl 3-aminobenzoate methanesulfonic acid salt, Aldrich, USA). Livers of three fish per cage were isolated, frozen in liquid nitrogen and stored at −80 °C for subsequent gene expression analysis of elongases (*elovl4* and *elovl5*) and transcription factors (*lxrα* and *srebp-1*). In addition, various tissues (eye, brain, testis, heart, liver, kidney, stomach, intestine and muscle) were isolated from nine croaker individuals fed diets with moderate n-3 LC-PUFA levels (1.05%) to investigate the tissue distribution of *elovl4* mRNA.

4.5 Primary hepatocyte isolation and incubation

The above feeding study provided preliminary data on the *in vivo*nutritional regulation of *elovl4* and *elovl5*. To confirm and support these data, an *in vitro* study was conducted to further investigate the role of certain transcription factors on the regulation of *elovl4* and *elovl5*elongases in *L. crocea*. Hepatocytes were isolated from five yellow croaker individuals (~50 g) starved for 24 h according to the published protocols[49] with slight modification[50, 51]. Briefly, croaker were anesthetized with MS 222 and the branchial arch cut followed by immersion in 70% ethanol for 3 min to sterilize the external surface. Liver tissues were excised aseptically and rinsed twice with phosphate buffered saline (PBS, pH 7.4 at 4 °C) supplemented with amphotericin-B (25 μg mL^{-1}), streptomycin (100 μg mL^{-1}) and penicillin (100 IU mL^{-1}). Thereafter, the liver was aseptically minced into 1 mm^3 pieces, followed by digestion in 0.25% sterile trypsin at room temperature for 15 min. Trypsin was neutralized with Dulbecco's Modified Eagle Medium (DMEM, Gibco, USA) containing 20% fetal bovine plasma (FBS, Invitrogen, USA). The cell suspension was collected and filtered through sterile 75 μm mesh, followed by centrifugation at low speed (100 g, 5 min) and the supernatant discarded. Isolated cells were washed in red blood cell lysis buffer (Beyotime Institute of Biotechnology, Haimen, China) for 2 min at 4 °C. Cell viability was evaluated using a hemocytometer under an inverted microscope after the cells were stained with 0.4% Trypan Blue. The hepatocytes were re-suspended in DMEM medium containing 1 mM glutamine, 20% FBS, penicillin (100 IU mL^{-1}) and streptomycin (100 μg mL^{-1}). Cells suspensions with more than 95% cell viability were used for the subsequent experiments.

Croaker primary hepatocyte suspensions were incubated with fatty acid (DHA or EPA) to confirm the influence of n-3 LC-PUFA on the transcription of the elongases (*elovl4* and *elovl5*) and transcription factors (*lxrα* and *srebp-1*) *in vivo*. Fatty acid (DHA and EPA, Cayman Chemical Co., USA) was supplemented to cells in the form of BSA/fatty acid complexes that were prepared at 10 mM concentration according to Ou *et al.*[52] and stored at −20 °C. Additionally, inhibitors/agonists of transcription factors were used to clarify the role of the transcription factors on the regulation of *elovl4* and *elovl5* elongases in *L. crocea*. GW3965 HCl (Selleckchem, Shanghai, China) was used as an LXRα agonist whereas FGH10019 (MCE, USA) was used as a SREBP-1 inhibitor, respectively. Cells were seeded in 6-well plates with a density of 2×10^6viable cells per well in DMEM/F12

(Gibco) containing 20% FBS, 100 U ml^{-1} penicillin and 100 μg ml^{-1} streptomycin, followed by incubation for 24 h. The hepatocytes were then washed and incubated for 1 h in FBS-free DMEM/F12 medium prior to incubation with EPA, DHA and the above inhibitors or agonists in triplicate wells. After incubation, cells were lysed in the wells and harvested for RNA extraction.

4.6 Real-time quantitative PCR (RT-qPCR) analysis

Specific primers for RT-qPCR were designed by Primer Premier 5.0 (Premier Biosoft) based on the cloned nucleotide sequences. The stability of β-actin was verified and confirmed and used as the reference gene53. The amplification was performed in a quantitative thermal cycler (Mastercycler EP Realplex, Eppendorf, Germany). The program was 95 °C for 2 min, followed by 40 cycles of 95 °C for 10 s, 57 °C for 10 s, and 72 °C for 20 s. At the end of each reaction, melting curve analysis of amplification products was carried out to confirm that a single PCR product was present in these reactions. The amplification efficiencies of the target and reference genes were determined from the slope of the log-linear portion of the calibration curve and PCR efficiency = $10^{(-1/\mathrm{Slope})} - 1$. The expression levels of the target genes were calculated following the $2^{-\Delta\Delta Ct}$ method described by Livak & Schmittgen[54].

4.7 Cloning of the *elovl4* and *elovl5* promoters

Genomic DNA was extracted from croaker liver using the SQ Tissue DNA Kit (OMEGA, USA) according to the manufacturer's instructions. Then, the genomic DNA was digested with four restriction enzymes (*Dra*I, *Eco*RV, *Pvu*II and *Stu*I), purified and ligated to Genome Walker Adaptors following the instructions of Universal Genome Walker 2.0 Kit user manual (Clontech, USA). The promoter region of elovl5 and elovl4 was amplified through a nested (two round) PCR combining the kit supplied primers Ap1 and Ap2 with Advantage 2 PCR Kit (Clontech, USA) and the gene specific primers designed on *elovl5* and *elovl4* cDNA (see Supplementary Table for primer details). The conditions for the primary round PCR were: 7 cycles of 25 s at 94 °C, 3 min at 72 °C, followed by 32 cycles of 25 s at 94 °C, 3 min at 65 °C, then with an additional 7 min at 67 °C after the final cycle. The conditions for the secondary round PCR were: 5 cycles of 25 s at 94 °C, 3 min at 72 °C, followed by 20 cycles of 25 s at 94 °C, 3 min at 65 °C, then with an additional 7 min at 67 °C after the final cycle. The PCR products were purified, cloned into pEASY-T1 and sequenced as described above.

4.8 Expression and reporter plasmids constructs

For expression plasmids, PCR fragments corresponding to the ORF of the croaker *srebp-1* (GenBank accession number: KP342262) were amplified with primers containing restriction sites for *Eco*RI and *Xho*I, respectively, and the ORF of croaker *lxrα* (GenBank accession number: XM_019273432) was amplified with primers containing restriction sites for *Cla*I and EcoRI (Supplementary Table). The DNA fragments were digested with corresponding restriction endonucleases (Takara) and ligated into a similarly restricted pCS2^{+} vector (Invitrogen, USA) to yield the expression plasmid constructs pCS2-srebp1and pCS-lxrα. The reporter plasmid Elovl5-Luc and Elovl4-Luc were engineered to contain a fragment of the large yellow croaker Elovl5 and Elovl4 promoters cloned into the *Kpn*I/*Xho*I sites of the pGL3 basic vector (Promega, USA) containing the coding sequences of firefly luciferase cDNA. The renilla luciferase plasmid pRLTK (Promega) was used as an internal control. Plasmids for transfection were prepared using TransGen Plasmid Mini Kit (Beijing, China) according to the manufacturer's instructions.

4.9 Cell culturing, transfection and luciferase assay

HEK 293 T cells were cultured in DMEM (Gibco) supplemented with 10% fetal bovine plasma (Invitrogen) at 37 °C in a humidified incubator under 5% CO_2. For DNA transfection, cells were seeded in 24-well plates until they were 90–100% confluent at the time of transfection. Then plasmids were transfected by using the Lipofectamine™ 2000 Reagent (Invitrogen) according to the manufacturer's instruction. Briefly, 0.6 μg expression plasmid, 0.2 μg reporter gene plasmid, 0.02 μg pRLTK renilla luciferase plasmid, and 2.0 μL Lipofectamine™ 2000 were co-transfected into the cells in each well in the 24-well plate. All assays were performed with three independent transfections. Firefly and renilla luciferase activities were measured using Dual Luciferase Reporter Assay System (Promega, USA) according to the manufacturer's instructions. Briefly, after 48 h transfection, HEK293T cells in the 24-well plates were washed twice with 100 μL PBS, then lysed with 100 μL 1x passive lysis buffer (PLB) at room temperature for 10 min. Cell lysate (20 μL) was transferred to a clean plate and 50 μL of luciferase assay reagent II and 1x Stop & Glo substrate were added in sequence, then firefly and renilla luciferase activities were measured using a InfiniTE200 plate reader (Tecan, Switzerland).

4.10 Statistical analysis

The results were presented as means ± standard error of the mean (SEM). Data from each treatment were subjected to one-way analysis of variance (ANOVA) and correlation analysis where appropriate using SPSS 19.0 for Windows. Tukey's multiple range test was chosen as a multiple comparison test with a significance level of 0.05. For the Elovl4 functional characterization, the saturated VLC-FA profiles from yeast expressing the *elovl4* were compared to the control by a Student's *t*-test ($P < 0.05$).

REFERENCES

1. Calder, P. C. Immunomodulation by omega-3 fatty acids. *Prostag. Leukotr. Ess.* **77**, 327–335 (2007).
2. Fleith, M. & Clandinin, M. T. Dietary PUFA for preterm and term infants: review of clinical studies. *Crit. Rev. Food Sci.* **45**, 205–229 (2005).
3. Nugent, A. P. The metabolic syndrome. *Nutr. Bull.* **29**, 36–43 (2004).
4. Sprague, M., Dick, J. R. & Tocher, D. R. Impact of sustainable feeds on omega-3 long-chain fatty acid levels in farmed Atlantic salmon, 2006–2015. *Sci. Rep. UK* 621892 (2016).
5. Bell, J. G. *et al.* Substituting fish oil with crude palm oil in the diet of Atlantic salmon (*Salmo salar*) affects muscle fatty acid composition and hepatic fatty acid metabolism. *J. Nutr.* **132**, 222–230 (2002).
6. Henriques, J., Dick, J. R., Tocher, D. R. & Bell, J. G. Nutritional quality of salmon products available from major retailers in the UK: content and composition of n-3 long-chain PUFA. *Brit. J. Nutr.* **112**, 964–975 (2014).
7. Monroig, Ó., Navarro, J. C. & Tocher, D. R. Long-chain polyunsaturated fatty acids in fish: recent advances on desaturases and elongases involved in their biosynthesis. *Avances en Nutrición Acuícola.* **11**, 257–283 (2011a).
8. Tocher, D. R. Omega-3 long-chain polyunsaturated fatty acids and aquaculture in perspective. *Aquaculture* **449**, 94–107 (2015).
9. Jakobsson, A., Westerberg, R. & Jaccobsson, A. Fatty acid elongases in mammals: their regulation and roles in metabolism. *Prog. Lipid Res.* **45**, 237–249 (2006).
10. Nugteren, D. H. The enzymic chain elongation of fatty acids by rat-liver microsomes. *BBA-Lipid. L. Met.* **106**, 280–290 (1965).
11. Castro, C. L. F., Tocher, D. R. & Monroig, Ó. Long-chain polyunsaturated fatty acid biosynthesis in chordates: Insights into the evolution of Fads and Elovl gene repertoire. *Prog. Lipid Res.***62**, 25–40 (2016).
12. Kiron, V., Satoh, S., Takeuchi, T. & Yoshizaki, G. Cloning and over-expression of a masu salmon (*Oncorhynchus masou*) fatty acid elongase-like gene in zebrafish. *Aquaculture* **282**, 13–18 (2008).
13. Kabeya, N. *et al.* Modification of the n-3 HUFA biosynthetic pathway by transgenesis in a marine teleost, nibe croaker. *J. Biotechnol.* **172**, 46–54 (2014).
14. Kuah, M. K., Jaya-Ram, A. & Shu-Chien, A. C. The capacity for long-chain polyunsaturated fatty acid synthesis in a carnivorous vertebrate: functional characterisation and nutritional regulation of a Fads2 fatty acyl desaturase with Δ4 activity and an Elovl5 elongase in striped snakehead (*Channa striata*). *BBA-Mol. Cell Biol. L.* **1851**, 248–260 (2015).
15. Li, S. *et al.* Molecular cloning and functional characterization of a putative *elovl4* gene and its expression in response to dietary fatty acid profiles in orange-spotted grouper *Epinephelus coioides*. *Aquac. Res.* **48**, 537–552 (2017).

16. Li, S. *et al.* Molecular cloning, functional characterization and nutritional regulation of the putative elongase Elovl5 in the orange-spotted grouper (*Epinephelus coioides*). *Plos One* **11**, e0150544 (2016).
17. Morais, S., Mourente, G., Martinez, A., Gras, N. & Tocher, D. R. Docosahexaenoic acid biosynthesis via fatty acyl elongase and Δ4-desaturase and its modulation by dietary lipid level and fatty acid composition in a marine vertebrate. *BBA-Mol. Cell Biol. L.***1851**, 588–597 (2015).
18. Ren, H. T., Yu, J. H., Xu, P. & Tang, Y. K. Influence of dietary fatty acids on muscle fatty acid composition and expression levels of Δ6 desaturase-like and elovl5-like elongase in common carp (*Cyprinus carpio var.* jian). *Comp. Biochem. Physiol. B* **163**, 184–192 (2012).
19. Yamamoto, Y. *et al.* Cloning and nutritional regulation of polyunsaturated fatty acid desaturase and elongase of a marine teleost, the nibe croaker *Nibea mitsukurii*. *Fisheries Sci.* **76**, 463–472 (2010).
20. Zuo, R., Mai, K., Xu, W., Dong, X. & Ai, Q. Molecular cloning, tissue distribution and nutritional regulation of a fatty acyl *elovl5*-like elongase in large yellow croaker, *Larimichthys crocea*. *Aquac. Res.***47**, 445–459 (2016).
21. Sampath, H. & Ntambi, J. M. Polyunsaturated fatty acid regulation of genes of lipid metabolism. *Annu. Rev. Nutr.* **25**, 317–340 (2005).
22. Horton, J. D., Goldstein, J. L. & Brown, M. S. SREBPs: activators of the complete program of cholesterol and fatty acid synthesis in the liver. *J. Clin. Invest.* **109**, 1125–1131 (2002).
23. Cha, J. Y. & Repa, J. J. The liver X receptor (LXR) and hepatic lipogenesis The carbohydrate-response element-binding protein is a target gene of LXR. *J. Biol. Chem.* **282**, 743–751 (2007).
24. Qin, Y., Dalen, K. T., Gustafsson, J. Å. & Nebb, H. I. Regulation of hepatic fatty acid elongase 5 by LXRα–SREBP-1c. *BBA-Mol. Cell Biol. L.* **1791**, 140–147 (2009).
25. Monroig, Ó. *et al.* Expression and role of Elovl4 elongases in biosynthesis of very long-chain fatty acids during zebrafish *Danio rerio* early embryonic development. *BBA-Mol. Cell Biol. L.* **1801**, 1145–1154 (2010).
26. Duan, Q. *et al.* Replacement of dietary fish oil with vegetable oils improves the growth and flesh quality of large yellow croaker (*Larimichthys crocea*). *J. Ocean U. China* **13**, 445–452 (2014).
27. Tocher, D. R. Fatty acid requirements in ontogeny of marine and freshwater fish. *Aquac. Res.* **41**, 717–732 (2010).
28. Guillou, H., Zadravec, D., Martin, P. G. & Jacobsson, A. The key roles of elongases and desaturases in mammalian fatty acid metabolism: Insights from transgenic mice. *Prog. Lipid Res.* **49**, 186–199 (2010).
29. Monroig, Ó. *et al.* Evolutionary functional elaboration of the Elovl2/5 gene family in chordates. *Sci. Rep.* **6**, 20510 (2016).
30. Cook, H. W. & McMaster, R. C. R. Fatty acid desaturationand chain elongation in eukaryotes. In (Eds) Vance, D. E., Vance, J. E., *Biochemistry of Lipids, Lipoproteins and Membranes*, Elsevier, Amsterdam, pp. 181–204 (2004).
31. Carmona-Antoñanzas, G., Monroig, Ó., Dick, J. R., Davie, A. & Tocher, D. R. Biosynthesis of very long-chain fatty acids (C > 24) in Atlantic salmon: Cloning, functional characterisation, and tissue distribution of an Elovl4 elongase. *Comp. Biochem. Physiol. B* **159**, 122–129 (2011).
32. Monroig, Ó. *et al.* Elongation of long-chain fatty acids in rabbitfish *Siganus canaliculatus*: Cloning, functional characterisation and tissue distribution of Elovl5-and Elovl4-like elongases. *Aquaculture* **350**, 63–70 (2012).
33. Fonseca-Madrigal, J. *et al.* Diversification of substrate specificities in teleostei Fads2: characterization of Δ4 and Δ6Δ5 desaturases of *Chirostoma estor*. *J Lipid Res.* **55**, 1408–1419 (2014).
34. Agbaga, M. P., Mandal, M. N. A. & Anderson, R. E. Retinal very long-chain PUFAs: new insights from studies on ELOVL4 protein. *J Lipid Res.* **51**, 1624–1642 (2010).
35. Zadravec, D. *et al.* ELOVL2 controls the level of n-6 28:5 and 30:5 fatty acids in testis, a prerequisite for male fertility and sperm maturation in mice. *J Lipid Res.* **52**, 245–25 (2011).

36. Sprecher, H. Metabolism of h ghly unsaturated n-3 and n-6 fatty acids. *BBA-Mol. Cell Biol. L.* **1486**, 219–231 (2000).
37. Kabeya, N. *et al.* Polyunsaturated fatty acid metabolism in a marine teleost, Nibe croaker *Nibea mitsukurii*: Functional characterization of Fads2 desaturase and Elovl5 and Elovl4 elongases. *Comp. Biochem. Physiol. B* **188**, 37–45 (2015).
38. Monroig, Ó., Webb, K., Ibarra-Castro, L., Holt, G. J. & Tocher, D. R. Biosynthesis of long-chain polyunsaturated fatty acids in marine fish: Characterization of an Elovl4-like elongase from cobia *Rachycentron canadum* and activation of the pathway during early life stages. *Aquaculture* **312**, 145–153 (2011b).
39. Gregory, M. K., See, V. H. L., Gibson, R. A. & Schuller, K. A. Cloning and functional characterisation of afatty acyl elongase from southern bluefin tuna (*Thunnus maccoyii*). *Comp. Biochem. Physiol. B* **155**, 178–185 (2010).
40. Wang, S. *et al.* Investigating long-chain polyunsaturated fatty acid biosynthesis in teleost fish: Functional characterization of fatty acyl desaturase (Fads2) and Elovl5 elongase in the catadromous species, Japanese eel *Anguilla japonica*. *Aquaculture* **434**, 57–65 (2014).
41. Lewis, M. J. *et al.* Targeted dietary micronutrient fortification modulates n-3 LC-PUFA pathway activity in rainbow trout (*Oncorhynchus mykiss*). *Aquaculture* **412**, 215–222 (2013).
42. Vagner, M. & Santigosa, E. Characterization and modulation of gene expression and enzymatic activity of delta-6 desaturase in teleosts: a review. *Aquaculture* **315**, 131–143 (2011).
43. Postic, C., Dentin, R., Denechaud, P. D. & Girard, J. Chrebp, a transcriptional regulator of glucose and lipid metabolism. *Annu. Rev. Nutr.* **27**, 179–92 (2007).
44. Minghetti, M., Leaver, M. J. & Tocher, D. R. Transcriptional control mechanisms of genes of lipid and fatty acid metabolism in the Atlantic salmon (*Salmo salar*) established cell line, shk-1. *BBA-Mol. Cell Biol. L.* **1311**, 194–202 (2011).
45. Saitou, N. & Nei, M. The neighbor-joining method: a new method for reconstructing phylogenetic trees. *Mol. Biol. Evol.* **4**, 406–425 (1987).
46. Agaba, M. K. *et al.* Cloning and functional characterisation of polyunsaturated fatty acid elongases from marine and freshwater teleost fish. *Comp. Biochem. Physiol. B* **142**, 342–352 (2005).
47. Li, M. *et al.* Characterization of two Δ5 fatty acyl desaturases in abalone (*Haliotis discus hannai* Ino). *Aquaculture* **416**, 48–56 (2013).
48. Monroig, Ó., Tocher, D. R., Hontoria, F. & Navarro, J. C. Functional characterisation of a Fads2 fatty acyl desaturase with Δ6/Δ8 activity and an Elovl5 with C16, C18 and C20 elongase activity in the anadromous teleost meagre (*Argyrosomus regius*). *Aquaculture***412**, 14–22 (2013).
49. Liebel, S. *et al.* Cellular responses of *Prochilodus lineatus*hepatocytes after cylindrospermopsin exposure. *Toxicol. In Vitro***25**, 1493–1500 (2011).
50. Zhang, Q. *et al.* miR-17 is involved in the regulation of LC-PUFA biosynthesis in vertebrates: Effects on liver expression of a fatty acyl desaturase in the marine teleost *Siganus canaliculatus*. *BBA-Mol. Cell Biol. L.* **1841**, 934–943 (2014).
51. Zhuo, M. Q. *et al.* Regulation of insulin on lipid metabolism in freshly isolated hepatocytes from yellow catfish (*Pelteobagrus fulvidraco*). *Comp. Biochem. Physiol. B* **177**, 21–28 (2014).
52. Ou, J. *et al.* Unsaturated fatty acids inhibit transcription of the sterol regulatory element-binding protein-1c (SREBP-1c) gene by antagonizing ligand-dependent activation of the LXR. *P. Natl. Acad. Sci. USA* **98**, 6027–6032 (2001).
53. Zheng, J. L. *et al.* Antioxidant defenses at transcriptional and enzymatic levels and gene expression of Nrf2-Keap1 signaling molecules in response to acute zinc exposure in the spleen of the large yellow croaker *Pseudosciaena crocea*. *Fish Shellfish Immun.***52**, 1–8 (2016).
54. Livak, K. J. & Schmittgen, T. D. Analysis of relative gene expression data using real-time quantitative PCR and the $2^{-\Delta\Delta CT}$ method. *Methods* 25, 402–408 (2001).

CHAPTER 10

Tissue Distribution of the MERS-Coronavirus Receptor in Bats

W. Widagdo[1], Lineke Begeman[1], Debby Schipper[1], Peter R. van Run[1], Andrew A. Cunningham[1], Nils Kley[3], Chantal B. Reusken[1], Bart L. Haagmans[1] & Judith M. A. van den Brand[1]

[1]*Department of Viroscience, Erasmus MC, Rotterdam, The Netherlands*

[2]*Institute of Zoology, Zoological Society of London, Regents Park, London, United Kingdom*

[3]*Institute for Novel and Emerging Infectious Diseases, Friedrich Loeffler Institute, Greifswald, Mecklenburg-Vorpommern, Germany*

ABSTRACT

Middle East respiratory syndrome coronavirus (MERS-CoV) has been shown to infect both humans and dromedary camels using dipeptidyl peptidase-4 (DPP4) as its receptor. The distribution of DPP4 in the respiratory tract tissues of humans and camels reflects MERS-CoV tropism. Apart from dromedary camels, insectivorous bats are suggested as another natural reservoir for MERS-like-CoVs. In order to gain insight on the tropism of these viruses in bats, we studied the DPP4 distribution in the respiratory and extra-respiratory tissues of two frugivorous bat species (Epomophorus gambianus and Rousettus aegyptiacus) and two insectivorous bat species (Pipistrellus pipistrellus and Eptesicus serotinus). In the frugivorous bats, DPP4 was present in epithelial cells of both the respiratory and the intestinal tract, similar to what has been reported for camels and humans. In the insectivorous bats, however, DPP4 expression in epithelial cells of the respiratory tract was almost absent. The preferential expression of DPP4 in the intestinal tract of insectivorous bats, suggests that transmission of MERS-like-CoVs mainly occurs via the fecal-oral route. Our results highlight differences in the distribution of DPP4 expression among MERS-CoV susceptible species, which might influence variability in virus tropism, pathogenesis and transmission route.

1. INTRODUCTION

Middle East respiratory syndrome coronavirus (MERS-CoV) emerged in the human population in 2012 and has been causing multiple outbreaks of human disease, mainly in the Arabian Peninsula1. The dromedary came[l] (Camelus dromedarius) has been shown to be the reservoir host for primary human infections[2-8], although other susceptible animals[9-11], including bats[12-13], are suspected also to be hosts for this virus. MERS-like-CoVs have been sequenced from bat samples, mainly from insectivorous bats, but they have not yet been successfully isolated[14,21]. Screening of over 5000 insectivorous bats from Ghana, Ukraine, Romania, Germany, and the Netherlands showed that MERS-CoV-like viruses were detected in 24.9% of Nycteris bats and 14.7% of Pipistrelle bats17.

MERS-CoV uses dipeptidyl peptidase-4 (DPP4) as its receptor to infect its target cells, including bat cells[22]. Analysis of DPP4 sequences from different bat species and in-vitro infection studies with various bat cell lines suggested that multiple bat species are susceptible to MERS-CoV[12-23]. MERS-like-CoVs probably also use DPP4 as indicated by studies on the Tylonycteris bat CoV HKU4, one of the MERS-like-CoVs[21]. HKU4 uses DPP4 to infect both bat and human cells *in vitro*[24, 25]. It is known that DPP4 is differently distributed in the respiratory tract of humans and other susceptible livestock animals, including dromedary camels[9, 26]. DPP4 expression in the nasal epithelium of the camel, llama, and pig allows them to develop upper respiratory tract infection upon intranasal inoculation with MERS-CoV[2, 9, 26,] while in humans, DPP4 is exclusively expressed in the lower respiratory tract epithelium, which is in line with acute pneumonia being the main clinical outcome of MERS-CoV infection[26, 27]. Additionally, the absence of DPP4 expression in the upper respiratory tract epithelium of sheep renders this tissue to be non-susceptible *in-vivo*[9]. These data indicate that the localization of DPP4 expression in tissues reflects MERS-CoV susceptibility and tropism in vivo. The localization of DPP4 expression in bat tissues, however, has not been studied, unlike that in other MERS-CoV susceptible species[9, 26.]

Our study aimed to understand the tropism of MERS-like-CoVs in bats by mapping the distribution of DPP4 expression in tissues from four bat species. DPP4 immunohistochemistry staining was performed on tissues collected from two widespread insectivorous bat species in Europe and Asia, the common pipistrelle bat (Pipistrellus pipistrellus) and the serotine bat (Eptesicus serotinus)[28, 29]; and two common frugivorous bat species in Africa, i.e. the Gambian epauletted fruit bat (Epomophorus gambianus) and the Egyptian fruit bat (Rousettus aegyptiacus)[30, 31]. These four bat species were chosen based on their interactions with humans as they roost and forage in the human habitat or serve as a human food source[28-31]. We show that DPP4 localization varies not only among MERS-CoV susceptible species[9, 26] but also between bat species, which may imply variability in MERS-like-CoVs tropism, pathogenesis, and transmission route.

2. RESULTS

Immunohistochemistry to detect DPP4 was performed on nose, lung, intestine, kidney, salivary gland, and liver tissues of different bat species: common pipistrelle bat, serotine bat, Gambian epauletted fruit bat (further referred as Gambian fruit bat), and Egyptian fruit bat. The assay was replicated two-three times for each tissue. We have used the same technique to map DPP4 localization in the respiratory tract tissues of human, dromedary camel, sheep, horse, pig, and llama[9, 26]. The antibody used in this study recognizes bat DPP4 as was demonstrated in transfection experiments using cloned Pipistrelle bat DPP4[22]. Hematoxylin and eosin staining on subsequent slides of the same tissues from the bats did not show significant histological changes.

DPP4 was not found in the nasal olfactory epithelial cells of common pipistrelle bat, serotine bat Gambian fruit bat, or Egyptian fruit bat (Fig. 1). In the nasal tissues of common pipistrelle bat, DPP4 was not detected in the respiratory epithelial cells lining the nasal cavity, but was detected in the epithelial cells lining the ducts of the submucosal glands in this species. In the serotine bat and Gambian fruit bat, multifocal DPP4 expression was detected in a limited number of nasal respiratory epithelial cells. In contrast, in the nasal tissues of the Egyptian fruit bat, DPP4 was prominently detected at the apical surface of the respiratory epithelial cells lining the nasal cavity as well as in glandular and ductular epithelial cells of the submucosal glands. In the lungs of the common pipistrelle and serotine bat, DPP4 was found in the endothelial cells of the capillaries but not in the bronchial, bronchiolar or alveolar epithelial cells. In the Gambian and Egyptian fruit bat, DPP4 was detected in the bronchial, bronchiolar and alveolar epithelial cells as well as in endothelial cells of small blood vessels (Fig. 1).

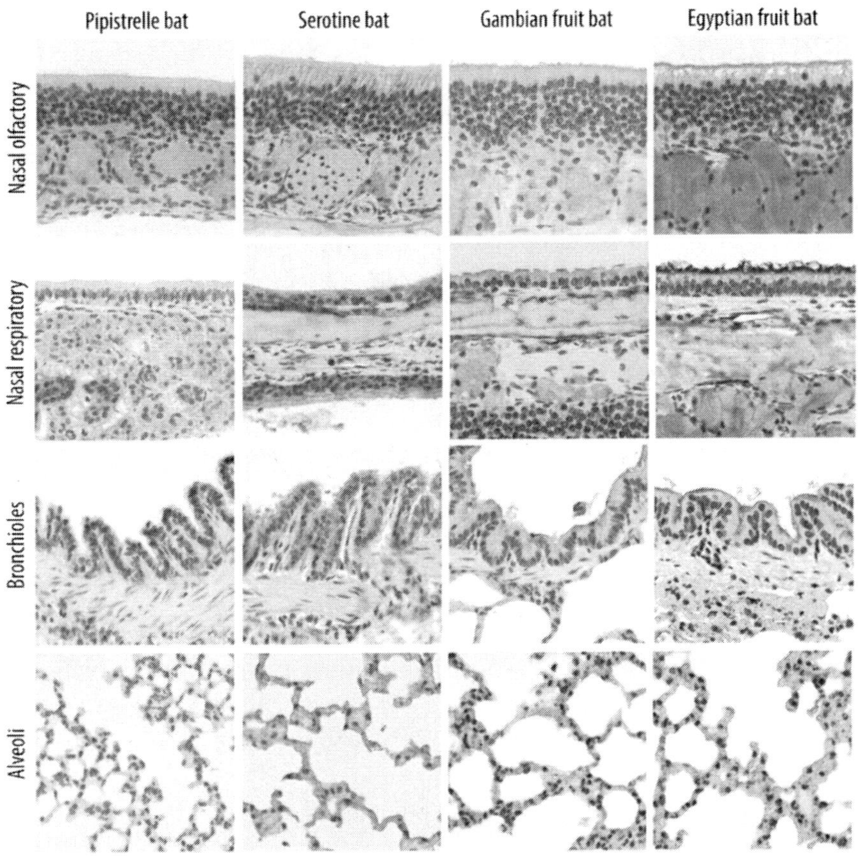

Figure 1 DPP4 expression in the respiratory tract tissues of common pipistrelle bat, serotine bat, Gambian epauletted fruit bat, and Egyptian fruit bat. DPP4 (indicated in red) is expressed in the nasal, bronchiolar and alveolar epithelium of the fruit bats, with limited expression in the epithelium lining the nasal cavity of serotine bats, and not detected in the epithelium lining the respiratory tract of the common pipistrelle bats. Original magnification x400 for all images.

In the intestinal tissues of all four bat species, DPP4 was prominently expressed on the apical surface of both small and large intestinal epithelial cells (Fig. 2).

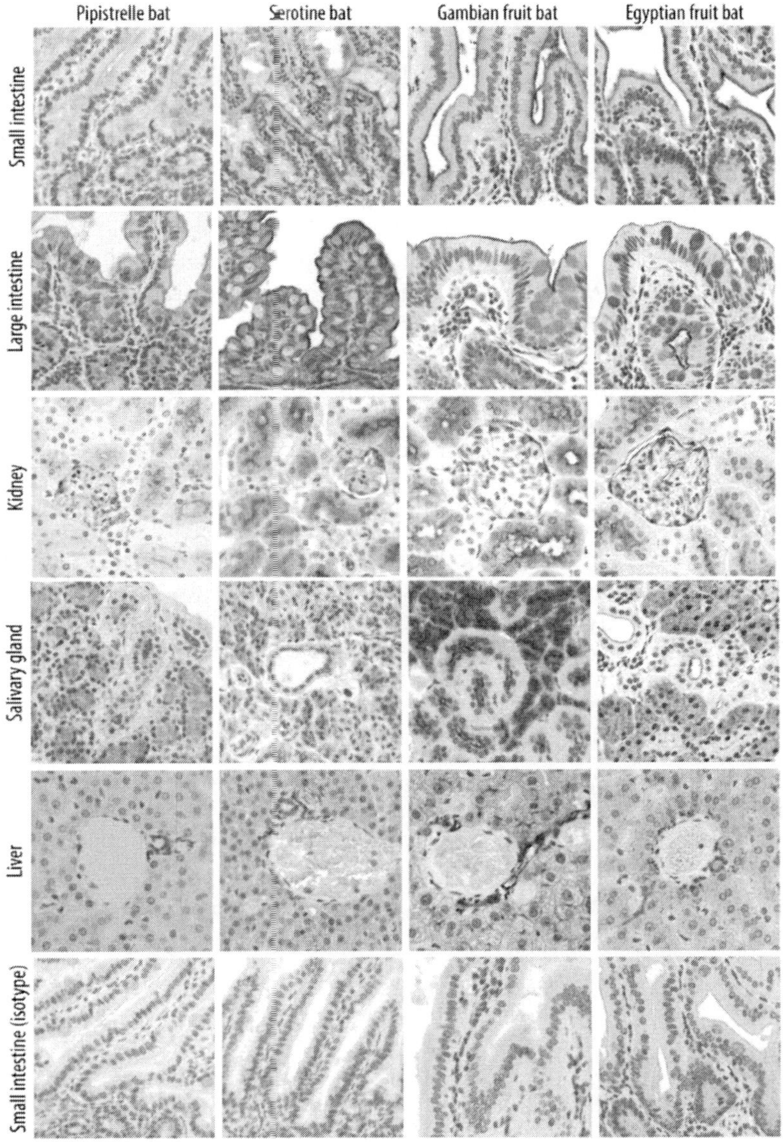

Figure 2 DPP4 expression in the intestine, kidney, salivary gland, and liver tissues of the common pipistrelle bat, serotine bat, Gambian epauletted fruit bat, and Egyptian fruit bat. In all four bat species, DPP4 (indicated in red) is detected on the apical surface of the intestinal epithelium, proximal tubular epithelium of the kidney, and in the salivary glands. Normal goat serum is used as isotype control for each tissue and showed no background signal. Only isotype control staining of the small intestines is shown. Original magnification x400 for all images.

In the kidney of all four bat species, DPP4 was found in glomerular cells, parietal squamous epithelial cells of the Bowman's capsule, and in the proximal tubular epithelial cells. In the salivary gland of common pipistrelle bat, DPP4 was only detected in the ductular epithelial cells, while in the serotine bat, it was detected in a limited number of glandular epithelial cells. In the Gambian and Egyptian fruit bat, it was detected in both the glandular and ductular epithelial cells of the salivary gland. In the liver of the common pipistrelle bat and serotine bat, DPP4 was present in a limited number of endothelial cells lining the sinusoids. In contrast, in the liver of the Gambian and Egyptian fruit bat, DPP4 was detected in the bile duct epithelial cells, in the endothelial cells of the hepatic arteries, and in the endothelial cells of the sinusoids (Fig. 2). Variation in DPP4 signal and localization were occasionally observed between animals within the same species. In one common pipistrelle bat, the paranasal sinus and pharynx were examined and showed a limited number of DPP4 positive epithelial cells. The results of the DPP4 immunohistochemistry staining were scored qualitatively and summarized in Table 1.

In general, our results showed that DPP4 was prominently expressed in the intestine and the respiratory tract tissues of the frugivorous bats, i.e. the Gambian and the Egyptian fruit bat. However, it is limitedly expressed in the respiratory tract tissues of the insectivorous bats, i.e. the common pipistrelle bat and the serotine bat.

Table 1: Overview of DPP4 expression in the tissues of the common pipistrelle bat, serotine bat, Gambian epauletted fruit bat and Egyptian fruit bat. +, positive detection; +/−, only expressed in a limited number of cells; −, undetected.

	Common pipistrelle bat	Serotine bat	Gambian fruit bat	Egyptian fruit bat
Nose				
✓Nasal respiratory epithelial cells	−	+/−	+/−	+
✓Nasal olfactory epithelial cells	−	−	−	−
✓Submucosal glands	+	+/−	+/−	+
Lung				
✓Bronchiolar epithelial cells	−	−	+/−	+
✓Alveolar epithelial cells	−	−	+/−	+
✓Endothelial cells of the capillaries and small blood vessels	+/−	+/−	+	+
Intestine				
✓Small intestinal epithelial cells	+	+	+	+

✓Large intestinal epithelial cells	+	+	+	+
Kidney				
✓Glomerular cells	+	+	+	+
✓Parietal squamous epithelial cells of the bowman capsule	+	+	+	+
✓Proximal tubular epithelial cells	+	+	+	+
Salivary gland				
✓Glandular epithelial cells	—	+/—	+/—	+
✓Ductular epithelial cells	+	—	+	+
Liver				
✓Hepatocytes	—	—	—	—
✓Bile ductular epithelial cells	—	—	+	+
✓Endothelial cells of the hepatic vein	—	—	—	—
✓Endothelial cells of the hepatic artery	—	—	+	+
✓Endothelial cells of the sinusoids	—	+/—	+	+

In the common pipistrelle bat, DPP4 was not detected in the nasal respiratory, nasal olfactory, bronchiolar, or alveolar epithelium, but was abundant on the apical surface of the epithelium lining the small and large intestine. We compared these findings to our previous results on dromedary camel and human tissues[26].

In dromedary camels, DPP4 is strongly detected in the nasal respiratory, tracheal, and bronchial epithelium, while there is limited expression in the alveolar epithelium (Fig. 3). In humans, it is not found in the nasal epithelium and is present mainly in the alveolar epithelium. Additionally, we performed DPP4 staining on intestinal tissues of dromedary camels obtained from a previous study[8]. We found that DPP4 was expressed mainly on the apical surface of the small intestinal epithelium (data not shown), similar to what has been reported for humans[32,35] (Fig. 3).

3. DISCUSSION

The tissue distribution of the MERS-CoV receptor, DPP4, has previously been studied in humans, dromedary camels, and other livestock animals[9, 26]. Here, we show that DPP4 is differentially expressed among bat species, especially between insectivorous and frugivorous bats. It is strongly detected in the intestine of the common pipistrelle bat, the serotine bat, the Gambian fruit bat and the Egyptian fruit bat. It is also prominent in the respiratory tract epithelium of the Gambian and Egyptian fruit bat, but expression is limited in that of the common pipistrelle and serotine bat. Given the essential role of DPP4 in the entry of MERS-CoV into cells, these results suggest that MERS-like-CoVs are not likely able to replicate in the respiratory tract in these two insectivorous bats. This is in line with our previous report on MERS-CoV infection experiment in sheep, showing that the lack of DPP4 in the respiratory tract of the sheep was associated with restricted MERS-CoV replication in these animals upon intranasal inoculation[9]. Rather, in these two insectivorous bats, MERS-like-CoVs may preferentially replicate in the gastrointestinal tract. This is partly supported by the fact that viral genomes of MERS-like-CoVs were mainly obtained from faecal and intestinal tissue samples of insectivorous bats[14,20, 36]. This intestinal tropism indicates that these viruses transmit mainly through the fecal-oral route.

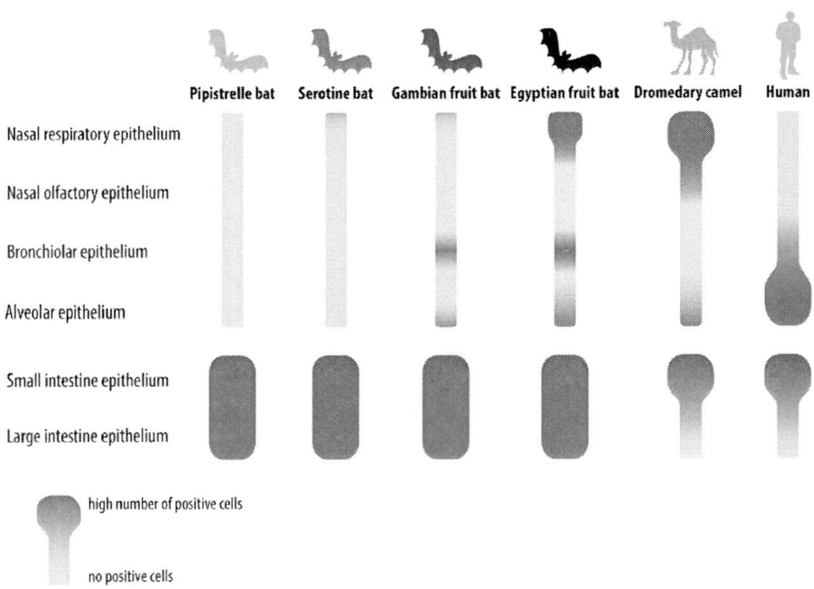

Figure 3 Different distribution of DPP4 expression in the lining respiratory tract and intestinal epithelium of the common pipistrelle bat, serotine bat, Gambian fruit bat, Egyptian fruit bat, dromedary camel, and human. In the common pipistrelle bat and serotine bat, DPP4 is limitedly detected in the respiratory tract and mainly expressed in the intestinal epithelium. In the Gambian and Egyptian fruit bat, DPP4 is found both in the respiratory tract and in the intestinal epithelium. In the dromedary camel, it is expressed in the upper respiratory tract and small intestine epithelium. In the human, it is predominantly expressed in the lower respiratory tract and small intestine epithelium.

Therefore, future screening of MERS-like-CoVs from insectivorous bats, particularly the common pipistrelle bat, might focus on fecal material, rectal swabs, or intestinal tissues, rather than throat or nasal swabs.

Prominent DPP4 expression in both respiratory tract and intestinal epithelium of the Gambian fruit bat and the Egyptian fruit bat suggests that MERS-CoV is able to replicate in both the respiratory tract and intestine of the fruit bats. These results are in line with the fact that MERS-CoV was able to replicate in the lungs of Jamaican fruit bat (Artibeus jamaicensis) upon intranasal and intraperitoneal inoculation[37]. Interestingly, viral RNA could be detected in the rectal swabs of these animals up to day 9 p.i. and infectious virus was also isolated in the duodenum of one of the bats at day 28 p.i.[37]. These data suggest that MERS-CoV infects and replicates in the intestine of these bats, not only in the respiratory tract. MERS-CoV infection in these bats is likely mediated by DPP4, since hamster BHK cells, a non-susceptible cell line, could be infected by MERS-CoV when modified to express Jamaican fruit bat's DPP437. DPP4 expression in the intestine and respiratory tract of these Jamaican fruit bats, however, was not investigated. Since the Jamaican fruit bat is a new world fruit bat, unlike the Gambian fruit bat and the Egyptian fruit bat, which are old world fruit bats, their genetic difference might influence the variation in DPP4 expression among these species. In contrast to the fruit bats, where DPP4 is expressed throughout the respiratory tract, DPP4 is rarely detected in the respiratory tract tissues of insectivorous bats. This limited DPP4 expression in insectivorous bats might significantly restrict the replication of MERS-like-CoVs in these tissues and minimize the possibility of transmission of these viruses from the respiratory tract.

The limited DPP4 expression in the respiratory tract of the two insectivorous bat species, particularly the common pipistrelle bat, is different from what has been reported for dromedary camels and humans. In humans, DPP4 is merely expressed in the lower respiratory tract, while in the dromedary camels, it is detected in the upper respiratory tract epithelium[26]. This renders humans to develop pneumonia upon MERS-CoV infection, while camels develop upper respiratory tract infection[2, 38, 39]. In the intestine of both dromedary camels and humans, DPP4 is mainly present in the apical surface of the small intestine epithelium[32,33,34,35]. MERS-CoV has been isolated from faecal samples of a naturally infected dromedary camel, which suggests that this virus is able to replicate in the intestinal tract of this species[40]. However, in dromedary camels, the chance of detecting MERS-CoV RNA in faecal samples is much lower than from nasal swabs[40]. We also observed that low amounts of viral RNA are detectable in rectal swabs taken from MERS-CoV- inoculated dromedary camels[2]. While MERS-CoV has not yet been isolated from human faecal samples, low amounts of viral RNA could be detected in stool samples of MERS patients[41], and several MERS patients have also been reported to suffer from diarrhoea[42,43,44]. These observations suggest that MERS-CoV replicates in the intestine of both dromedary camels and humans although only to a limited extent. It is currently unclear what factors restrain MERS-CoV replication in the intestinal tract of dromedary camels and humans. The human intestinal tract is protected by a mucus layer, commensal microorganisms, multiple innate and adaptive immune cells45. Also, adenosine deaminase (ADA), a natural antagonist of DPP4 that can inhibit MERS-CoV infection *in vitro*[10], has also been found in the human intestine. The amount of ADA in the human intestine is four times higher compared to that in the lung[46.] The presence of DPP4 in the intestinal tract of bats suggests an intestinal tropism of MERS-like-CoVs. We also detected DPP4 in the salivary glands and kidneys in all of the bats. In vitro, MERS-CoV has also been shown to replicate in primary kidney cell culture derived from common pipistrelle bat[13]. However, there has been no further evidence supporting the susceptibility of these two tissues in vivo, nor have there been any reports of MERS-like-CoVs isolated from these two tissues or from bat urine samples. Whether these viruses are transmitted through bat saliva or urine, therefore, is currently unclear.

In general, our study describes the variation in DPP4 distribution among four bat species, with notable differences between insectivorous and frugivorous bats. More importantly, the tissue distribution of DPP4 in insectivorous bats, believed to be one of the natural hosts for MERS-like-CoVs, is different to that in dromedary camels and humans. Our results indicate intestinal tropism of MERS-like-CoVs in the insectivorous bats we examined. The existence of a co-receptor that might influence MERS-like-CoVs tropism and replication in these bats, however, could not be disregarded. CEACAM5 is recently reported as an attachment factor that facilitates entry of MERS-CoV *in-vitro*[47]. Whether CEACAM5 plays an important role in-vivo, particularly in bats, remains to be investigated. *In-vivo* infection experiments are necessary to confirm our findings, but such studies are ethically and technically challenging. Nevertheless, our data are relevant for future monitoring and surveillance of MERS-like-CoVs in insectivorous bats, particularly in the common pipistrelle bat[14, 17], as well as for future efforts to better understand the pathogenesis and transmission of MERS-like-CoVs in their natural host.

4. MATERIALS AND METHODS

Common pipistrelle and serotine bats were found stranded and severely wounded on different occasions, and admitted to an official local bat shelter in the Netherlands. The animals were euthanized by veterinarians due to ethical reasons using officially approved methods. The Gambian fruit bats and three of four Egyptian fruit bats used in this study originated from free-ranging populations in Ghana. The bats were sampled for an unrelated study and this study was approved by the Ethics Committee of the Zoological Society of London (ref. WLE715) and the council for scientific and industrial research in Accra, Ghana. One of the Egyptian fruit bats was obtained from the captive colony at the Friederich Loeffler Institute, Riems, Germany. It had been euthanized due to reasons not related to this study. All methods were performed in accordance with the relevant guidelines and regulations.

After euthanasia, the bats were necropsied and tissues were collected. Parts of the lung, intestine, salivary gland, liver, and kidney were obtained from nine common pipistrelle bats, seven serotine bats, three Gambian fruit bats, and four Egyptian fruit bats. Parts of the noses were obtained from five common pipistrelle bats, six serotine bats, three Gambian fruit bats, and three Egyptian fruit bats. These tissues were all fixed in 10% formalin and embedded in paraffin. The noses were decalcified in 10% EDTA for 9 days before being embedded in paraffin. The formalin fixed paraffin embedded tissues were sectioned (4 μm), deparaffinized, and subsequently hydrated. Citric acid buffer 10 mM pH 6 was used to retrieve antigens. Blocking with normal rabbit serum 5% was performed prior to staining with polyclonal goat IgG anti-DPP4 (R&D systems, Abingdon, UK) in 5 μg/ml concentration. Normal goat serum (MP Biomedicals, Santa Ana, CA, USA) in equal concentration was used as negative control. DPP4 staining was performed at 4 °C overnight. Secondary antibody rabbit anti-goat IgG labeled with peroxidase were applied subsequently in 1:200 dilution for 1 hour at room temperature (Dako, Glostrup, Denmark). The red signal was revealed with 3-amino-9-ethyl-carbazole (Sigma-Aldrich, St. Louis, Missouri, USA) before counterstaining with hematoxylin.

Dromedary camel intestinal tissues were obtained from three different animals sacrificed at day 14 post infection with MERS-CoV in a previous MERS-CoV vaccination experiment[2]. Two of these animals were vaccinated beforehand, while one was not. MERS-CoV was not detected in the intestinal tissues of these animals using PCR, virus titration or immunohistochemistry detecting nucleoprotein of MERS-CoV. Information on DPP4 expression in human respiratory and intestinal tissues was derived from the previous studies[26, 33-35].

ACKNOWLEDGEMENTS

This study is supported by a TOP project Grant (91213066) and by the Zoonoses in the night project (50-52200-98-308) both funded by ZonMW. The E. gambianus and three of four R. aegyptiacus used in this study originated from a study in Ghana in collaboration with Richard Suu-Ire, from the Forestry Commission, Accra. We would like to thank Anne Buschmann-Balkema for her assistance in preparing the R. aegyptiacus tissues that come from the Friederich Loeffler Institute, Riems, Germany. We would like to thank Stichting Vleermuisopvang Oss for providing the tissues of P. pipistrellus. We thank Brigitta M Laksono for her advice on the schematic figure.

REFERENCES

1. Zumla, A., Hui, D. S. & Perlman, S. Middle East respiratory syndrome. Lancet 386, 995–1007, doi:10.1016/S0140-6736(15)60454-8 (2015).
2. Haagmans, B. L. et al. An orthopoxvirus-based vaccine reduces virus excretion after MERS-CoV infection in dromedary camels. Science 351, 77–81, doi:10.1126/science.aad1283 (2016).
3. Raj, V. S. et al. Isolation of MERS coronavirus from a dromedary camel, Qatar, 2014. Emerg Infect Dis 20, 1339–1342, doi:10.3201/eid2008.140663 (2014).
4. Reusken, C. B. et al. Middle East respiratory syndrome coronavirus neutralising serum antibodies in dromedary camels: a comparative serological study. Lancet Infect Dis 13, 859–866, doi:10.1016/S1473-3099(13)70164-6 (2013).
5. Reusken, C. B. et al. Middle East Respiratory Syndrome coronavirus (MERS-CoV) serology in major livestock species in an affected region in Jordan, June to September 2013. Euro Surveill 18, 20662, doi:10.2807/1560-7917.ES2013.18.50.20662 (2013).
6. Meyer, B. et al. Antibodies against MERS coronavirus in dromedary camels, United Arab Emirates, 2003 and 2013. Emerg Infect Dis 20, 552–559, doi:10.3201/eid2004.131746 (2014).
7. Alraddadi, B. M. et al. Risk Factors for Primary Middle East Respiratory Syndrome Coronavirus Illness in Humans, Saudi Arabia, 2014. Emerg Infect Dis 22, 49–55, doi:10.3201/eid2201.151340 (2016).
8. Haagmans, B. L. et al. Middle East respiratory syndrome coronavirus in dromedary camels: an outbreak investigation. Lancet Infect Dis 14, 140–145, doi:10.1016/S1473-3099(13)70690-X (2014).
9. Vergara-Alert, J. et al. Livestock Susceptibility to Infection with Middle East Respiratory Syndrome Coronavirus. Emerg Infect Dis 23, 232–240, doi:10.3201/eid2302.161239 (2017).
10. Raj, V. S. et al. Adenosine deaminase acts as a natural antagonist for dipeptidyl peptidase 4-mediated entry of the Middle East respiratory syndrome coronavirus. J Virol 88, 1834–1838, doi:10.1128/JVI.02935-13 (2014).
11. Reusken, C. B. et al. MERS-CoV Infection of Alpaca in a Region Where MERS-CoV is Endemic. Emerg Infect Dis 22, 1129–1131, doi:10.3201/eid2206.152113 (2016).
12. Cai, Y. et al. CD26/DPP4 cell-surface expression in bat cells correlates with bat cell susceptibility to Middle East respiratory syndrome coronavirus (MERS-CoV) infection and evolution of persistent infection. PLoS One 9, e112060, doi:10.1371/ journal.pone.0112060 (2014).
13. Eckerle, I. et al. Replicative Capacity of MERS Coronavirus in Livestock Cell Lines. Emerg Infect Dis 20, 276–279, doi:10.3201/eid2002.131182 (2014).
14. Reusken, C. B. et al. Circulation of group 2 coronaviruses in a bat species common to urban areas in Western Europe. Vector Borne Zoonotic Dis 10, 785–791, doi:10.1089/vbz.2009.0173 (2010).
15. Memish, Z. A. et al. Middle East respiratory syndrome coronavirus in bats, Saudi Arabia. Emerg Infect Dis 19, 1819–1823, doi:10.3201/eid1911.131172 (2013).

16. Ithete, N. L. et al. Close relative of human Middle East respiratory syndrome coronavirus in bat, South Africa. Emerg Infect Dis 19, 1697–1699, doi:10.3201/eid1910.130946 (2013).
17. Annan, A. et al. Human betacoronavirus 2c EMC/2012-related viruses in bats, Ghana and Europe. Emerg Infect Dis 19, 456–459, doi:10.3201/eid1903.121503 (2013).
18. Yang, L. et al. MERS-related betacoronavirus in Vespertilio superans bats, China. Emerg Infect Dis 20, 1260–1262, doi:10.3201/eid2007.140318 (2014).
19. Wacharapluesadee, S. et al. Group C betacoronavirus in bat guano fertilizer, Thailand. Emerg Infect Dis 19, 1349–1351, doi:10.3201/eid1908.130119 (2013).
20. Kim, H. K. et al. Detection of Severe Acute Respiratory Syndrome-Like, Middle East Respiratory Syndrome-Like Bat Coronaviruses and Group H Rotavirus in Faeces of Korean Bats. Transbound Emerg Dis 63, 365–372, doi:10.1111/tbed.12515 (2016).
21. van Boheemen, S. et al. Genomic characterization of a newly discovered coronavirus associated with acute respiratory distress syndrome in humans. MBio 3, e00473-12–e00473-12, doi:10.1128/mBio.00473-12 (2012).
22. Raj, V. S. et al. Dipeptidyl peptidase 4 is a functional receptor for the emerging human coronavirus-EMC. Nature 495, 251–254, doi:10.1038/nature12005 (2013).
23. Cui, J., Eden, J. S., Holmes, E. C. & Wang, L. F. Adaptive evolution of bat dipeptidyl peptidase 4 (dpp4): implications for the origin and emergence of Middle East respiratory syndrome coronavirus. Virol J 10, 304, doi:10.1186/1743-422X-10-304 (2013).
24. Wang, Q. et al. Bat origins of MERS-CoV supported by bat coronavirus HKU4 usage of human receptor CD26. Cell Host Microbe 16, 328–337, doi:10.1016/j.chom.2014.08.009 (2014).
25. Yang, Y. et al. Receptor usage and cell entry of bat coronavirus HKU4 provide insight into bat-to-human transmission of MERS coronavirus. Proc Natl Acad Sci USA 111, 12516–12521, doi:10.1073/pnas.1405889111 (2014).
26. Widagdo, W. et al. Differential expression of the MERS-coronavirus receptor in the upper respiratory tract of humans and dromedary camels. J Virol. 90, 4838–4842, doi:10.1128/JVI.02994-15 (2016).
27. Who Mers-Cov Research, G. State of Knowledge and Data Gaps of Middle East Respiratory Syndrome Coronavirus (MERS-CoV) in Humans. PLoS Curr 5, doi: 10.1371/currents.outbreaks.0bf719e352e7478f8ad85fa30127ddb8 (2013).
28. Hutson, A. et al. Eptesicus serotinus, http://dx.doi.org/10.2305/IUCN.UK.2008.RLTS.T7911A12867244.en (2008)
29. Hutson, A. et al. Pipistrellus pipistrellus, http://dx.doi.org/10.2305/ IUCN.UK.2008.RLTS.T17317A6968203.en (2008).
30. Korine, C. Rousettus aegyptiacus, http://dx.doi.org/10.2305/IUCN.UK.2016-2.RLTS.T29730A22043105.en (2016).
31. Tanshi, I. & Fahr, J. Epomophorus gambianus, e.T7903A22122670 (2016).
32. Uhlen, M. et al. Proteomics. Tissue-based map of the human proteome. Science 347, 1260419–1260419, doi:10.1126/science.1260419 (2015).
33. Gorvel, J. P. et al. Expression of sucrase-isomaltase and dipeptidylpeptidase IV in human small intestine and colon. Gastroenterology 101, 618-625, doi:S0016508591003001 [pii], 101, 618–25 (1991).
34. Quaroni, A. et al. Expression and different polarity of aminopeptidase N in normal human colonic mucosa and colonic tumors. Int J Cancer 51, 404–411, doi:10.1002/(ISSN)1097-0215 (1992).
35. Darmoul, D. et al. Regional expression of epithelial dipeptidyl peptidase IV in the human intestines. Biochem Biophys Res Commun 203, 1224-1229, doi:S0006291X84723138 [pii], 203, 1224-9 (1994).
36. Corman, V. M. et al. Rooting the phylogenetic tree of middle East respiratory syndrome coronavirus by characterization of a conspecific virus from an African bat. J Virol 88, 11297–11303, doi:10.1128/JVI.01498-14 (2014).
37. Munster, V. J. et al. Replication and shedding of MERS-CoV in Jamaican fruit bats (Artibeus jamaicensis). Sci Rep 6, 21878, doi:10.1038/srep21878 (2016).

38. Ng, D. L. et al. Clinicopathologic, Immunohistochemical, and Ultrastructural Findings of a Fatal Case of Middle East Respiratory Syndrome Coronavirus Infection in the United Arab Emirates, April 2014. Am J Pathol 186, 652–658, doi:10.1016/j.ajpath.2015.10.024 (2016).
39. Hocke, A. C. et al. Emerging human middle East respiratory syndrome coronavirus causes widespread infection and alveolar damage in human lungs. Am J Respir Crit Care Med 188, 882–886, doi:10.1164/rccm.201305-0954LE (2013).
40. Hemida, M. G. et al. MERS coronavirus in dromedary camel herd, Saudi Arabia. Emerg Infect Dis 20, 1231–1234, doi:10.3201/eid2007.140571 (2014).
41. Drosten, C. et al. Clinical features and virological analysis of a case of Middle East respiratory syndrome coronavirus infection. Lancet Infect Dis 13, 745–751, doi:10.1016/S1473-3099(13)70154-3 (2013).
42. Tsiodras, S. et al. A case of imported Middle East Respiratory Syndrome coronavirus infection and public health response, Greece, April 2014. Euro Surveill 19, 20782, doi:10.2807/1560-7917.ES2014.19.16.20782 (2014).
43. Assiri, A. et al. Epidemiological, demographic, and clinical characteristics of 47 cases of Middle East respiratory syndrome coronavirus disease from Saudi Arabia: a descriptive study. Lancet Infect Dis 13, 752–761, doi:10.1016/S1473-3099(13)70204-4 (2013).
44. Guery, B. et al. Clinical features and viral diagnosis of two cases of infection with Middle East Respiratory Syndrome coronavirus: a report of nosocomial transmission. Lancet 381, 2265–2272, doi:10.1016/S0140-6736(13)60982-4 (2013).
45. Peterson, L. W. & Artis, D. Intestinal epithelial cells: regulators of barrier function and immune homeostasis. Nat Rev Immunol 14, 141–153, doi:10.1038/nri3608 (2014).
46. Edwards, Y. H., Hopkinson, D. A. & Harris, H. Adenosine deaminase isozymes in human tissues. Ann Hum Genet 35, 207–219, doi:10.1111/ahg.1971.35.issue-2 (1971).
47. Chan, C. M. et al. Carcinoembryonic Antigen-Related Cell Adhesion Molecule 5 Is an Important Surface Attachment Factor That Facilitates Entry of Middle East Respiratory Syndrome Coronavirus. J Virol 90, 9114–9127, doi:10.1128/JVI.01133-16 (2016).

CHAPTER 11

The Habitats, Burrowing Behavior, Physiology Adaptation and Life Cycle of Spadefoot Toads (Pelobates syriacus, Boettger, 1869) at the Southern Limit of Its Distribution in Israel

Gad Degani[1,2]

[1]*Faculty of Science and Technology, Tel Hai Academic College, Tel-Hai, Israel*

[2]*MIGAL-Galilee Research Institute, Kiryat Shmona, Israel*

ABSTRACT

The present study describes the habitats, life cycle, larvae growth, burrowing behavior and terrestrial adaptation of Spadefoot toads in Israel based on observations and data collected during more than 30 years in northern Israel. The distribution area in Israel is from the north in the Upper Galilee and Golan Heights (annual rainfall range of 500 - 1000 mm) to the southern coastal plain (annual rainfall of about 250 mm). Among the 51 different breeding places of amphibians, only ponds where water was available for a few months were used by Spadefoot toads and metamorphosed populations were found around these ponds. The larvae underwent metamorphosis during the summer and autumn in northern Israel, and during the spring in central and southern Israel. A negative correlation exists between the percentage of toads burrowing and soil moisture levels, with greater burrowing behavior occurring under dry conditions. The plasma concentration increased during burrowing by electrolytes and urea accumulations. The burrowing behavior helped Spadefoot toads survive in this area at the southern border of its distribution.

Keywords: Breeding Places, Burrowing Behavior, Larvae, Life Cycle, Pelobates syriacus, Plasma Concentration

1. INTRODUCTION

The knowledge of the biology, ecology and genetics of Spadefoot toads (Pelobates syriacus) in Israel at the southern border of its distribution is important for understanding their adaptation to this area. This species is found in the southeastern Balkans, eastern to southeastern Transcaucasia, northern Iran, and south of the Levant [1] . Spadefoot toads are studied extensively in Europe. The populations are found around breeding places where the substrates of terrestrial habitats include sandy soils, heath lands and deciduous woodlands with loamy soils, while the spawning biotopes include a variety of permanent or semi-permanent ponds [1] [2] . Spadefoot toads were 7.4 years old on average (maximum age recorded 12), despite the fact that sexual maturity (i.e., the age of the youngest adults observed) was reached at a similar age [1] [2] . The size of the egg clutches varied from less than 1000 to almost 6000 eggs, with an average value of 2800 eggs per clutch [2] .

The distribution, habitat and breeding place selection of ponds are related to water availability during the spring and early summer in Israel [3] , larval growth and complete metamorphosis [4] [5] , burrowing behavior [6] , physiology adaptations to various environments [7] , and biology [8] [9] . The burrowing toad P. syriacus spends 8 - 10 months of the year being buried in soil (Warburg, 1971, 1972). During the summer, the soil dries out, and the osmotic relationship between the toad and the surrounding soil changes. At a certain point, water begins to pass from the toad to the soil (or the air) and the toad becomes dehydrated.

Spadefoot toads are highly specialized and have a narrow ecological niche that selects lower levels of diversity [10] . The genetic variation among their populations was studied in the southern area of the Mediterranean shoreline [11] [12] . Nevo [13] suggested that Pelobates was best explained by the environmental variability model based on selection for homozygosity as an adaptive strategy in the relatively constant and narrow subterranean niche. Although several aspects of Spadefoot toads in Israel have been studied, the life cycle of Spadefoot toads at the southern border of its distribution in Israel has not been described in detail. Due to their secretive behavior (nocturnal activity, weak underwater breeding call and cryptic coloration), their life histories are not well documented, and overall there is a scarcity of information regarding this genus. Moreover, this genus, Pelobates, is distributed in different climates, which affects life cycle, and some basic biological characteristics are similar and might help to complete the information of the species in Israel, Pelobates syriacus.

The purpose of the present study is to describe the habitats, life cycle, larvae growth, burrowing behavior and terrestrial adaptation of Spadefoot toads in Israel based on observations and data collected during more than 30 years in northern Israel.

2. MATERIALS AND METHODS

2.1. Study Area

The study was carried out over four consecutive decades (1979-2014). Different aquatic habitats were dispersed over an area of approximately 1400 km^2 in northern Israel. The habitats included springs and streams that were stable water bodies where water was available year-round, rock pool holes, which were filled by rainwater and whose hydroperiods were long (about 200 d/yr), and ponds, which were flooded during the autumn when rainfall began, and gradually dried out between the late winter months and early summer.

2.2. Sampling

The water body was examined by nets (pore size 450 μm) from a depth of approximately 10 - 30 cm, as described previously by [3] . Rain pools and ponds in Israel that were filled by rain held the water for 3 - 4 months before drying up; large ponds dried up after a period of 6 - 12 years. The elevation of the breeding places is between 0 to 1000 m above sea level. The breeding places that were examined are presented in Table 1 and Figure 1.

2.3. Biotic and Abiotic Water Parameters

Water quality testing during larval growth, which was described in detail in the chapter on water quality, included dissolved oxygen (%), water temperature (°C), pH, electrical conductivity (µs/cm), ammonium concentration (mg/l), nitrate concentration (mg/l), chlorophyll a concentration (mg/l), water volume (m3) and aquatic invertebrates-number of taxa and biomass (µg/l). Water parameters were measured at a depth of 10 cm every two weeks during the period in which the pools were filling up. Temperature (°C) and level of oxygen dissolved in water (dissolved oxygen concentration, mg/L−1 and oxygen saturation, %) were obtained by a hand-held oxygen meter (WTW, Oxi330 set, Germany) in situ, and one liter of water was sampled for the laboratory water quality tests, including pH (WTW, pH315i, Germany, electrical conductivity (EC, mS/cm−1) corrected to 25°C (WTW, Multiline P4, Germany) [14] , chlorophyll a concentration (mg/L−1) extracted from GF/F filters in 90% ethanol [14] , and ammonium concentrations (NH4, mg/L−1) carried out using kits based on color changes. Color changes were measured using a dedicated spectrophotometer (NOVA 60, Merck; Germany) [14] .

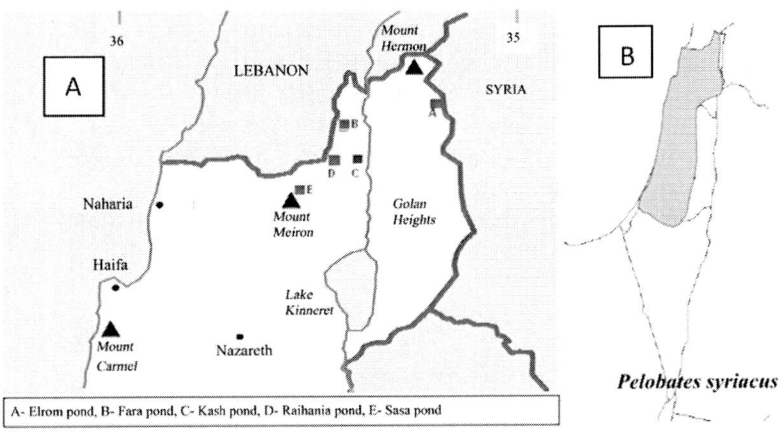

Figure 1. The study area: (A) Samples taken in different northern breeding sites of amphibians (Table 1); (B) Area in Israel where Pelobates syriacus are distributed.

2.4. Buried Behavior Exterminate

Fifty juvenile P. syriacus were collected from beneath stones in the Sasa and Dovev pond area of northern Israel during the summer (August and September) and maintained in an aquarium measuring 100 × 30 × 20 cm at the Migal Laboratory in northern Israel. In the experimental room, the temperature was maintained at 26°C and the aquarium was half full of moist soil. A light regime of 12 hours light and 12 hours darkness was maintained. Soil moisture level was determined according to [15] . Fully saturated soil was considered to have a moisture level of 100% and was determined in a 500 ml glass flask half full of soil and filled with water above the soil line for 1 hr. The glass flask was turned upside down for 10 minutes in order to allow all the free water to flow out. The flask was subsequently dried in a stove for 24 hrs at a temperature of 105°C. Full saturation was calculated according to Degani (1988).

2.5. Terrestrial Adaption

Ten juvenile P. syriacus were maintained in a 70% soil aquarium measuring 100 × 30 × 20 cm at the Migal Laboratory in northern Israel for three months, and after that period, blood samples were taken. The control group was maintained in the aquarium with 4 cm water for seven days and then the samples were taken. Blood samples were taken from acclimatized juveniles (tap water and burrowing juveniles) by heart puncture using micropipettes washed with lithium heparin. The blood samples were centrifuged immediately for 15 min at 300 g, and the plasma was either analyzed immediately or frozen for analysis later. Sample preparation was according to [16] . Sodium, potassium and chloride in plasma were determined using methods described by [16] . Urea was determined using a method described by [17] .

2.6. Statistical Analysis

The range of biotic and abiotic parameters of aquatic habitats was calculated, and average water parameters were analyzed by one-way analysis of variance (ANOVA), with the level of significance between groups set at $p < 0.05$ (ANOVA) (Tukey test).

Table 1. Aquatic systems examined in northern Israel; Si—Salamandra infra immaculata; Tv—Triturus vittatus vittatus; Hs—Hyla savignyi; Bv—Bufo viridis; Ps—Pelobates syriacus; Rb—Rana bedriaga.

Water Body	Type of Body	Galilee	Larvae of Spadefoot toads	Larvae of Amphibia
Nimrod Pond	Winter Pond	Northern Golan		*Si, Bv*
Nimrod Spring	Continuous Spring	Northern Golan		*Bv*
Barn Pond	Winter Pond	Hula Valley		*Bv*
Shunit Spring	Seasonal Spring	Hula Valley		*Bv*
Pond North KS	Winter Pond	Hula Valley		*Bv*
Pond South KS	Winter Pond	Hula Valley		*Bv*
Nahalin Spring	Seasonal Spring	Naftali Heights		*Bv, Hs*
Hula NR	Permanent Pond	Hula Valley		*Bv, Rb*
Sal Pond	Winter Pond	Central Golan		*Rb, Hs*
Katzarin Pond	Winter Pond	Central Golan		*Tv*
Paras Pond	Winter Pond	Central Golan		*Tv, Bv, Hs*
Surman Pond	Winter Pond	Central Golan		*Bv, Hs*
Oil Pipe Pond	Winter Pond	Central Golan		*Bv, Hs, Rb*
Gamla Pond	Winter Pond	Central Golan		*Bv, Hs*
Tel Bazak Pond	Winter Pond	Central Golan		*Bv, Hs*
Darach Pond	Winter Pond	Central Golan		*Bv*
Sbina Pond	Winter Pond	Central Golan		*Bv, Hs, Rb*
Bageria Pond	Winter Pond	Southern Golan		*Rb*
Irsim Pond	Winter Pond	Southern Golan		*Rb*
Irism Pond North	Winter Pond	Southern Golan		
Nov Pond North	Winter Pond	Southern Golan		*Rb*
Hispin Pond	Winter Pond	Southern Golan		*Bv*
Shachar Pond	Winter Pond	Northern Israel		*Tv, Bv, Hs*
Kash Pond	Winter Pond	Northern Israel		*Bv, Hs*
Dalton Pond 1	Winter Pond	Upper Galilee		*Bv, Hs*
Dalton Pond 2	Winter Pond	Upper Galilee		*Bv, Hs*
Gush Halav Pond	Winter Pond	Upper Galilee		*Si, Bv*
Bechania Pond	Winter Pond	Upper Galilee		
Sasa Pond	Winter Pond	Upper Galilee	*Ps*	*Si, Tv, Hs, Bv*
Humama Spring	Continuous Spring	Upper Galilee		*Si*
Zakin Spring	Continuous Spring	Upper Galilee		*Si*
Bet Jan Reservoir	Permanent Pond	Upper Galilee		*Si*
Fasuta Pond	Winter Pond	Western Galilee		*Si*
Menachem Pond	Winter Pond	Western Galilee		*Tv, Hs*

Shetula Pond	Winter Pond	Western Galilee		
Maron Pond	Winter Pond	Western Galilee		*Si, Hs*
Sumara Pond	Winter Pond	Western Galilee		*Si*
Tel Arad Pond	Winter Pond	Western Galilee		*Si, Hs, Tv*
Sasa Rock Pool	Rock Pool (al. year)	Upper Galilee		*Si*
Tel Dan	Continuous Stream	Northern Galilee		*Si*
Balad Spring				*Si, Hs*
Navoraya Spring				*Si, Hs, Rb*
Maalot Rock Pool				*Si*
Manof Pool				*Tv, Hs,* *Bv, Ps Si, Bv*
Kash Pool			*Ps*	*Si, Tv, Hs, Bv, Rb*
Dovev Pool	Winter Pond	Upper Galilee		*Tv, Hs, Bv, Ps,* Si
Matityahu Pool				
Lehavot Pool				*Hs, Rb*
Fara Pool		Northern Galilee	*Ps*	*Si, Tv, Hs, Bv, Rb*
Raihaniya Pool	Winter Pond	Upper Galilee	*Ps*	*Bv*
Elrom Pond	Winter Pond	Northern Golan	*Ps*	

3. RESULTS

Spadefoot toads (Pelobates syriacus) are found in northern Israel from the north in the Upper Galilee and Golan Heights (annual rainfall range of 500 - 1000 mm) (Figure 1), to the southern coastal plain (annual rainfall of about 250 mm). Among the 51 different breeding places of amphibians, only certain ponds (Table 1) are used by Spadefoot toads and metamorphosed populations were found around those ponds. Water in the breeding places of Spadefoot toads is available for more a few months and is sometimes available all year round, the time for the larvae to grow and complete metamorphosis (Table 2). Spadefoot toads can breed several times during the warm season, usually when there is a drop in barometric pressure, two or more inches of rainfall, and in the darkness of night.

The life cycle of Spadefoot toads is presented in Figure 2. Males are the first to arrive at the pond, sending out a mating call. This call consists of a short, explosive "wonk," similar to the call of a crow. The females then arrive, and breeding occurs underwater. Breeding time in Israel varies according to location and geographical and ecological conditions. It takes place in March-April in northern Israel, Upper Galilee and the Golan Heights, and earlier in the lower locations and in central Israel. In central Israel, the breeding period is during the winter. The amplexus of Spadefoot toads involves the male holding the female above the back legs, they swim together, and the female lays 2 to 4 lies of eggs (Figure 2). The number of eggs differs, and ranges between 2000 to 10,000 per female. Larvae growth and complex metamorphosis take place in northern Israel during the summer and early autumn until October. Larvae growth in various breeding places is presented in Figure 3.

There is no information on adult terrestrial activity of Spadefoot toads. Considering the spatial distribution of rain pools and the dispersal ability of juvenile adults of only about 2 - 4 km, it seems that most of the population is isolated.

The results of this experiment show that a negative correlation exists between the percentage of toads burrowing and soil moisture levels. In soil having a low moisture level (40%), a relatively high percentage of toads (73%) was buried compared to a low percentage (9%) at a high soil moisture level (100%) (Figure 4 and Figure 5), and the concentration of plasma in the soil was higher (Figure 6). During burrowing in the soil, the urea in the plasma increased dramatically compared to other components.

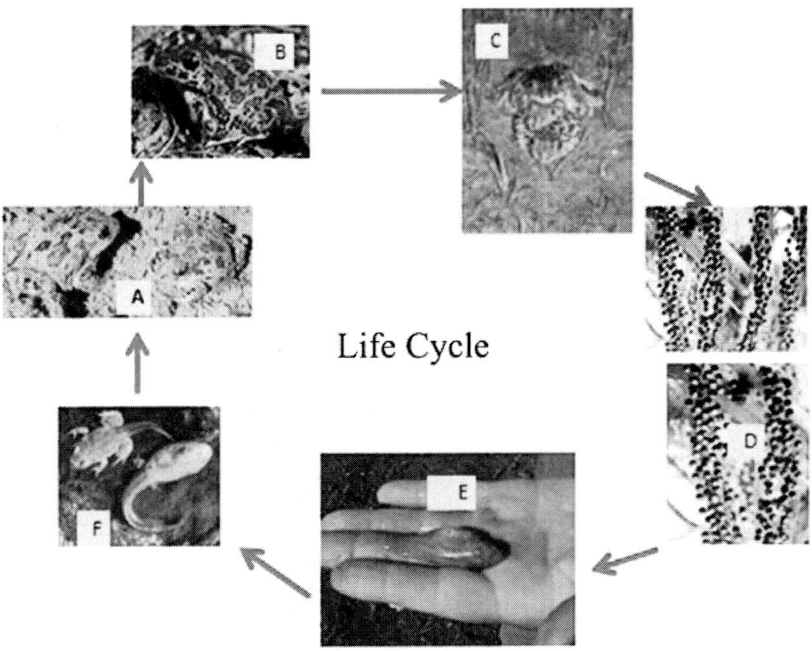

Figure 2. Life cycle of P. syriacus: juveniles (A) and mature females (B); Amplexuses (Cogălniceanu et al., 2013) (C); Eggs (D); Tadpole (E); Metamorphosis (F).

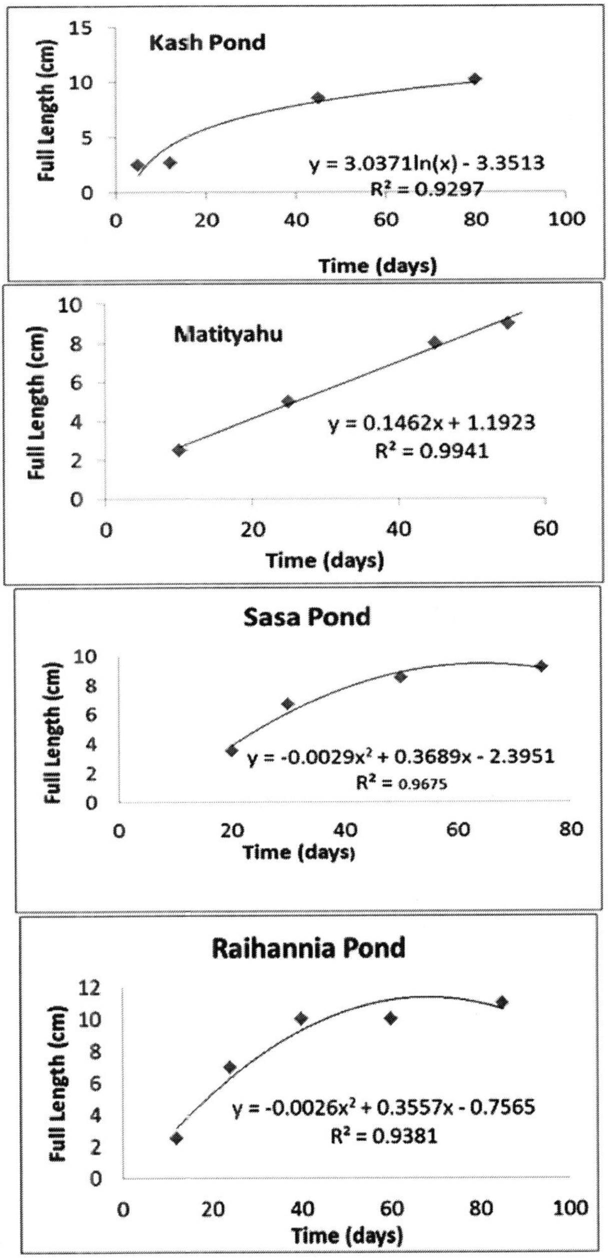

Figure 3. Growth rates of larvae of Pelobates syriacus. The Y axis shows larval size in mean of full length (cm), and the X axis shows time periods in the number of days that larvae had spent in the water body until they reached metamorphosis.

4. DISCUSSION

Spadefoot toads (Pelobates syriacus) are highly specialized burrowing and nocturnal species having a narrow ecological niche among amphibians in Israel, and all of them are found in relatively extreme conditions [6] . Spadefoot toad tadpoles are the largest among Israel's amphibian species [14] [18] , and the larval period can last several months until metamorphosis [4] [18] , as is supported by the present study. However, the selection of breeding places included only winter ponds where water was available for a few months, as was found in Europe [1] [2] and in the present study. According to previous studies at the southern border of its distribution of larvae growth and complete metamorphosis, we suggested that the selection of ponds where water was available for a few months should be correlated to the phenomena of this species (Pelobates syriacus)—large tadpoles and long periods to complete metamorphosis, as was found in the present study and previous studies [5] , and to the ecological conditions and water quality of the breeding places, e.g., the water temperature must be above 20°C and most of the time above 25°C for Spadefoot toad tadpoles to complete metamorphosis after a relatively long growth period (more than three months) [4] [5] [18] .

Table 2. Mean values and standard deviation (SD) of biotic and abiotic water quality parameters measured in breeding sites of Pelobates syriacus.

Site	Taxa (no.)	Mean Biomass (µg/l)	Rangechlor[a] (mg/l)	Mean hyd (days)	Mean Max. Volume (m^3)	Residential/ Agricultural Areas
Kash	6	1.2E+05	1.7 - 222.4	172 ± 8	2335 (444)	naa
Dovev	8	3.3E+03	1.9 - 89.2	183 ± 16	278 (31)	naa
Matityahu	10	5.9E+04	0.7 - 43.1	210 ± 4	4839 (847)	naa
Lehavot	6	2.3E+04	1.0 - 65.1	130 ± 19	10272 (2291)	naa
Sasa	7	1.7E+05	1.3 - 347.7	215 ± 18	11463 (1559)	naa
Fara	8	9.6E+05	0.9 - 581.7	297 ± 54	1109 (107)	naa
Raihania	7	3.2E+05	0.6 - 237.0	271 ± 18	7956 (390)	nra/naa

chlor. [a]chlorophyll *a*; hyd—hydroperiods; ay—all year; nra—near residential areas; naa—near agricultural areas.

Due to their secretive behavior (nocturnal activity, weak underwater breeding call and cryptic coloration), their adaptation to semi-arid environments is not well-documented in central Israel [8] [19] and northern Israel [6] [7] [20] , and overall there is a scarcity of information regarding this genus. According to the results of the present study, the selection and adaptation to ecological conditions is not done according to the aquatic phase of Spadefoot toads but following metamorphosis. This hypothesis, which is supported by relatively many studies on the conditions of larvae growth of Spadefoot toads [14] , shows that the water analysis of larvae is suitable for other amphibians in Israel, and that the distribution of Spadefoot toads is in drier conditions [21] .

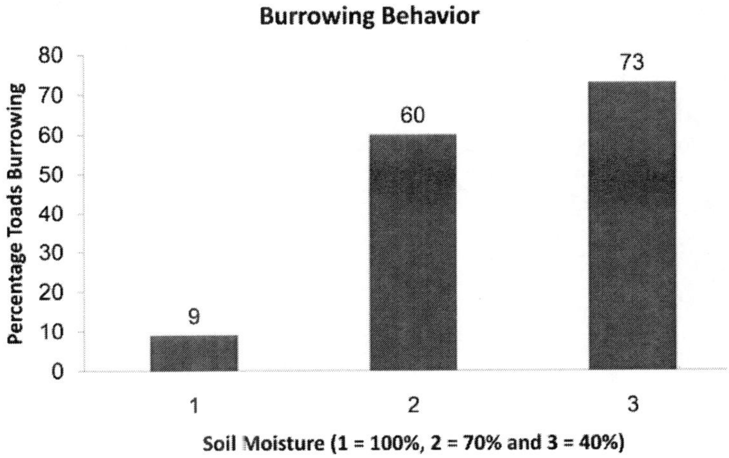

Figure 4. The effect of soil moisture on burrowing behavior of Spadefoot toads (Pelobates syriacus).

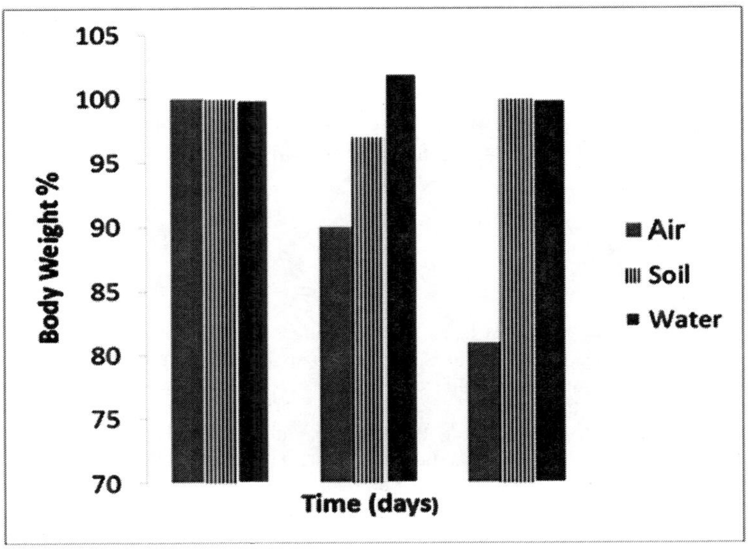

Figure 5. Spadefoot toads dehydrated under different conditions: in air, burrowed in soil and in water.

The burrowing behavior of metamorphosed Spadefoot toads preventing dehydration, as was found in the present study, is very important adaptation to terrestrial life. The accumulation of urea and electrolytes in the plasma during burrowing behavior seems to be a physiological adaptation, as was found in the present study, and is supported by previous research on this species [7] [22] . The only better adaptation to terrestrial life in amphibians in Israel is Green toad (Pseudepidalea viridis (Syn. Bufo viridis)) whose burrowing behavior is less than Spadefoot toads but has a higher adaptation to terrestrial life [17] [23] [24] . The other anuran in Israel, which is less adapted to terrestrial life, Hyla savygny [25] and Rana bedreagae [26] , exhibits no burrowing behavior and ophiology adaptation to terrestrial life.

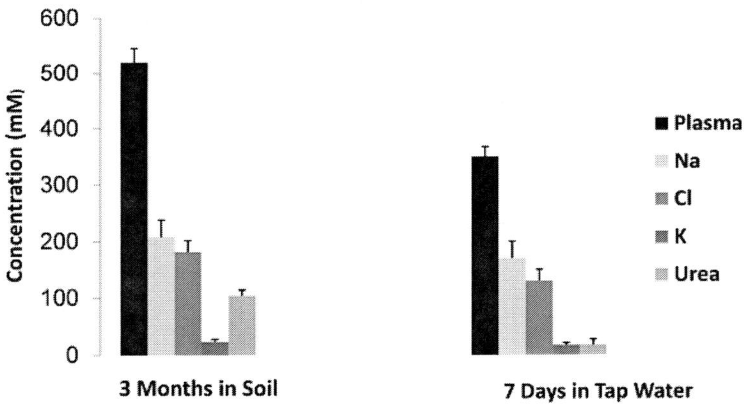

Figure 6. Changes in salinity and composition of plasma in juvenile Pelobates syriacus burrowed in soil and in water.

REFERENCES

1. Dzukic, G., Beskov, V., Sidorovska, V., Cogalniceanu, D. and Kaleziae, L.M. (2005) Historical and Contemporary Ranges of the Spadefoot Toads Pelobates spp. (Amphibia: Anura) in the Balkan Peninsula. Acta Zoologica Cracoviensia, 48A, 1-9. http://dx.doi.org/10.3409/173491505783995699
2. Cogălniceanu, D., Székely, P., Iosif, R., Székely, D. and Stănescu, F. (2013) Life History and Conservation of Spadefoot Toads (Genus Pelobates) in Romania. FrogLog, 21, 24-26.
3. Degani, G. and Kaplan, D. (1999) Distribution of Amphibian Larvae in Israeli Habitats with Changeable Water Availability. Hydrobiologia, 405, 49-56. http://dx.doi.org/10.1023/A:1003796820900
4. Degani, G. (1982) Amphibian Tadpole Interaction in Winter Pond. Hydrobiologia, 96, 3-7. http://dx.doi.org/10.1007/BF00006274
5. Degani, G. (1986) Growth and Behavior of Six Species of Amphibian Larvae in Winter Pond in Israel. Hydrobiologia, 140, 5-10. http://dx.doi.org/10.1007/BF00006723

6. Degani, G. and Carmali, D. (1988) Burrowing Behavior of Pelobates syriacus. Biology of Behaviour, 13, 22-29.
7. Degani, G., Goldenberg, S. and Warburg, M.R. (1983) Changes in Ion, Urea Concentrations and Blood Plasma Osmolarity of Pelobates syriacus Juveniles under Varying Conditions. Comparative Biochemistry and Physiology Part A: Physiology, 75, 619-623. http://dx.doi.org/10.1016/0300-9629(83)90430-9
8. Gafny, S. (1986) The Biology and Ecology of the Syrian Spadefoot Toad Pelobates syraicus syriacus in Israel. MSc Thesis, Tel Aviv University, Tel Aviv.
9. Gafny, S. (2004) Threatened Amphibians of Israel. In: Dolev, A. and Perevolotsky, A., Eds., Endangered Species in Israel, Red List of Threatened Animals Vertebrate, Israel Nature and Park Authority and the Society for the Preservation of Nature, Pub, Jerusalem, 55-68.
10. Shpun, S., Huffman, J., Nevo, E. and Katz, U. (1993) Is the Distribution of Pelobates syriacus Related to Its Limited Osmoregulatory Capacity? Comparative Biochemistry and Physiology Part A: Physiology, 105, 135-139. http://dx.doi.org/10.1016/0300-9629 (93)90185-7
11. Munwes, I., Geffen, E., Roll, U., Friedmann, A., Daya, A., Tikochinski, Y. and Gafny, S. (2010) The Change in Genetic Diversity down the Core-Edge Gradient in the Eastern Spadefoot Toad (Pelobates syriacus). Molecular Ecology, 19, 2675-2689. http://dx.doi.org/ 10. 1111/j.1365-294X.2010.04712.x
12. Degani, G. (2013) Genetic Variation among Various Populations of Spadefoot Toads (Pelobates syriacus, Boettger, 1869) at Breeding Sites in Northern Israel. Advances in Biological Chemistry, 3, 440-447. http://dx.doi.org/10.4236/abc.2013.35047
13. Nevo, E. (1976) Adaptive Strategies of Genetic Systems in Constant and Varying Environments. In: Karlin, S. and Nevo, E., Eds., Population Genetics and Ecology, Academic Press, New York, 141-158. http://dx.doi.org/10.1016/B978-0-12-398560-6.50011-0
14. Goldberg, T., Eviatar, E. and Degani, G. (2009) Breeding Site Selection According to Suitability for Amphibian Larval Growth under Various Ecological Conditions in the Semi-Arid Zone of Northern Israel. Ecologia Mediterranea, 35, 65-74.
15. Degani, G. and Warburg, M.R. (1980) The Response to Substrate Moisture of Juvenile and Adult Salamandra salamandra (L.) (Amphibia; Urodela). Biology Behaviour, 5, 281-229.
16. Degani, G. (1981) The Adaptation of Salamandra salamandra (L.) from Different Habitats to Terrestrial Life. British Journal of Herpetology, 6, 169-172. [Citation Time(s):2]
17. Degani, G. (1985) Urea Tolerance and Osmoregulation in Bufo viridis and Rana ridibunda. Comparative Biochemistry and Physiology Part A: Physiology, 82, 833-836. http:// dx.doi.org/10.1016/0300-9629(85)90491-8
18. Goldberg, T., Nevo, E. and Degani, G. (2012) Amphibian Larval in Various Water Bodies in the Semi-Arid Zone. Zoological Studies, 51, 345-361.
19. Gad, D., Thomsen, M.V. and Johannsen, H.G. (1999) [Spontaneous Rupture of the Left External Iliac Vein. A Complication of Deep Vein Thrombophlebitis?]. Ugeskrift for Læger, 161, 6639.
20. Degani, G. and Kaplan, D. (1999) Distribution of Amphibian Larvae in Israeli Habitats with Changeable Water Availability. Hydrobiologia, 404, 49-56. http://dx.doi.org/10.1023/ A:1003796820900
21. Munwes, I., Geffen, E., Roll, U., Friedmann, A., Daya, A., Tikochinski, Y. and Gafny, S. (2010) The Change in Genetic Diversity down the Core-Edge Gradient in the Eastern Spadefoot Toad (Pelobates syriacus). Molecular Ecology, 19, 2675-2689. http://dx.doi.org/10. 1111/j.1365-294X.2010.04712.x
22. Degani, G. (1986) Osmotic Stress and Osmoregulation of Tadpoles and Juveniles Pelobates syriacus. Comparative Biochemistry and Physiology Part A: Physiology, 83, 365-370. http://dx.doi.org/10.1016/0300-9629(86)90591-8
23. Degani, G. (1985) Osmoregulation in Red Blood Cells of Bufo viridis. Comparative Biochemistry and Physiology Part A: Physiology, 81, 451-453. http://dx.doi.org/10. 1016/0300-9629(85)90163-X
24. Degani, G., Silanikove, N. and Shkolnik, A. (1984) Adaptation of Green Toad (Bufo viridis) to Terrestrial Life by Urea Accumulation. Comparative Biochemistry and Physiology Part A: Physiology, 77, 585-587. http://dx.doi.org/10.1016/0300-9629(84)90233-0

25. Degani, G., Nagar, R. and Yom-Din, S. (2012) Molecular DNA Variation in Hyla felixarabica. Herpetol. Romanica, 6, 51-67.
26. Degani, G., Goldberg, T. and Yom-Din, S. (2013) The Ecology and Variation in DNA of Rana bedreagae from Various Breeding Site in North Israel. Res. Open. J. Anim. Sci, 1, 1-14.

CHAPTER 12

Spicule formation in calcareous sponges: Coordinated expression of biomineralization genes and spicule-type specific genes

Oliver Voigt[1], Maja Adamska[2], Marcin Adamski[2], André Kittelmann[1], Lukardis Wencker[1] & Gert Wörheide[1,3,4]

[1]*Department of Earth and Environmental Sciences, Palaeontology and Geobiology, Ludwig-Maximilians-Universität München, Richard-Wagner-Str. 10, 80333 Munich, Germany.*

[2]*Research School of Biology, ANU College of Medicine, Biology and Environment, The Australian National University, Canberra, 46 Sullivans Creek Road, Acton ACT 2601, Australia.*

[3]*GeoBio-Center, Ludwig-Maximilians-Universität München, Richard-Wagner-Str. 10, 80333 München, Germany.*

[4]*Bayerische Staatssammlung für Paläontologie und Geologie, Richard-Wagner-Str. 10, 80333 München, Germany.*

ABSTRACT

The ability to form mineral structures under biological control is widespread among animals. In several species, specific proteins have been shown to be involved in biomineralization, but it is uncertain how they influence the shape of the growing biomineral and the resulting skeleton. Calcareous sponges are the only sponges that form calcitic spicules, which, based on the number of rays (actines) are distinguished in diactines, triactines and tetractines. Each actine is formed by only two cells, called sclerocytes. Little is known about biomineralization proteins in calcareous sponges, other than that specific carbonic anhydrases (CAs) have been identified, and that uncharacterized Asx-rich proteins have been isolated from calcitic spicules. By RNA-Seq and RNA *in situ* hybridization (ISH), we identified five additional biomineralization genes in *Sycon ciliatum:* two bicarbonate transporters (BCTs) and three Asx-rich extracellular matrix proteins (ARPs). We show that these biomineralization genes are expressed in a coordinated pattern during spicule formation. Furthermore, two of the ARPs are spicule-type specific for triactines and tetractines (ARP1 or *SciTriactinin*) or diactines (ARP2 or *SciDiactinin*). Our results suggest that spicule formation is controlled by defined temporal and spatial expression of spicule-type specific sets of biomineralization genes.

1. INTRODUCTION

By the process of biomineralization many animal groups produce mineral structures like skeletons, shells and teeth. Biominerals differ in shape considerably from their inorganic mineral counterparts[1]. In order to build specific skeletal structures, organisms have to control the biomineralization process. This control involves proteins with different functions. For calcium carbonate biominerals, which are the most widespread minerals formed by animals[2], the directional transport and accumulation of inorganic ions to the calcification site is achieved by specialized transporters, i.e. by bicarbonate transporters (BCTs) or Ca2+-transporters e.g., refs [3] and [4]. Linked to bicarbonate transport and pH-regulation is the catalytic activity of carbonic anhydrases (CAs), which catalyse the reversible reaction of CO2 to bicarbonate[5]. Specialized CAs are key biomineralization proteins in calcium carbonate producing animals[6]. In addition, proteins of the skeletal organic matrix (SOM) have been identified by means of proteomics, transcriptomics and genomics[7]. Skeletal proteomes comprise mostly secreted proteins, and often include acidic proteins with high proportions of the amino acids aspartic acid or glutamic acid[7,8]. These acidic SOM proteins presumably interact with the calcium carbonate crystals and thereby can influence the growth and shape of biominerals[9]. However, little is known how the expression of biomineralization genes is coordinated and influences the biomineral shape.

Calcareous sponges (Porifera, class Calcarea) are an ideal system to address this question. Their calcite spicules are relatively simple structures, which can be distinguished by the number of their rays (actines) in monaxonic diactines (initially growing in two directions), three-rayed triactines, and four-rayed tetractines[10] (Fig. 1A). They are produced by only a few specialized cells, the sclerocytes, often within just a few days[11,12]. The spicules are growing within an extracellular space, sealed by septate junctions between the membranes of the sclerocytes[13,14], and are surrounded by an organic sheath that is secreted by the sclerocytes[14]. Each spicule is formed by two (diactines), six (triactines) or seven (tetractines) sclerocytes, of which one (termed founder cell) promotes tip growth, and the other, at least in some species, thickens the spicule (the thickener cell)[15,16] (Fig. 1B,C). Each founder and thickener cell pair originates from the division of a precursor cell; in case of triactine sclerocytes, these precursors form triplets before they divide[14,15,16]. At least in diactines, based on spicule staining experiments[11] and TEM observations[14] it was suggested that during initial stages of spicule formation the two sclerocytes contribute equally to tip elongation, before one starts functioning as a thickener cell.

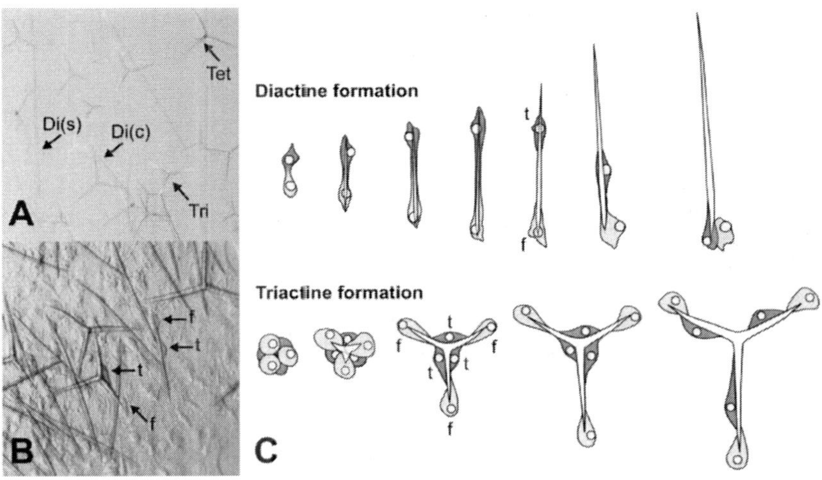

Figure 1: Spicule types and spicule formation in *S. ciliatum*.(A) Isolated spicules (fluorescence of calcein staining overlayed, showing spicule growth during 18 h12): Di(s) = oscular slender diactines, Di (c) = curved diactines of the distal end of radial tubes, Tri = triactines of the radial tubes and the atrial skeleton, Tet = tetractines of the atrial skeleton. (B) In vivo formation of spicules by sclerocytes (f = founder cell, t = thickener cell). (C) Movement of founder (f) and thickener (t) cells during diactine and triactine formation. (A) and (C) modified from ref. 12, (C) partially redrawn from ref. 16.

Little is known about biomineralization genes in calcareous sponges; only two specific CAs have been identified12,17,18. Furthermore, Asx (aspartic acid or asparagine)- rich proteins (ARPs) were extracted from spicules of different species, but have been only characterized by their amino acid composition19,20. We performed our study on the widespread calcareous sponge Sycon ciliatum, a model species for developmental biology with a sequenced genome21,22,23. The spicule formation by sclerocytes in this species has been documented by light microscopy15 and electron microscopy13,14. Sycon ciliatum has four spicule types (Fig. 1A), which can be readily distinguished and occur in specific body parts: (1) long, slender diactines (also called trichoxea), which form a palisade-like ring structure around the osculum; (2) curved diactines, which are restricted to the distal end of the radial tubes; (3) triactines, which form the atrial skeleton and the walls of the radial tubes; and (4) tetractines, which occur in the atrial skeleton (Fig. 1A). Triactines and tetractines with their three-rayed basal system form a scaffolding support for the tissues of the radial tubes (including the innermost layer of the water-propelling and filtering choanocytes), and the central cavity. Diactines, which protrude from the sponge body at the tips of the radial tubes and around the osculum, may serve as mechanical protection against blockage of influx and efflux openings.

A previous study found that spicule formation and the expression of two biomineralization genes, the carbonic anhydrases SciCA1 and SciCA2, is increased in the apical part of S. ciliatum sponges, where new radial tubes and the slender diactines of the osculum are built12. By RNA-Seq analysis we identified additional key biomineralization genes of calcareous sponges and studied their temporal and spatial expression patterns by RNA in situ hybridisation (ISH) to understand how they interact in the spicule formation process.

2. RESULTS AND DISCUSSION

2.1 Identification and expression patterns of biomineralization candidate genes

We identified new additional genes involved in biomineralization in Sycon ciliatum by screening RNA-Seq data of apically overexpressed genes22,24 for potential candidates, focussing on bicarbonate transporters and secreted, Asx-rich, proteins (ARPs). Bicarbonate transporters of the solute carrier 4 (SLC4) family are known to be involved in carbon transport and pH regulation25, and a specific variant has been shown to be a key biomineralization gene in scleractinian corals4. ARPs appear to be a major component of the spicule matrix proteome of calcareous sponges, as revealed by analyses of amino acid composition from proteins isolated from the spicules of various species19,20.

Among the apically overexpressed transcripts, we identified two SLC4 proteins and three ARPs with signal peptides (ARP1-3). Partial or complete coding sequences were PCR-amplified, cloned and sequenced. The sclerocyte-specific expression of all five genes was verified by in situ hybridization (ISH), confirming their expected involvement in biomineralization (Fig. 2). To further interpret the expression patterns in the absence of the calcitic spicules, which dissolve during the ISH procedure, double ISH was performed with two different colour detections with combinations of probes for the five new genes and the previously studied carbonic anhydrases SciCA1 and SciCA212.

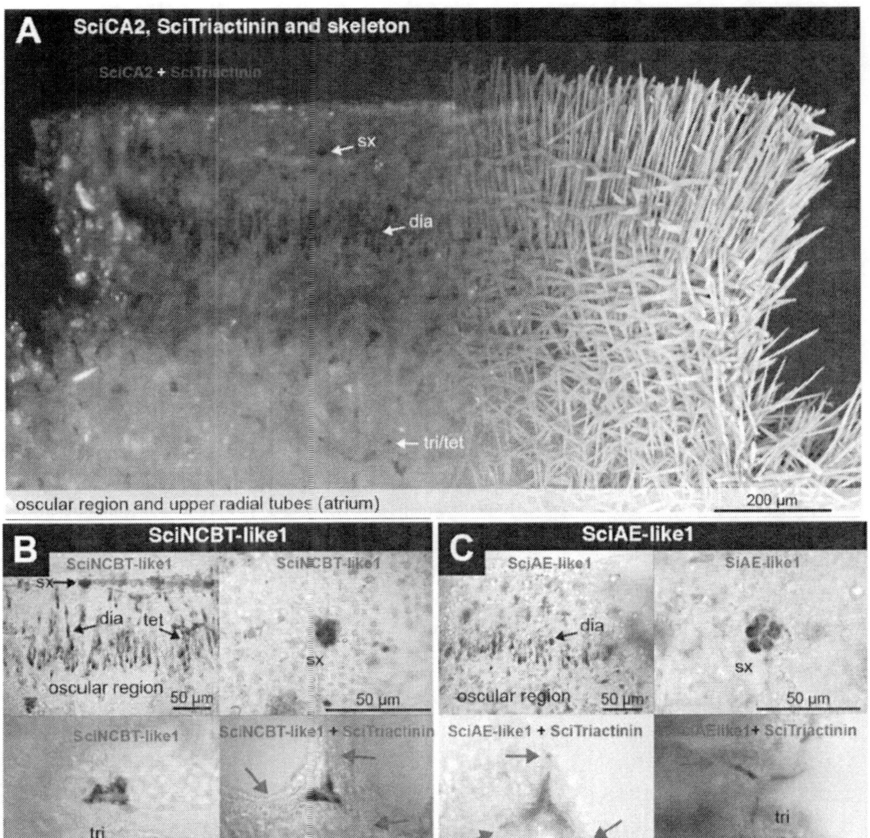

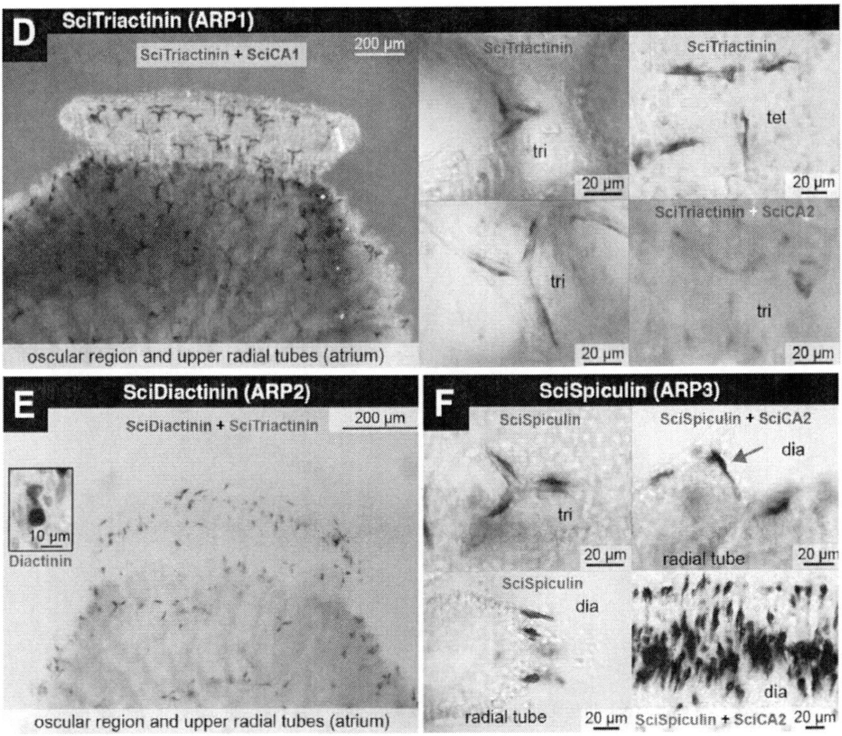

Figure 2: Expression patterns of biomineralization genes. (A) Overview over oscular region (atrial side) with *SciCA2* expression (blue), *SciTriactinin* expression (red) and an overlay of a µCT image to show the position of the dissolved spicules. (**B**) *SciNCBT-like1* expression in diactine, triactine and tetractine sclerocytes. Double-ISH with SciTriactinin-specific probes suggests expression in founder cells at later stages of spicule formation. (**C**) *SciAE-like1* expression in diactine, triactine and tetractine sclerocytes. Double-ISH with SciTriactinin-specific probes suggests expression in founder cells at later stages of spicule formation. (**D**) *SciTriactinin* expression is specific to triactine and tetractine thickener cells. *SciCA1* is only expressed in early stages, *SciTriactinin* in later stages. Double ISH with *SciCA2* reveals that at later stages of triactine and tetractine formation *SciCA2* expression only occurs in founder cells. (**E**) Expression of *SciDiactinin* in diactine forming sclerocytes. Inset: two close diactine-forming sclerocytes of early diactine formation. (**F**) *SciSpiculin* expression in thickener cells of triactines (tetractines not shown) and diactines. Double ISH with *SciCA2* reveals that founder cells of diactines are not expressing *SciSpiculin*, but *SciCA2* (radial tubes and oscular diactines). Abbreviations: dia = diactines, sx = sextet of sclerocytes, early stage of triactine (and tetractine) formation (compare Fig. 1C), tri = triactines; tet = tetractines.

2.2 Expression patterns of sclerocyte-specific *S. ciliatum* SLC4 proteins

Phylogenetic analyses (Fig. 3) of SLC4 proteins revealed that one (scigt008985) of the identified, potentially sclerocyte-specific SLC4-proteins of S. ciliatum belongs to the group of Na+/HCO3− co-transport proteins (NCBT-like), and that the other one (scigt015021) falls in the group of HCO3−/Cl− anion exchange proteins (AE-like). We therefore termed these Sycon ciliatum SLC4 proteins SciNCBT-like1 (scigt008985), and SciAE-like1 (scigt015021).

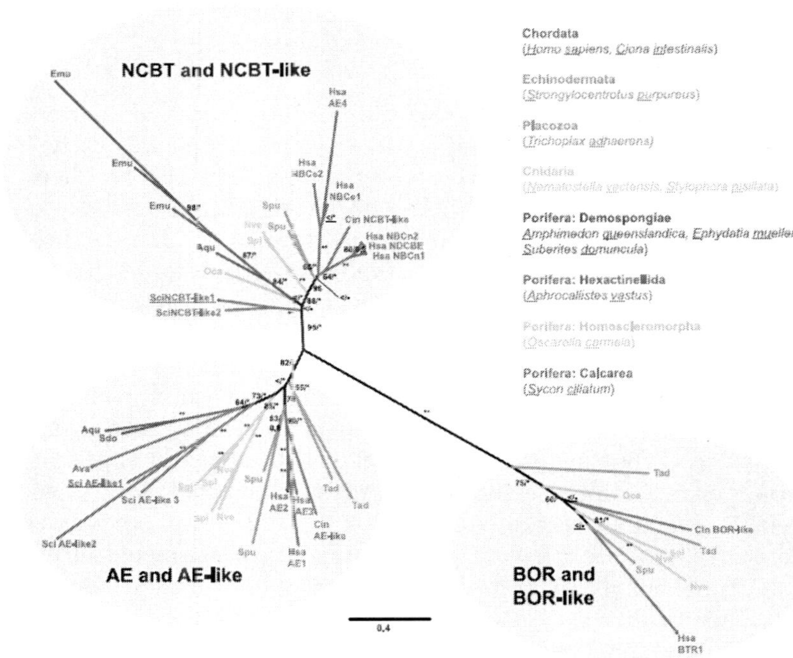

Figure 3: Phylogeny of SLC4 proteins. Maximum Likelihood tree (Phyml[48], LG + I + G + F), bootstrap values (BS) of 200 replicates and posterior probability (PP) of Bayesian analysis (MrBayes[49], LG + I + G + F, 5 million generations, burnin = 25% of sampled trees) are provided at the nodes (BS/PP; "*"= 100 BS or PP > 0.96; "** " = BS of 100 and PP > 0.97; " <=" support values below 50/0.5, "−" = node not present in Bayesian analysis), value on scale bar = substitutions/site. Biomineralization-specific proteins of S. ciliatum and Stylophora pistillata (SLC4γ) are underlined. The two biomineralization SLC4-proteins of S. ciliatum belong to the NCBT-like and the AE-like group, respectively.

Additional SLC4 proteins of S.ciliatum also belong to the two SLC4 groups and accordingly were termed SciNCBT-like2 (scigt018445), SciAE-like2 (scigt016671) and SciAE-like3 (scigt026034).

SciNCBT-like1 and SciAE-like1 showed similar expression patterns. They were expressed in founder and thickener cells of all spicule types, similar to the S. ciliatum carbonic anhydrase SciCA212 (Fig. 2A–C). Both are expressed in regions of increased spicule formation and expressing cells form an oscular ring (Fig. 2B,C), and are more abundant in the upper radial tubes (Suppl. Figure 1). Expression occurred in sclerocytes of diactines, triactines and tetractines. In the latter two, expression occurred in all six cells of the initial sextet (Fig. 2B,C). Double ISH with ARP1 revealed further details (see below).

2.3 Expression patterns and properties of ARPs

In contrast to SciNCBT-like1 and SciAE-like1, the expression patterns of the three ARPs were more specific: ARP1 (scigt005329) was exclusively expressed in founder cells of tri- and tetractines (Fig. 1D); we therefore termed this protein SciTriactinin. ARP2 (scigt017205) was expressed mostly in cells found in the oscular region, in which oscular diactines are formed, and in the distal end of radial tubes, where curved diactines are built (Fig. 1E). On several occasions, ARP2 expression occurred in two close sclerocytes (Fig. 2E, inset). When detected together in double ISH with SciCA2, a marker of active sclerocytes[12], only a small fraction of active sclerocytes expressed ARP2 (Suppl. Figure 1). In our view, these ISH patterns suggest expression only in a short time during spicule formation in early-stage diactine sclerocytes. Because no expression in triactine- or tetractine-specific sclerocytes was detected, we named this protein SciDiactinin. Finally, ARP3 (scigt005329) was expressed in thickener cells of all spicule types in later stages of spicule formation (Fig. 1F, Suppl. Figure 1). Accordingly, we termed this protein SciSpiculin, in reference to Haeckel's name for unidentified organic components in calcareous sponge spicules[26].

SciTriactinin, SciDiactinin and SciSpiculin are short proteins (with 143, 158 and 418 amino acids, respectively, Fig. 4), with an N-terminal signal peptide and a high content of aspartic acid, which makes them highly acidic (isoelectric points 3.6–3.8). Additionally, serine is a frequent amino acid in these proteins. Several O-linked glycosylation sites are predicted by Glyco EP[27] in all three ARPs, but only SciTriactinin has three potential N-linked glycosylation sites. Despite a short, shared motif (ADPPTP) found near the C-terminus of SciTriactinin and SciDiactinin, the three ARPs are not particularly similar to each other. Spiculin is characterized by a 39 amino acid repeat motif, which was present in eight complete (five in the genomic sequence, see Methods), and one partial copy in the cDNA sequence (Fig. 4). Previous reports about high Asx and serine content in proteins isolated from the intraspicular matrix of several calcareous sponge species suggested that acidic proteins are a major component of the spicule matrix proteome[19,20]. Therefore, we propose that SciTriactinin, SciDiactinin and SciSpiculin are important intraspicular matrix proteins. This proposal is supported by (1) the higher expression of these genes in the top body part of S. ciliatum, where increased spicule formation occurs; (2) the sclerocyte-specific expression of the ARPs; and (3) the presence of signal peptides, and therefore their potential secretion into the extracellular space of spicule formation.

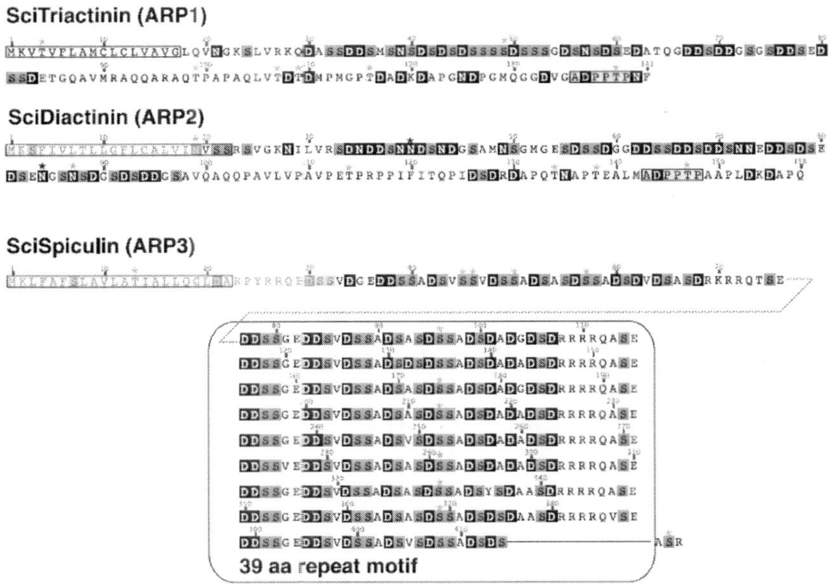

Figure 4: Amino acid sequences of the ARPs SciTriactinin (ARP1), SciDiactinin (ARP2) and SciSpiculin (ARP3). Aspartic acid and asparagine residues are highlighted by white letters on black background, serine by grey boxes. N-terminal signal peptides are marked by lined boxes, a short shared motif of *SciTriactinin* and *SciDiactinin* is marked by a yellow. Potential glycosylation sites are labelled with *(blue = O-linked glycosylation sites, black: N-linked glycosylation sites). Grey letters show parts that were not sequenced from cDNA due to position of the forward primers. For SciTriactinin and SciDiactinin, protein predictions from transcriptome data are presented, the SciSpiculin sequence is provided from combined clone sequence and transcriptome data.

2.4 Temporal and spatial expression of biomineralization genes during spicule formation

The expression levels in different body parts (top, middle bottom, Fig. 5A) were studied by RNA-Seq, using the available datasets22. The expression profiles of SciNCBT-like1 and SciAE-like1 were similar to that of SciCA1 und SciCA212 regarding their apical overexpression and maximum expression levels (Fig. 5B). Of the remaining SLC4 proteins, SciNCBT-like2, SciAE-like2 had equal expression levels in all body parts, and expression levels of SciAE-like3 were much lower (Fig. 5B). Maximal expression levels of the three ARPs were lower compared to the sclerocyte-specific CAs and BCTs. All were significantly higher expressed in apical parts in comparison to middle body parts, and, with exception of SciDiactinin, to bottom body parts.

Double ISH of combinations of biomineralization gene probes provided additional insight into the temporal and spatial expression in different stages of spicule formation: the results are summarized in Fig. 5C. SciNCBT-like1, SciAE-like1 and SciCA1 and SciCA2 are expressed in all sclerocytes of all spicule types in the initial spicule formation stages (SciCA2 expression begins later12). At later stages, when the founder and the thickener cells become separated, the expression of these genes is restricted to the founder cells. At this stage, we did not observe expression of SciCA1 (Fig. 2D). The expression of SciNCBT-like1 (Fig. 2B), SciAE-like1 (Fig. 2C) in founder cells in these later spicule formation stages was less frequently observed than the expression of SciCA2 (Fig. 2A,D); therefore, their expression likely ceases earlier. In the case of the SLC4 transporters, it can be assumed that these transmembrane transporters remain functional for a certain amount of time after their formation; so their production may not be necessary until the very end of the spicule growth. SciSpiculin is expressed in thickener cells of all spicule types in later spicule formation stages, again, after the separation of founder and thickener cell (Fig. 2F). In contrast, SciDiactinin and SciTriactinin are spicule type-specific. SciDiactinin is expressed in both, founder and prospective thickener cells, in initial diactine stages of diactines (oscular and curved diactines, Fig. 2E). SciTriactinin is specific to triactine and tetractine thickener cells, and expression begins approximately with the separation of founder and thickener cells. In summary, of the seven biomineralization genes observed here, five are expressed in sclerocytes of all spicule types (SciCA1, SciCA2, SciNCBT-like1, SciAE-like1 and SciSpiculin), and two are spicule type-specific (SciDiactinin and SciTriactinin). Furthermore, during initial spicule formation stages, the expression of founder and (prospective) thickener cells of one spicule type is identical.

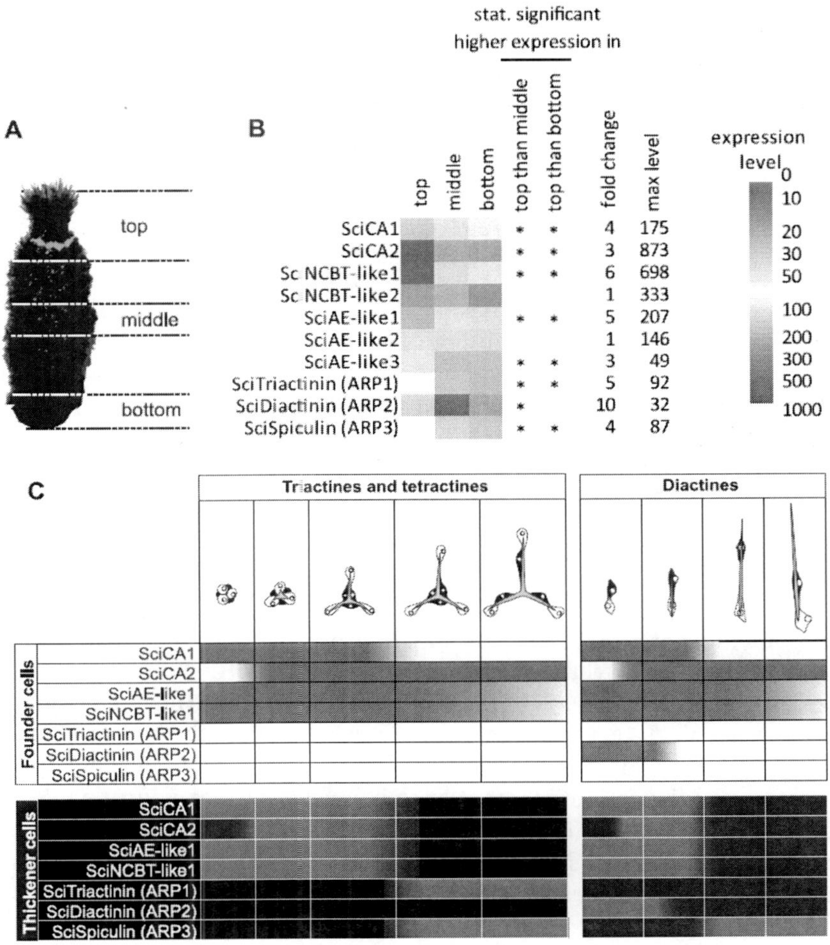

Figure 5: Spatial and temporal expression of seven biomineralization genes. (A) Schematic view of S. ciliatum body parts that were compared in the RNA-Seq analysis. The green colour depicts spicule formation, which is increased in apical regions, and was deduced from calcein-staining experiments[12]. (B) Expression levels of biomineralization genes and remaining SLC4 proteins in top, middle and bottom parts of S. ciliatum, the colour scale is from blue (lowest) through white (medium) to red (highest). Expression levels were calculated with expected_count from RSEM package[42], normalized between datasets with the DESeq package[43] and then log 10 transformed. Statistically significantly (padj ≤ 0.1) overexpressed genes in top vs. middle or top vs. bottom comparisons are marked by asterisks. (C) Summary of biomineralization gene expression in founder cells and (prospective) thickener cells of different spicule types, based on observations of the double ISH

experiments. In both, tri- and tetractines on the one hand, and diactines on the other hand, the founder cells and prospective thickener cells show initially identical expression patterns. The most dramatic change of expression occurs in later stages in thickener cells, which of the seven genes only expresses SciSpiculin (all spicule types) and SciTriactinin (only triactine- and tetractine-specific thickener cells).

In later stages, the expression of the biomineralization genes changes, especially in the thickener cells, which no longer express the sclerocyte-specific CAs and SLC4 genes, but begin to produce the ARPs SciSpiculin and/or SciTriactinin. These observations are consistent with previous reports that in contrast to other species of calcareous sponges, the thickener cells of S. ciliatum do not appear to deposit additional calcite on the spicule[14,15]; therefore CA activity and bicarbonate transport are unnecessary in thickener cells. The missing thickening activity suggests another role for the thickener cells, which involves the expression of SciSpiculin and SciTriactinin (see below).

2.5 Evolution of SLC4 and ARP proteins

Our phylogenetic analyses of SLC4 proteins revealed that sponge proteins occur in all three SLC4 groups (NCBT and NCBT-like, AE and AE-like, BOR and BOR-like, Fig. 3). SLC4 proteins of the BOR-like group are missing in calcareous sponges (and other sponges with the exception of the homoscleromorph Oscarella, Fig. 3). Because genomic data is not available for Hexactinellida, it cannot be excluded that additional SLC4-like transporters (NCBT and NCBT-like or BOR and BOR-like) are present in this sponge class. Nonetheless, our phylogenetic analyses (Fig. 3) confirm that sponges possess more SLC4-like transporters than only the previously reported AE-like protein of the demosponge Suberites domuncula28. In the clade "NCBT and NCBT-like" and "AE and AE-like", many lineages, including calcareous sponges, show lineage-specific gene duplications: In addition to the two sclerocyte-specific SLC4 genes, of the three additional SLC4 proteins that are encoded in the S. ciliatum genome, one is placed in the NCBT and NCBT-like group (SciNCBT-like2), and is closest related to SiNCBT-like1 (Fig. 3). The other two were found to be of the AE and AE-like group and form a clade with SciAE-like1; within the clade, SciAE-like3 is the sister group to a clade of SciAE-like1 and SciAE-like2 (Fig. 3). Because Sycon ciliatum NCBT-like and AE-like BCTs are each monophyletic, a lineage-specific diversification of both SLC4 groups within Calcarea can be suggested. The biomineralizing SLC4γ from scleractinians also belongs to the AE-like SLC4-proteins, but is not especially closely related to the SciAE-like1 protein (Fig. 3). These transporters were likely independently recruited for the process of biomineralization in Calcarea and Scleractinia, possibly following lineage-specific duplications in both lineages. Similar observations were reported for the evolution of CAs[12].

The evolution of the ARPs is more obscure. No conserved domains could be identified in the ARPs. BLAST searches against transcriptome of the closely related species Sycon coactum29 found one significant hit for SciTriactinin (S. coactum contig_18526), which has a sequence similarity of 47% and similar aspartic acid composition (S. ciliatum: 18.2%, S. coactum 17.2%) and serine contents (S. ciliatum: 16.8%, S. coactum: 18.6%) and may represent a true homolog of SciTriactinin (Suppl. Figure 2). It has an additional potential ORF that would encode 141 additional N-terminal amino acids if it would get translated (see legend of Suppl. Figure 2). This 141 amino acid sequence lacks any known protein domains and shows no homology by BLAST searches. An incomplete transcript (coding for 100 N-terminal amino acids) was found as significant BLAST-hit for SciSpiculin (S. coactum contig_22784). It contained two copies of a 30 amino acid repeat (Suppl. Figure 2). Possibly, the incompleteness is due to assembly problems and more repeats are present in the mature protein similar to the eight complete copies of the 39 amino acid motif in SciSpiculin. The similarity in the first 100 amino acid positions of SciSpiculin and the potential S. coactum homolog is 45%, their aspartic acid content is similar (S. ciliatum: 21.0%, S. coactum: 23.0%), while the S. coactum serine content is higher (S. ciliatum: 28.0%, S. coactum: 39%). No BLAST hits were found for SciDiactinin in S. coactum. BLAST searches of the ARPs (neglecting the signal peptides) against the transcriptome of the more distantly related calcareous sponge Leucosolenia complicata[22] failed to provide any hits for any of the three ARPs (with maximum E-value cut-off of 10). Therefore, ARPs appear to be either evolving so fast that homology is obscured rapidly (e.g. in Diactinin even between very closely related species), or they represent lineage-specific innovations.

2.6 Potential function of biomineralization genes

Potential function of SciAE-like1 and SciNCBT-like1 Although SLC4 proteins can be assigned to the groups AE-like, NCBT-like or BOR-like based on their phylogenetic affinities, the function and stoichiometry of transport of only a few members of each group are known, excluding for example sponge proteins[28]. Therefore, neither the direction nor the mode of transport (Na^+-independent Cl- cotransport for SciAE-like1, or Na^+-coupled for SciNCBT-like1) can be deduced for the two sclerocyte-specific SLC4 proteins in S. ciliatum. However, it is reasonable to assume that the proteins are involved in the guided transport of bicarbonate to the calcification site through the sclerocyte, i.e., trafficking bicarbonate from the mesohyl into the sclerocyte and/or trafficking bicarbonate that is formed within the sclerocyte through the activity of SciCA1 to the intercellular space of calcification[12]. This interconnected function of the sclerocyte-specific CAs and SLC4 proteins would also explain the striking similarity in their expression profiles in body parts (Fig. 5B).

Potential function ARPs Acidic proteins with a high aspartic acid content have been found in the organic matrices of carbonate skeletons in many animals, including stony corals e.g., refs 30 and 31 and coralline demosponges[32]. Important functions of these proteins in the biomineralization process have been suggested[8]. For example, aspartic acid residues in these proteins have the ability to bind Ca^{2+} ions, and some can interact with specific crystal faces of growing biominerals, thereby influencing the crystal shape. Depending on the conditions, inhibition or promotion of crystallisation has been reported for acidic skeletal organic matrix (SOM) proteins8. The presence of Asx-rich-protein extracts of calcareous sponge spicules has been found to influence the shape of calcite crystal formation in in vitro experiments[20]. It has also been proposed that differences in crystal texture among spicule types of calcareous sponges are influenced by the acidic SOM proteins[33]; accordingly, specialized proteins were suggested to interact with specific crystal faces, inhibiting their growth, and thus influencing the preferred direction of crystal growth, which differed among triactines, curved diactines and oscular diactines. We therefore suggest that the spicule-type specific ARPs SciTriactinin and SciDiactinin are involved in the development of the different crystallographic growth patterns between diactines and triactines/tetractines.

Because SciDiactinin was expressed in the early stages of spicule formation, it presumably plays a role in the initial nucleation process of diactine spicules. However, our results cannot explain the previously reported differences between curved diactines and oscular diactines[33]. In contrast to the crystal texture of curved diactines, that of oscular diactines did not differ considerably from that of synthetic calcite, which was attributed to a lack or a low concentration of intraspicular proteins[33]. Yet, a difference in protein abundance might exist between the slender and curved diactines, potentially due, for example, to faster growth rates (about two times) of the former[12]. Also, species-specific differences between S. ciliatum and other species may exist: The ARPs are highly specific to S. ciliatum, and we found only two recognizable orthologs in the transcriptome of Sycon coactum, and none in the transcriptome of Leucosolenia complicata. The previous study[33] mentioned above investigated a different species (Sycon sp. from the Mediterranean), but it is known that genus Sycon is polyphyletic, such that its member species may be only distantly related[34].

Because SciTriactinin and SciSpiculin are only expressed in thickener cells in late spicule formation stages, they cannot act in the earlier stages. Observations on the organization of triactines from Clathrina sp. may be relevant: they possess a calcite core containing Asx-rich proteins, which is surrounded by a phase of amorphous calcium carbonate (ACC) stabilized by Glx-rich proteins and itself is covered by a thin calcitic sheath[20]. Although ACC has not been reported from other calcareous sponges[35], it is difficult to detect, and it was speculated that stabilized ACC may be more widespread in calcareous sponges than is currently recognized[36], potentially even as a transition stage in spicule maturation. Provided that the spicules in S. ciliatum show an identical organisation, SciTriactinin and SciSpiculin may be involved in the formation of the outermost thin calcitic sheath. In such a scenario, additional Asx-rich proteins from the calcitic core of triactines and tetractines of S. ciliatum could be expected, similar to the findings in Clathrina sp[20]. Mineralogical studies on the fine structure of newly formed spicules and the identification of additional ARPs, and Glx-rich proteins of a potential ACC layer could provide further insight.

3. CONCLUSION

Spicule formation is a highly dynamic process that requires the concerted temporal regulation of gene expression in the sclerocytes involved to build the complex architecture of the calcareous sponge skeleton. The expression of the seven biomineralization genes studied here in the prospective founder and thickener cells of each spicule type is identical in the initial stages of spicule formation. In later stages of spicule formation, expression of founder and thickener cells differentiate from each other. This observation is consistent with the fact that each thickener and founder cell pair develops from a single precursor cell with subsequent spatio-temporal diversification15. Of the seven biomineralization genes analysed, the two biomineralizing CAs and the two biomineralizing SLC4 genes and the ARP SciSpiculin (ARP3) provide a common genetic ground pattern for the formation of all spicule types of S. ciliatum. In contrast, the ARPs SciDiactinin (ARP2) and SciTriactinin (ARP1) are spicule type-specific modifications in the genetic biomineralization toolkit and present evidence for genetic determination of biomineral shape in calcareous sponges. Our results highlight that genetic control over the biomineralization is essential in the formation of different biomineral shapes as observed even in such simple biominerals as calcitic sponge spicules, which are formed by only a few cells.

4. METHODS

4.1 Identification of biomineralization genes

Sycon ciliatum sponges were collected in Norway, tissue fixed for RNA extraction and RNA in situ hybridization as described before[12,22,24]. Previous studies provided transcriptomes of different life-cycle stages and body parts and provided lists of genes with higher expression in apical body parts[22,24], in which biomineralization is increased[12]. From this list, two bicarbonate transporters of the SLC4 family were identified. ARPs were identified by selecting apically overexpressed genes with Asx-contents larger than 20% and with a signal peptide. While SciTriactinin (ARP1) and SciDiactinin (ARP2) were complete transcripts, the transcriptome assembly of SciSpiculin (ARP3) did not yield the C-terminal stop codon, probably due to the presence of a 117 bp repeat motif (coding for 39 amino acids), and we therefore identified the corresponding ORF on the genomic scaffold 29508[22] to design 5′ primers.

4.2 Cloning, sequencing and sequence analysis

Primers (Suppl. Table 1) for each of the target genes were designed using the primer3 as implemented in Geneious R8 (http://www.geneious.com)37. SciTriactinin and SciDiactinin reverse primers were designed to introduce a T7 recognition site for RNA antisense probe generation from PCR products. A pool of cDNA from different life stages was used as template for PCRs. PCR-products of all templates were cloned into the pCR4 vector (Invitrogen), clones were prepared for sequencing with vector-specific primers using the BigDye Terminator sequencing kit v.3.1 (Applied Biosystems). Bidirectional sequencing was performed at the Sequencing Service at the LMU Biozentrum on an ABI 3730 capillary sequencer (Applied Biosystems). Forward and reverse sequences were assembled in Geneious R8 (http://www.geneious.com)[37]. All sequences have been submitted to the European Nucleotide Archive (accession codes LT674110- LT674121, http://www.ebi.ac.uk/ena/data/view/ LT674110-LT674121). Alignments of genomic and amplified sequences are available in the Open Data LMU repository (http://dx.doi.org/ 10.5282/ubm/data.97). Cloning of SciTriactinin (ARP1) yielded two versions of which one had a six base pair (two amino acid) insertion compared to the transcriptome sequence (scigt017205). The sequenced SciDiactinin (ARP2) fragment did not cover the complete 5′ coding region of the gene. For further analyses, the predicted gene sequences from the genome were used. Sequencing of SciSpiculin (ARP3) cDNA revealed three additional 117 bp direct repeats compared to the genomic sequence. We believe that the genomic assembly probably failed to assemble the 8 × 117 bp repeat region of the gene, which may also be the reason for the incompleteness of the transcriptomic sequence (see above). For further analyses, the clone sequence was complemented with the 5′ end of the transcriptomic and genomic sequence, which was not amplified with our primers. Amino acid composition and isoelectric point of ARPs were determined in Geneious R8 (http://www.geneious.com)[37]. Signal peptides of ARPs were detected with signalP 4.1[38], potential glycosylation sites were predicted with GlycoEP (http://www.imtech.res.in/raghava/glycoep)27. BLAST searches[39] against GenBank databases and searches in pfam[40,41] were conducted, but for the ARPs yielded no significant similarities to known proteins or domains. BLAST was also used to identify ARPs in the transcriptome of Sycon coactum (https://era.library.ualberta.ca/files/bjh343s467#.WE53UKKLS1s)[29]. Bicarbonate transporters SciNCBT-like1 and Sci-AE-like1 were unambiguously homologous to other SLC4 proteins, to which they could be aligned (see below).

4.3 RNA *in situ* hybridization and RNA-Seq

Antisense RNA probes of all five genes were generated by in vitro transcription using T7 or T3 RNA polymerase and plasmids or PCR products (for SciTriactinin, SciDiactinin) with introduced T7 sites in the reverse primers (Suppl. Table 1). Probes were labeled using the digoxigenin (DIG) or fluorescein RNA labelling kit (Roche). RNA antisense probes for SciCA1 and SciCA2 were available from a previous study[12]. Fixed tissues of S. ciliatum (small sponges or parts of larger sponges) were used in ISH experiments, which were performed according to previously published protocols[12,21]. For double ISH, two probes labelled with either DIG or fluorescein were applied, and the first probe was detected with NBT/BCIP and the second with Fast Red (Roche). Whole mount ISH experiments were documented with Leica M165F or Leica DMLB microscope. To increase the depth of field, multi-focus images were combined with Helicon Focus 4.2.9 (HeliconSoft).

In detail RNA-Seq analysis of the expression of the seven biomineralization genes and the remaining SLC4 genes was performed using existing transcriptomic RNA-seq datasets from top, middle and bottom body section of S. ciliatum sponges[22] available at ArrayExpress (http://www.ebi.ac.uk/arrayexpress) under accession number E-MTAB-2430. Expression levels were calculated with expected_count from RSEM package[42], normalized between datasets with the DESeq package[43] and then log 10 transformed. Statistically significantly ($padj \leq 0.1$) overexpression of genes was determined in comparisons top vs. middle or top vs. bottom.

4.3 Phylogenetic analysis of SCL4 proteins

Additional SCL4 proteins of S. ciliatum and other phyla were identified by BLAST[39] from available transcriptomic or genomic data (Suppl. Table 2). Protein sequences were aligned with MUSCLE[44] implemented in Seaview[45]. Gblocks[46] was used to select conserved sites suitable for the phylogenetic analyses. The best fitting model for Maximum Likelihood (ML) analysis and Bayesian inference (LG + I + G + F) was determined under the Akaike Information Criterion (AIC) with Prottest3[47]. ML likelihood analysis including a 200 replicate bootstrap analysis was performed with PhyML 3[48]. Bayesian inference was conducted in MrBayes 3.2.6[49] (5 million generations, sampling every 200th tree and discarding the first 25% of sampled trees as burnin to calculate the consensus tree). Sufficient parameter sampling of the analysis was confirmed by inspection of the parameter files in tracer v1.6 (http://tree.bio.ed.ac.uk/software/tracer/). The SLC4 alignment (including sequence identifiers and information about sites included in the analyses) is available via the Open Data LMU repository (http://dx.doi.org/10.5282/ubm/data.97).

ACKNOWLEDGEMENTS

Kasia· Sluzek generated some of the RNA ISH probes used in this study. Ana Sofia Ortega Arbulu, Carolin Gut and Laura Leiva carried out some of the double RNA ISH experiments during LMU-funded student projects. O.V. thanks LMU for financial support by the studi_forscht@Geo program.

REFERENCES

1. Lowenstam, H. A. *Minerals formed by organisms*. *Science* **211**, 1126–1131 (1981).
2. Murdock, D. J. E. & Donoghue, P. C. J. *Evolutionary Origins of Animal Skeletal Biomineralization*. *Cells Tissues Organs* **194**, 98–102 (2011).
3. Barott, K. L., Perez, S. O., Linsmayer, L. B. & Tresguerres, M. Differential localization of ion transporters suggests distinct cellular mechanisms for calcification and photosynthesis between two coral species. Am. J. Physiol. Regul. Integr. Comp. Physiol. 309, R235–46 (2015).
4. Zoccola, D. et al. Bicarbonate transporters in corals point towards a key step in the evolution of cnidarian calcification. Sci. Rep. 5, 9983 (2015).
5. Tripp, B. C., Smith, K. & Ferry, J. G. Carbonic anhydrase: new insights for an ancient enzyme. J. Biol. Chem. 276, 48615–48618 (2001).
6. Le Roy, N., Jackson, D. J. & Marie, B. The evolution of metazoan α-carbonic anhydrases and their roles in calcium carbonate biomineralization. Front. Zool. 11, 75 (2014).

7. Marin, F., Bundeleva, I., Takeuchi, T., Immel, F. & Medakovic, D. Organic matrices in metazoan calcium carbonate skeletons: Composition, functions, evolution. J. Struct. Biol. 196, 98–106 (2016).
8. Marin, F. & Luquet, G. In Handbook of biomineralization(ed. Bäuerlein, E.) 1, 273–290 (Wiley-VCH Verlag GmbH, 2007).
9. Addadi, L. & Weiner, S. Interactions between acidic proteins and crystals: stereochemical requirements in biomineralization. Proc. Natl. Acad. Sci. USA. 82, 4110–4114 (1985).
10. Manuel, M. Phylogeny and evolution of calcareous sponges. Can. J. Zool. 84, 225–241 (2006).
11. Ilan, M., Aizenberg, J. & Gilor, O. Dynamics and growth patterns of calcareous sponge spicules. Proceedings of the Royal Society of London B Biological Sciences 263, 133–139 (1996).
12. Voigt, O., Adamski, M., Sluzek, K. & Adamska, M. Calcareous sponge genomes reveal complex evolution of α-carbonic anhydrases and two key biomineralization enzymes. BMC Evol. Biol. 14, 230 (2014).
13. Ledger, P. W. Septate junctions in the calcareous sponge Sycon ciliatum. Tissue Cell 7, 13–18 (1975).
14. Ledger, P. W. & Jones, W. C. Spicule formation in calcareous sponge Sycon ciliatum. Cell Tissue Res. 181, 553–567 (1977).
15. Woodland, W. Memoirs: Studies in spicule formation: I.–The development and structure of the spicules in Sycons: with remarks on the conformation, modes of disposition and evolution of spicules in calcareous sponges generally. Q. J. Microsc. Sci. 49, 231–282 (1905).
16. Minchin, E. A. Materials for a monograph of the Ascons. II: – The formation of spicules in the genus Leucosolenia, with some notes on the histology of the sponges. Q. J. Microsc. Sci. 52, 301–355 (1908).
17. Müller, W. E. G. et al. Common genetic denominators for Ca++ -based skeleton in Metazoa: role of osteoclast-stimulating factor and of carbonic anhydrase in a calcareous sponge. PLoS One 7, e34617 (2012).
18. Müller, W. E. G. et al. The enzyme carbonic anhydrase as an integral component of biogenic Ca-carbonate formation in sponge spicules. FEBS Open Bio 3, 357–362 (2013).
19. Aizenberg, J., Ilan, M., Weiner, S. & Addadi, L. Intracrystalline macromolecules are involved in the morphogenesis of calcitic sponge spicules. Connect. Tissue Res. 34, 255–261 (1996).
20. Aizenberg, J., Lambert, G., Addadi, L. & Weiner, S. Stabilization of amorphous calcium carbonate by specialized macromolecules in biological and synthetic precipitates. Adv. Mater. 8, 222–226 (1996).
21. Fortunato, S. et al. Genome-wide analysis of the sox family in the calcareous sponge Sycon ciliatum: multiple genes with unique expression patterns. Evodevo 3, 14 (2012).
22. Fortunato, S. A. et al. Calcisponges have a ParaHox gene and dynamic expression of dispersed NK homeobox genes. Nature 514, 620–623 (2014).
23. Fortunato, S. A. V., Vervoort, M., Adamski, M. & Adamska, M. Conservation and divergence of bHLH genes in the calcisponge Sycon ciliatum. Evodevo 7, 23 (2016).
24. Leininger, S. et al. Developmental gene expression provides clues to relationships between sponge and eumetazoan body plans. Nat. Commun. 5, 3905 (2014).
25. Romero, M. F., Chen, A.-P., Parker, M. D. & Boron, W. F. The SLC4 family of bicarbonate (HCO3–) transporters. Mol. Aspects Med. 34, 159–182 (2013).
26. Haeckel, E. Die Kalkschwämme. Eine Monographie in zwei Bänden Text und einem Atlas mit 60 Tafeln Abbildungen. 1–3 (Verlag von Georg Reimer, 1872).
27. Chauhan, J. S., Rao, A. & Raghava, G. P. S. In silico platform for prediction of N-, O- and C-glycosites in eukaryotic protein sequences. PLoS One 8, e67008 (2013).
28. Parker, M. D. & Boron, W. F. The divergence, actions, roles, and relatives of sodium-coupled bicarbonate transporters. Physiol. Rev. 93, 803–959 (2013).
29. Riesgo, A., Farrar, N., Windsor, P. J., Giribet, G. & Leys, S. P. The analysis of eight transcriptomes from all Porifera classes reveals surprising genetic complexity in sponges. Mol. Biol. Evol. 31, 1102–1120 (2014).

30. Puverel, S. et al. Soluble organic matrix of two Scleractinian corals: partial and comparative analysis. Comp. Biochem. Physiol. B Biochem. Mol. Biol. 141, 480–487 (2005).
31. Drake, J. L. et al. Proteomic analysis of skeletal organic matrix from the stony coral Stylophora pistillata. Proc. Natl. Acad. Sci. USA. 110, 3788–3793 (2013).
32. Germer, J., Mann, K., Wörheide, G. & Jackson, D. J. The skeleton forming proteome of an early branching metazoan: A molecular survey of the biomineralization components employed by the coralline sponge Vaceletia sp. PLoS One 10, e0140100 (2015).
33. Aizenberg, J. et al. Morphogenesis of calcitic sponge spicules - a role for specialized proteins interacting with growing crystals. FASEB J. 9, 262–268 (1995).
34. Voigt, O., Erpenbeck, D. & Wörheide, G. Molecular evolution of rDNA in early diverging Metazoa: first comparative analysis and phylogenetic application of complete SSU rRNA secondary structures in Porifera. BMC Evol. Biol. 8, 69 (2008).
35. Smith, A. M., Berman, J., Key, J. M. M. & Winter, D. J. Not all sponges will thrive in a high-CO2 ocean: Review of the mineralogy of calcifying sponges. Palaeogeogr. Palaeoclimatol. Palaeoecol. 392, 463–472 (2013).
36. Sethmann, I. & Wörheide, G. Structure and composition of calcareous sponge spicules: a review and comparison to structurally related biominerals. Micron 39, 209–228 (2008).
37. Kearse, M. et al. Geneious Basic: an integrated and extendable desktop software platform for the organization and analysis of sequence data. Bioinformatics 28, 1647–1649 (2012).
38. Petersen, T. N., Brunak, S., von Heijne, G. & Nielsen, H. SignalP 4.0: discriminating signal peptides from transmembrane regions. Nat. Methods 8, 785–786 (2011).
39. Altschul, S. F., Gish, W., Miller, W., Myers, E. W. & Lipman, D. J. Basic Local Alignment Search Tool. J. Mol. Biol. 215, 403–410 (1990).
40. Bateman, A. et al. The Pfam protein families database. Nucleic Acids Res. 32, D138–41 (2004).
41. Finn, R. D. et al. Pfam: clans, web tools and services. Nucleic Acids Res. 34, D247–51 (2006).
42. Li, B. & Dewey, C. N. RSEM: accurate transcript quantification from RNA-Seq data with or without a reference genome. BMC Bioinformatics 12, 323 (2011).
43. Anders, S. & Huber, W. Differential expression analysis for sequence count data. Genome Biol. 11, R106 (2010).
44. Edgar, R. C. MUSCLE: a multiple sequence alignment method with reduced time and space complexity. BMC Bioinformatics 5, 113 (2004).
45. Gouy, M., Guindon, S. & Gascuel, O. SeaView version 4: a multiplatform graphical user interface for sequence alignment and phylogenetic tree building. Mol. Biol. Evol. 27, 221–224 (2010).
46. Castresana, J. Selection of conserved blocks from multiple alignments for their use in phylogenetic analysis. Mol. Biol. Evol. 17, 540–552 (2000).
47. Darriba, D., Taboada, G. L., Doallo, R. & Posada, D. ProtTest 3: fast selection of best-fit models of protein evolution. Bioinformatics 27, 1164–1165 (2011).
48. Guindon, S. et al. New algorithms and methods to estimate maximum-likelihood phylogenies: assessing the performance of PhyML 3.0. Syst. Biol. 59, 307–321 (2010).
49. Ronquist, F. & Huelsenbeck, J. P. MrBayes 3: Bayesian phylogenetic inference under mixed models. Bioinformatics 19, 1572–1574 (2003).

CHAPTER 13

Automated measurement of cattle surface temperature and its correlation with rectal temperature

HongXiang Kou[1,2], YiQiang Zhao[3] , Kang Ren[4], XiaoLi Chen[1], YongQiang Lu[4] , Dong Wang[1*]

[1]*The Key Laboratory for Farm Animal Genetic and Utilization of Ministry of Agriculture of China, Institute of Animal Science, Chinese Academy of Agricultural Science, Beijing, China,*

[2]*College of Animal Science and Technology, Jilin Agricultural University, Changchun, China,*

[3]*State Key Laboratory for Agro-biotechnology, China Agricultural University, Beijing, China,*

[4]*Animal Husbandry Station of Beijing, Beijing, China*

ABSTRACT

The body temperature of cattle varies regularly with both the reproductive cycle and disease status. Establishing an automatic method for monitoring body temperature may facilitate better management of reproduction and disease control in cattle. Here, we developed an Automatic Measurement System for Cattle's Surface Temperature (AMSCST) to measure the temperature of metatarsus by attaching a special shell designed to fit the anatomy of cattle's hind leg. Using AMSCST, the surface temperature (ST) on the metatarsus of the hind leg was successively measured during 24 hours a day with an interval of one hour in three tested seasons. Based on ST and rectal temperature (RT) detected by AMSCST and mercury thermometer, respectively, a linear mixed model was established, regarding both the time point and seasonal factors as the fixed effects. Unary linear correlation and Bland-Altman analysis results indicated that the temperatures measured by AMSCST were closely correlated to those measured by mercury thermometer (R^2 = 0.998), suggesting that the AMSCST is an accurate and reliable way to detect cattle's body temperature. Statistical analysis showed that the differences of STs among the three seasons, or among the different time points were significant ($P<0.05$), and the differences of RTs among the different time points were similarly significant ($P<0.05$). The prediction accuracy of the mixed model was verified by 10-fold cross validation. The average difference between measured RT and predicted RT was about 0.10 ± 0.10°C with the association coefficient of 0.644, indicating the feasibility of this model in measuring cattle body temperature. Therefore, an automated technology for accurately measuring cattle body temperature was accomplished by inventing an optimal device and establishing the AMSCST system.

1. INTRODUCTION

As an important marker of both the health and physiological status of livestock, body temperature has been used to evaluate the regularity of oestrus and ovulation [1,2], pregnancy [3], parturition [4] and disease occurrence [5,6] in cows. However, manual measurement of body temperature is costly and time-consuming. It increases the potential risk of disease transmission and is impracticable to measure temperature over the course of a full diurnal cycle as well [7–11]. It is urgent, therefore, to establish a simple, precise, efficient and automated method for cattle body temperature detection.

The rapid development of wireless telemetry technology laid a foundation for the automatic determination of body temperature. Early in 1974, Bligh proposed the use of wireless telemetry technology to measure animal body temperature [12]. This led to an expansion in the research of automated monitoring technology to detect dairy cow's body temperature using infrared and implantable measurement devices. It is very difficult to accurately measure bovine body temperature in a non-contacted way due to its thick body hair. Vaginal implantation of wireless sensors made good progress in remote body temperature measurement of cows achieving an accuracy of 96.61% [13]. However, it has not yet been applied in the management of reproduction, because of lacking the supporting automated technology. Morais [14] and Miranda [15] reported the automatic detection of body temperature via implanting the temperature sensor into the cows' vulvar muscles. Unfortunately, this approach cannot be used in dairy cows due to the vaginal damage. Automatic measurement of reticulo-rumen boluses temperature was attempted [5,16] but failed to return accurate body temperature due to the influence of water drinking and rumen fermentation. The measurement of dairy cow body temperature via automated detection technology based on infrared thermography (IRT) suggested that the eye, vulvar and the highest udder surface temperature (ST) are good proxies for rectal temperature (RT) [17,18]. Concerning the fact that IRT can be influenced by wind speed and direct sunlight exposure [19], it may be difficult to get accurate body temperature through analyzing IRT image [18,20]. The objective of this study was to establish a technology to automatically measure the body temperature of dairy cattle. We invented a specific device according to the anatomy of cattle's hind leg, using which the ST on the metatarsus was successfully detected. The establishment of our Automatic Measurement System for Cattle's Surface Temperature (AMSCST) based on thermistor sensor (Fig 1) provides a highly reliable and automatic cattle body temperature detection system.

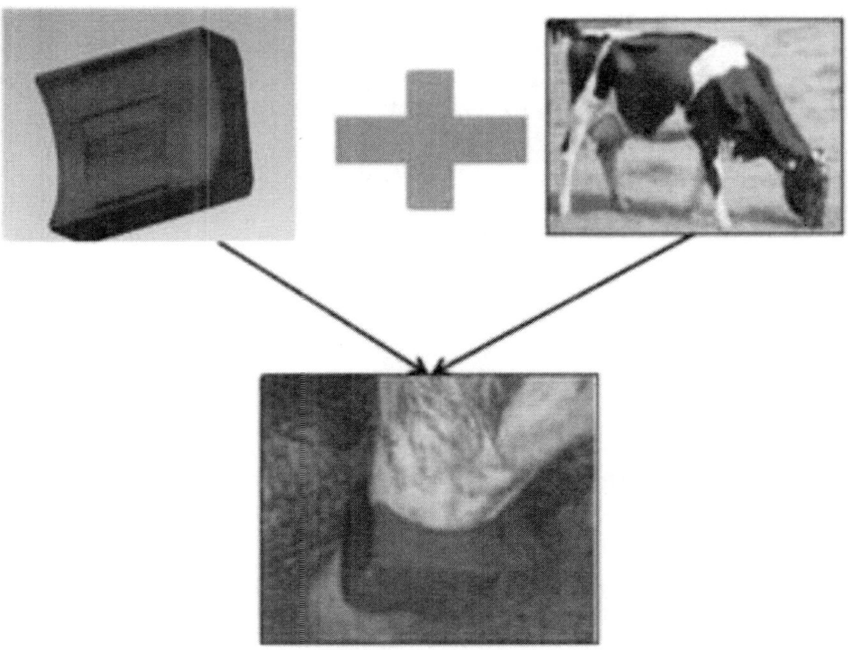

Figure 1. Installation diagram of the data detection device of AMSCST.

2. MATERIALS AND METHODS

In the present experiment, animal care and samples collection procedures were approved and conducted under established standard of the Institute of Animal Science, Chinese Academy of Agricultural Sciences, Beijing, China.

2.1 Equipment

In this experiment, we used a commercially available mercury thermometer, veterinary mercury thermometers, as well as the AMSCST including seven sets of data detection devices, one for each of the seven animal, (animal physiological state detector in vitro: length 80 mm, height 74 mm, thickness 24 mm; patent number: CN201520869032.5) as depicted in Fig 1, a data collector, and one computer system.

The ST was measured using our AMSCST from the metatarsus of the hind leg with sparse hair. The temperature sensor is mounted inside the metatarsus where the muscle and blood vessels are rich. A shell shape was designed to fix the temperature sensor appropriate to the cattle anatomy. The temperature sensor was placed in close contact with the muscle via the shell and mounted on a concave nest positioned centrally on the side of the shell. This design minimizes the influence of environment factors such as wind, leading to more accurate temperature measurement.

2.2 Testing and correcting the temperature measuring equipment

AMSCST was tested by placing 7 sets data detection devices of the AMSCST into the calorstat prior to the experiment to calibrate them. Firstly the temperature of the calorstat was increased from 30°C to 40°C using a steady increase rate of 1°C per 30 minutes, and then we automatically measured its temperature every 30 minutes using AMSCST. The temperature was also concurrently measured using a conventional mercury thermometer fixed to the inside of the glass door of the calorstat. This testing experiment was repeated three times. After the validation of its accuracy and reliability, AMSCST was applied to the cattle temperature detection.

2.3 Animal feeding and management

Seven healthy Simmental cows, which were 60 to 90 days postpartum, multiparous and non-pregnant, were used as experiment animals. All cattle were housed in a free stall barn, fed five times daily (at 2:00, 6:00, 12:00, 18:00, 22:00) on an automatic total mixed rations (TMR) diet with drinking water freely available. Each cow was attached with one set of data detection device.

2.4 Collection of the ST and RT

Both ST and RT were measured for periods during the winter (February 2nd to 8th), summer (June 2nd to 8th) and autumn (October 2nd to 8th) in 2015 in a cattle farm in Shijiazhuang city, Hebei province of China. Since weather in spring was considered similar to that in autumn, no measurement of ST and RT in spring in this study.

Carrying and adapting the data detection device. The data detection devices were attached to the metatarsus of testing cattle's hind legs 3 days prior to the formal experiment to let the cattle get used to the device. ST was measured every 1 hour starting at 20:00 by data detection devices, and data were collected every four hours by the data collector from data detection devices.

Formal experiment. ST of cattle was continuously measured for 3 days using the AMSCST. Meanwhile RT was measured every 2 hours starting at 20:00 by a veterinary mercury thermometer. The testing cattle were held still for no less than 10 minutes prior to measuring the RT. The thermometer was removed 5 minutes after inserting into the rectum and the temperature was recorded. Each cow was continuously measured for 3 days and 12 RT data were collected per day.

2.5 Statistical analysis

The t-test was used for the data obtained by the laboratory test section to find the significant difference. Relationship between the temperature data recorded by AMSCST and by mercury thermometer was evaluated using both Pearson correlation and linear regression. A Bland-Altman test [21] was used to test the consistency between the two measurements.

For each individual tested cow, 24 STs were obtained by AMSCST daily at an interval of 1 hour, while 12 RTs were collected per day at an interval of 2 hours by manual measurement using mercury thermometer. To match ST data set, the missing 12 RTs were arbitrarily generated by averaging two adjacent RTs since the difference of RTs between two adjacent time points was negligible. Totally, there were 24 STs and RTs data points for each cattle per day, and a total of 1512 paired STs and RTs data in all 3 tested seasons were obtained. We used the linear mixed model to estimate parameters for the fixed effects in regression. We also trained a linear mixed model to predict RT using ST adjusting for time and season. Only season or the season and time point are included as fixed effect in model 1 and model 2, respectively. In each model, we considered the variation between animals as well as covariation within animals for repeated measurements.

The models were established as follows:

$$\text{model 1}: Y = a + b_0X_0 + b_sX_S + b_iX_i + e; \tag{1}$$

$$\text{model 2}: Y = a + b_0X_0 + b_sX_S + b_tX_t + b_iX_i + e. \tag{2}$$

In the two models, Y represents RT, a represents intercept, X_0 represents ST, b_0 is the parameter vector of X_0, X_s represents seasonal design matrix, b_s is the nonrandom parameter vector of X_s, that is fixed effects; X_t represents time design matrix, b_t is the nonrandom parameter vector of X_t, that is fixed effects. X_i represents the individual design matrix, b_i is the random parameter vector of X_i, that are random effects; e is the vector of residual random errors. The outcome parameters are listed in the schedules. Determination coefficient (R^2) is a measure of the reduction in variance when predictor variables have been added into the model. In SAS proc mixed, determination coefficient cannot be obtained directly, so we estimated residual from the full model(A) and model with intercept only (B), then used (B-A)/B to obtain the determination coefficient. RT was predicted based on ST by using the optional mixed liner model. The prediction accuracy was evaluated by comparing Pearson correlation coefficient, in a 10-fold cross validation approach [22]. For the training set, randomly picked 90% of the total 1512 data points were used to build the regression model in each run, and then the estimated parameters were used to predict RTs in the left 10% data points. The process was repeated 10 times. All analyses were carried out using SAS program.

3. RESULTS

3.1 AMSCST test

The pre-cattle laboratory test analysis demonstrated a close relationship between AMSCST and mercury thermometer measurement, with temperatures sourced from both devices synchronously increasing from 30°C to 40°C. The t-test analysis showed that there was no significant difference among AMSCST devices. The strong relationship between the temperatures detected by AMSCST and mercury thermometer (R^2 = 0.998) was shown in Fig 2A, and the agreement between the temperature data measured by the seven sets of devices of the calorstat was shown in Fig 2B. The average difference between the temperature detected by different devices is 0.00 ± 0.07°C (95% CI: -0.22 to 0.22°C, Fig 2B). These results suggested the feasibility and reliability of AMTSCST which provides rather accurate and stable data in comparison with the conventional manual method.

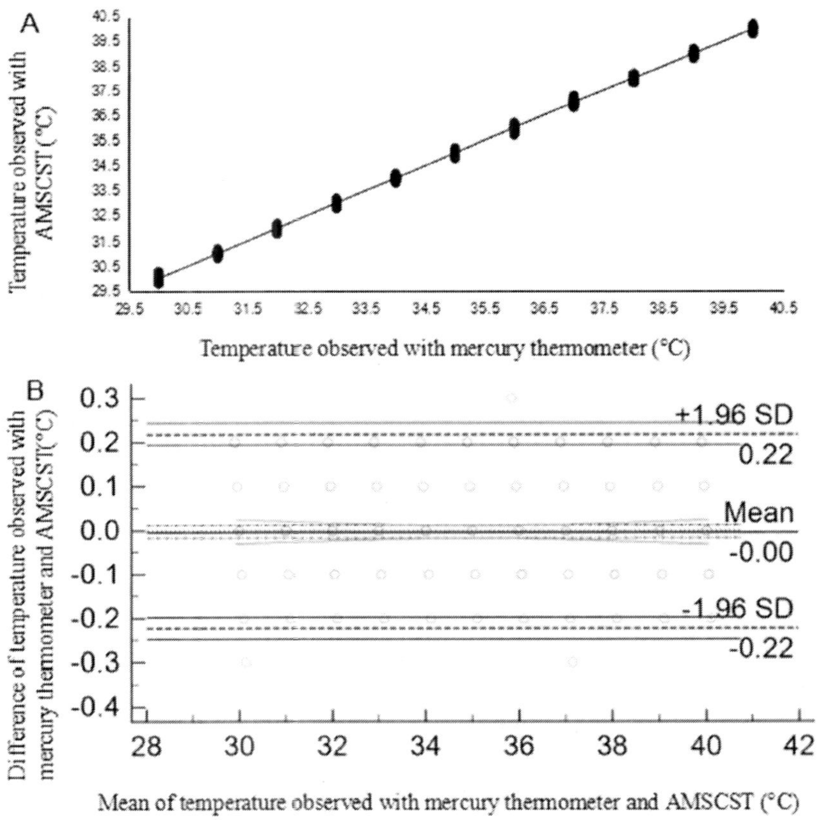

Figure 2. Consistency of temperature measured by AMSCST and mercury thermometer. A The linear curve of the measured temperature of AMSCST and mercury thermometer; B. Bland and Altman plot showing agreement between measurements obtained by the mercury thermometer (as the reference) and AMSCST.

Table 1. The mean ST and RT in different days and different seasons (°C).

Measurement day	ST			RT		
	Winter	Summer	Autumn	Winter	Summer	Autumn
First day	34.31 ± 1.97^{A}	35.29 ± 1.83^{A}	34.28 ± 1.61^{A}	38.57 ± 0.23^{A}	38.44 ± 0.17^{A}	38.42 ± 0.20^{A}
Second day	34.16 ± 1.80^{A}	35.86 ± 0.99^{A}	34.43 ± 1.52^{A}	38.57 ± 0.24^{A}	38.52 ± 0.16^{A}	38.43 ± 0.20^{A}
Third day	34.22 ± 1.92^{A}	35.35 ± 1.20^{A}	34.29 ± 1.53^{A}	38.60 ± 0.28^{A}	38.47 ± 0.13^{A}	38.39 ± 0.19^{A}
Mean±SD	34.23 ± 1.90^{a}	35.50 ± 1.43^{c}	34.33 ± 1.55^{b}	38.58 ± 0.25^{d}	38.48 ± 0.16^{d}	38.41 ± 0.20^{d}

Note: A indicates there is no significant difference in the same column ($P > 0.05$);
a, b, c and d represent the differences in the same row, the same letter indicates there is no significant difference in the same row ($P > 0.05$), while the different letters indicate the group differs significantly from others in the same row ($P < 0.05$).

3.2 Analysis results of each factor

The ST and RT differences among days and seasons. There was no significant difference in RT within either days or seasons (*P*>0.05) (Table 1). This confirmed that the cow's RT is nearly constant under the normal physiological conditions. Similarly, the ST difference among days was not significant (*P*>0.05) suggesting that AMSCST performance was fairly stable. However, ST varies by season with significant difference (P<0.05). The highest ST appeared in summer, while the lowest appeared in winter.

ST and RT differences across different daily time points. Table 2 shows that ST was significantly different from the measurement time point and largely consistent with the expected regular diurnal variation. The lowest ST appeared from 5:00 to 7:00, then gradually increased to peak value from 13:00 to 15:00. Subsequently, ST gradually declined again to another trough level. Similarly, RT was different significantly at different time points with similar pattern to that of ST although the diurnal variation of RT was marginal compared with ST.

3.3 Regression analysis of ST to RT

Considering season as the only fixed effect. Linear mixed model 1 analysis was performed based on the ST and RT data, in which the seasonal factor was treated as the fixed effect as Eq 1.

The linear mixed model 1 was used to analyze the ST and RT data, and the estimated parameter values in this model were shown in S1 Table. The determination coefficient of the model is 0.399, which means that the model does not sufficiently fit.

Considering both season and time point as fixed effects. Since the determination coefficient of the previous model is 0.399, we then added time factor as fixed effect and obtained model 2 as Eq 2.

The estimated parameter values in model 2 are shown in S2 Table. The determination coefficient of model 2 was 0.562, thus providing a superior fit of the observed data.

Predicted RT. By employing the estimated parameters (in S2 Table) of season and time point in model 2, we obtained the predicted RT. The average difference between predicted RT and observed RT was 0.10°C. 90% (1360/1512) of the absolute differences were no larger than 0.2°C.

Table 2. Diurnal mean ST and RT (°C).

Time	ST			RT		
	Winter	Summer	Autumn	Winter	Summer	Autumn
0:00	33.81 ± 1.43^{CD}	34.90 ± 1.47^{C}	33.79 ± 1.20^{BC}	38.53 ± 0.17^{C}	38.46 ± 0.14^{CD}	38.36 ± 0.11^{C}
4:00	33.73 ± 2.04^{E}	35.26 ± 0.99^{C}	33.32 ± 0.84^{CD}	38.34 ± 0.22^{D}	38.42 ± 0.10^{D}	38.31 ± 0.08^{C}
8:00	33.35 ± 1.85^{DE}	34.62 ± 0.85^{C}	32.90 ± 0.96^{D}	38.36 ± 0.19^{D}	38.26 ± 0.09^{E}	38.11 ± 0.12^{D}
12:00	35.98 ± 1.03^{A}	37.15 ± 0.82^{A}	35.86 ± 0.78^{A}	38.83 ± 0.18^{A}	38.56 ± 0.11^{B}	38.47 ± 0.10^{B}
16:00	35.07 ± 1.51^{AB}	36.31 ± 1.46^{B}	35.79 ± 0.82^{A}	38.71 ± 0.19^{B}	38.64 ± 0.09^{A}	38.75 ± 0.14^{A}
20:00	34.40 ± 1.32^{BC}	35.17 ± 1.44^{C}	34.10 ± 0.90^{B}	38.53 ± 0.19^{C}	38.51 ± 0.09^{BC}	38.49 ± 0.09^{B}

Note: A, B, C, D and E represent the differences in the same column. The same letter indicates there is no significant difference in the same column ($P > 0.05$), while different letters indicate the group differs significantly from others in the same column ($P < 0.05$).

To evaluate the prediction accuracy of the above models, Pearson correlation coefficient between the predicted RT and measured RT was also calculated for the validation set upon 10-fold cross validation. The correlation coefficient of the model 2 is 0.644, which is higher than 0.569 from model 1.

4. DISCUSSION

Automated collection of body temperature data was pioneered using IRT to scan parts of human body. This process can assist in diagnosis a variety of diseases including breast cancer, local inflammation and vascular embolism [23–25]. However, the surface of the livestock such as cattle and horses is covered with thick hair, and hairless body parts such as eyes, udder and muzzle are not suitable for automated temperature detection because of their limited space for devices. Automated body temperature detection technology developed slowly, with few publications reporting the relationship between mastitis and udder ST measured by IRT [26,27]. Previoulsy, Metzner et al. measured udder ST using IRT, and reported its correlation with RT [18]. Because the udder ST can be markedly influenced by conditions such as mastitis, it is thus difficult to find an anatomical location for reliably monitoring the body temperature of dairy cows [27]. In our study, cattle's ST was detected by measuring of the inside of the metatarsus of the cattle's hind leg using our AMSCST. The design of shell shape of the AMSCST device greatly minimized the influence of environmental factors on body temperature, leading to more accurate and reliable measurement. Metzner [18] analyzed the relationship between RT and the udder ST's maximum value and arithmetic mean of dairy cows using IRT. The results showed a better linear relationship between RT and the maximum value of the udder ST than RT and the arithmetic mean of the udder ST. The equation was RT = 5.68 + 0.874 × Tmax (Tmax represents the maximum value of the udder ST), and the determination coefficient was 0.646. However, the determination coefficient using RT and the arithmetic mean of the udder ST was smaller at 0.432. Coupled with the high cost and special equipment required by IRT, it is difficult to apply this approach in practice. Suthar et al. investigated the relationship between vaginal temperature (VT) and RT, and reported a determination coefficient between RT and VT of 0.846 for cows of 135 ± 56 days in milk and 0.884 for cows of 3 ± 1 days in milk respectively, indicating a strong relationship between RT and VT [28]. In fact the measuring duration in his study was only 80 ± 10 minutes with an interval of 1 minute, which was not long enough to reflect the variation of the body temperature. Another disadvantage is that the vagina is often affected by inflammation, and thus it is difficult to monitor the normal temperature of dairy cows. Furthermore,

Suthar's study only collected and analyzed the temperature for a limited period of time in the daylight hours, and didn't consider the diurnal effect. The fitting curve may be difficult to reflect the regularity of cattle's body temperature.

In this study, we compared two liner mixed models where the effect of time point was considered or not. An optimal model which included both season and time point, gave out a determination coefficient of 0.562. This value is 30% higher than the determination coefficient of 0.432 between RT and the average temperature of udder ST reported by Metzner. Furthermore, our detection system realized automation. Although the determination coefficient obtained by Suthar is higher, our method was easier to apply in practice since no stress response to animals when measuring ST compared with measuring VT. The predicted RT using this optimized model provides a good approximation to the actual observed RT, with the average difference of 0.10 ± 0.10°C, and 90% absolute difference is no bigger than 0.2°C. This indicates that the ST measured using our AMTSCST can be used as a reasonable estimation of cattle body temperature. Our invention may be extended to study other physiological phenomena displaying regularity such as estrus, pregnancy and parturition, and it can provide aid in disease diagnosis as well.

5. CONCLUSIONS

This study described an automated technology for measuring cattle body temperature on the metatarsus of cattle's hind leg. This method may play an important role in the future study for heat detection, pregnancy and disease diagnosis.

ACKNOWLEDGMENTS

This work was supported by grants from the Beijing Innovation Team of Technology System in Dairy Industry (BAIC06-2016), National Natural Science Foundation of China (No. 31372296) and the Agricultural Science and Technology Innovation Program (ASTIP) (cxgc-ias-06).

REFERENCES

1. Lewis GS, Newman SK. 1984. Changes throughout oestrous cycles of variables that might indicate oestrus in dairy cows. J. Dairy Sci 67:146±152. https://doi.org/10.3168/jds.S0022-0302(84)81278-3 PMID: 6538581
2. Talukder S, Kerrisk KL, Ingenhoff L, Thomson PC, Garcia SC, Celi P. 2014. Infrared technology for estrus detection and as a predictor of time of ovulation in dairy cows in a pasture-based system. Theriogenology 81(7):925±935. https://doi.org/10.1016/ j.theriogenology.2014.01.009 PMID: 24560547
3. Gil Z, Kural J, Szarek J, Wierzchoś E. 2001. Increase in milk and body temperature of cows as a sign of embryo entry into the uterus. Theriogenology 56(4):685±697. PMID: 11572449
4. Wrenn TR, Bitman J, Sykes JF. 1958. Body Temperature Variations in Dairy Cattle during the Estrous Cycle and Pregnancy. J. Dairy Sci 41:1071±1076.
5. Timsit E, AssieÂ S, Quiniou R, Seegers H, Bareille N. 2011. Early detection of bovine respiratory disease in young bulls using reticulo-rumen temperature boluses. Veterinary journal 190(1):136±142.
6. Schaefer AL, Cook NJ, Bench C, Chabot JB, Colyn J, Liu T et al. 2012. The non-invasive and automated detection of bovine respiratory disease onset in receiver calves using infrared thermography. Research in veterinary science 93(2):928±935. https://doi.org/10.1016/ j.rvsc.2011.09.021 PMID: 22055252
7. Hahn GL, Eigenberg RA, Nienaber JA, Littledike ET. 1990. Measuring physiological responses of animals to environmental stressors using a microcomputer-based portable data logger. Journal of animal science 68(9):2658±2665. PMID: 2211395
8. Hicks LC, Hicks WS, Bucklin RA. 2001. Comparison of methods of measuring deep body temperatures of dairy cows. Livestock Environment VI, Proceedings of the 6th International Symposium:432±438.
9. Firk R, Stamer E, Junge W, Krieter J. 2002. Automation of oestrus detection in dairy cows: a review. Livestock Production Science 75:219±232.
10. Prendiville D, Lowe J, Earley B. 2002. Radiotelemetry systems for measuring body temperature. Beef Prod. No. Ser.:57.
11. Aalseth, E. 2005. Fresh cow management: What is important, what does it cost, and what does it return? 7th Western Dairy Management Conf., Reno, NV. Kansas State Universityand Cooperative Extension Service, Manhattan:1±12.
12. Bligh J, Heal JW. 1974. The use of radio-telemetry in the study of animal physiology. The Proceedings of the Nutrition Society 33(2):173±181. PMID: 4456381
13. Kyle BL, Kennedy AD, Small JA. 1998. Measurement of vaginal temperature by radiotelemetry for the prediction of estrus in beef cows. Theriogenology 49(8):1437± 1449. PMID: 10732008
14. Morais R, Valente A, Almeida JC, Silva AM, Soares S, Reis MJCS. 2006. Concept study of an implantable microsystem for electrical resistance and temperature measurements in dairy cows, suitable for estrus detection. Sensors and Actuators A 132:354±361.
15. Miranda N, Morais R, Dias M, Viegas C, Silva F, SeroÃdio C et al. 2009. Bioimplantable impedance and temperature monitor low power micro-system suitable for estrus detection. Procedia Chemistry 1:505±508.
16. Liang D, Wood CL, McQuerry KJ, Ray DL, Clark JD, Bewley JM. 2013. Influence of breed, milk production, season, and ambient temperature on dairy cow reticulorumen temperature. J. Dairy Sci 96(8):5072±5081. https://doi.org/10.3168/jds.2012-6537 PMID: 23769360
17. Hoffman G, Schmidt M, Ammon C, Rose-Meierhofer S, Burfeind O, HeuwieserWet al. 2013. Monitoring the body temperature of cows and calves using video recordings from an infrared thermography camera. Veterinary research communications 37(2):91±99. https://doi.org/10.1007/s11259-012-9549-3 PMID: 23264091

18. Metzner M, Sauter-Louis C, Seemueller A, Petzl W, Klee W. 2014. Infrared thermography of the udder surface of dairy cattle: characteristics, methods, and correlation with rectal temperature. Veterinary journal 199(1):57±62.
19. Church JS, Hegadoren PR, Paetkau MJ, Miller CC, Regev-Shoshani G, Schaefer AL et al. 2014. Influence of environmental factors on infrared eye temperature measurements in cattle. Research in veterinary science 96(1):220±226. https://doi.org/10.1016/j.rvsc.2013.11.006 PMID: 24290729
20. Jiang LJ, Ng EY, Yeo AC, Wu S, Pan F, Yau WY et al. 2005. A perspective on medical infrared imaging. Journal of Medical Engineering and Technology 29:257±267. https://doi.org/10.1080/03091900512331333158 PMID: 16287675
21. Bland MJ, Altman DG. 1986. Statistical methods for assessing agreement between two methods of clinical measurement. Lancet 327 307±310.
22. Kohavi R. 1995. A study of cross-validation and bootstrap for accuracy estimation and model selection. IJCAI: 1137±1145.
23. Gautherie M, Gros CM. 1980. Breast thermography and cancer risk prediction. Cancer 45(1):51±56. PMID: 7351006
24. Gallo LM. 1985. Quantitative infrared thermography to identify varicoceles as the cause of maleinfertility. Biomed Tech 30(11):284.
25. Hu L, Gupta A, Gore JP, Xu LX. 2004. Effect of forced convection on the skin thermal expression of breast cancer. Journal of biomechanical engineering 126(2):204±211. PMID: 15179850
26. Paulrud CO, Clausen S, Andersen PE, Bjerring M, Rasmussen MD. 2002. Infrared thermography to evaluate milking induced alterations in teat tissue fluid circulation. J. Dairy Sci 85:84.
27. Pezeshki A, Stordeur P, Wallemacq H, Schynts F, Stevens M, Boutet P et al. 2011. Variation of inflammatory dynamics and mediators in primiparous cows after intramammary challenge with Escherichia coli. Veterinary research 42:15. https://doi.org/10.1186/1297-9716-42-15 PMID: 21314974
28. Suthar V, Burfeind O, Maeder B, Heuwieser W. 2013. Agreement between rectal and vaginal temperature measured with temperature loggers in dairy cows. The Journal of dairy research 80(2):240±245. https://doi.org/10.1017/S0022029913000071 PMID: 23506659

CHAPTER 14

Effects of dietary methionine on breast muscle growth, myogenic gene expression and IGF-I signaling in fast- and slow-growing broilers

Chao Wen, Xueying Jiang, Liren Ding, Tian Wang & Yanmin Zhou

College of Animal Science and Technology, Nanjing Agricultural University, Nanjing, 210095, China. Correspondence

ABSTRACT

This study investigated the responses of fast- (FG) and slow- (SG) growing broilers to dietary methionine (Met) status. The broilers were subjected to low (LM, 0.38 and 0.28 g/100 g), adequate (AM, 0.51 and 0.42 g/100 g) and high (HM, 0.65 and 0.52 g/100 g) Met during 1–21 and 22–42 d, respectively. Compared with the LM diets, the AM and HM diets increased body weight gain only in the FG broilers. The HM diets increased breast muscle yield only in the FG broilers, although insulin-like growth factor-I (IGF-I) concentration was increased in both strains of broilers. The HM diets increased mRNA levels of myogenic regulatory factors (MRF4, Myf5) and myocyte enhancer factor 2 (MEF2A and MEF2B) in the FG broilers, and increased MEF2A and decreased myostatin mRNA level in the SG broilers. Extracellular signal-regulated kinase (ERK) phosphorylation of breast muscle was increased by the HM diets in both strains of broilers, but mechanistic target of rapamycin (mTOR) phosphorylation was increased by the AM and HM diets only in the FG broilers. These results reflect a strain difference in broiler growth and underlying mechanism in response to dietary Met.

1. INTRODUCTION

Methionine (Met) is an essential amino acid for protein synthesis and a precursor of S-adenosylmethionine, the methy[l] donor for DNA methylation1. As the first limiting amino acid in broiler diets, Met has been demonstrated to affect growth performance and breast meat yield in broilers[2]. Therefore, it is important to provide adequate Met to optimise broiler growth and production. Growth performance and breast muscle yield are often used to estimate Met requirement of broilers, which has been reported to depend on gender, age, nutrient density, rearing environment, etc[3, 4]. However, few studies have compared Met requirement between broiler strains differing in growth rate. The Met contents given in the NRC (1994)[5] are generally minimum levels that satisfy general productive activities of broilers without strain being considered. In general, fast-growing (FG) broilers have superior growth performance and breast muscle yield to slow-growing (SG) broilers[6, 7]. Thus, it can be implied that the responses of growth performance and breast muscle yield to dietary Met status may be different between FG and SG broilers, which needs further elucidation.

Skeletal muscle development is regulated by myogenic regulatory factors (including MyoD, Myf5, MyoG, and MRF4), myocyte enhancer factor 2 (including MEF2A, B, C, and D), myostatin (MSTN) and insulin-like growth factor-I (IGF-I) signaling pathways[8, 9]. The downstream targets of IGF-I include extracellular signal-regulated kinase (ERK), mechanistic target of rapamycin (mTOR) and forkhead box O (FoxO), which play an important role in cell proliferation and protein synthesis[10, 11]. Our laboratory recently reported that MSTN, mTOR and FoxO4 were involved in the enhancement of breast muscle growth by Met in broilers with lower but not high hatching weight[12], suggesting that the responses to dietary Met were dependent on hatching weight of chicks. However, whether these responses are affected by broiler strain remains unclear. The FG and SG broilers may have different Met requirements[13], so they may have different responses of breast muscle yield and associated gene and protein expression to dietary Met status. Therefore, the objective of this study was to compare the effects of dietary Met on growth performance, breast muscle yield and expression of myogenic genes and proteins associated with IGF-I signaling pathway in FG and SG chickens.

2. RESULTS

2.1 Growth performance

Higher body weight gain and feed intake as well as lower feed conversion ratio were observed ($P < 0.001$) in the FG broilers than in the SG broilers (Table 1). Compared with the LM diets, the AM and HM diets increased ($P < 0.05$) body weight gain and the AM diets increased ($P < 0.05$) FI in the FG broilers, but no difference was found in the SG broilers. There was no Met effect on feed conversion ratio within each strain.

Table 1: Effects of dietary methionine on growth performance of broilers from 1 to 42 d of age. Results are presented as means and pooled SEM (n = 6).

	Body weight gain (kg)	Feed intake (kg)	Feed conversion ratio
Strain			
FG	2.01	3.94	1.96
SG	1.14	2.68	2.36
Met			
LM	1.49^{b}	3.19^{b}	2.20
AM	1.62^{a}	3.40^{a}	2.14
HM	1.62^{a}	3.34^{a}	2.13
Strain × Met			
FG+LM	1.91^{b}	3.80^{b}	1.99^{b}
FG+AM	2.07^{a}	4.08^{a}	1.98^{b}
FG+HM	2.06^{a}	3.94^{ab}	1.91^{b}
SG+LM	1.07^{c}	2.71^{c}	2.41^{a}
SG+AM	1.17^{c}	2.58^{c}	2.31^{a}
SG+HM	1.17^{c}	2.74^{c}	2.34^{a}
SEM	0.02	0.03	0.02
P value			
Strain	<0.001	<0.001	<0.001
Met	0.004	0.010	0.173
Strain × Met	0.777	0.358	0.453

2.2 Breast muscle weight and insulin-like growth factor-I concentration

The FG broilers had higher ($P < 0.001$) absolute and relative breast muscle weight than the SG broilers, but the IGF-I concentration did not differ between the two strains of broilers (Table 2). Compared with the LM diets, the HM diets increased ($P < 0.05$) absolute and relative breast muscle weight only in the FG broilers, but the AM and HM diets increased ($P < 0.05$) IGF-I concentration of breast muscle in both strains of broilers.

Table 2: Effects of dietary methionine on breast muscle weight and insulin-like growth factor-I (IGF-I) concentration of broilers at 42 d of age.

	Breast muscle weight		
	Absolute (g)	Relative (g/kg BW)	IGF-I (ng/mg protein)
Strain			
FG	396	181.7	13.63
SG	132	97.5	14.10
Met			
LM	228[b]	130.3[b]	12.72[c]
AM	265[ab]	140.7[ab]	14.08[b]
HM	300[a]	147.8[a]	14.80[a]
Strain × Met			
FG+LM	341[b]	167.9[b]	12.33[d]
FG+AM	393[ab]	180.4[ab]	13.71[bc]
FG+HM	456[a]	196.9[a]	14.85[a]
SG+LM	116[c]	92.7[c]	13.10[cd]
SG+AM	137[c]	101.0[c]	14.45[ab]
SG+HM	144[c]	98.7[c]	14.75[ab]
SEM	9	2.6	0.14
P value			
Strain	<0.001	<0.001	0.103
Met	0.020	0.039	<0.001
Strain × Met	0.191	0.180	0.310

2.3 Gene expression

The main effects showed that the FG broilers had higher ($P < 0.05$) mRNA expression of MRF4, Myf5, MEF2B and MEF2D in breast muscle than the SG broilers (Table 3). Compared with the LM diets, the HM diets increased ($P < 0.05$) mRNA levels of MRF4, Myf5, MEF2A and MEF2B in the FG broilers, and increased MEF2A and decreased MSTN mRNA level in the SG broilers. There were no differences in mRNA levels of MyoD, MyoG or MEF2C among treatments.

Table 3: Effects of dietary methionine on relative mRNA levels in breast muscle of broilers at 42 d of age.

	MSTN	MyoD	MyoG	MRF4	Myf5	MEF2A	MEF2B	MEF2C	MEF2D
Strain									
FG	1.03	0.98	0.92	1.01	0.96	0.93	1.03	0.92	1.03
SG	1.21	0.72	1.04	0.65	0.64	0.83	0.64	0.69	0.54
Met									
LM	1.56^{a}	0.71	0.80	0.59^{b}	0.53^{b}	0.64^{b}	0.65^{b}	0.62	0.56
AM	1.02^{b}	0.85	1.01	0.81^{ab}	0.78^{b}	0.94^{a}	0.75^{b}	0.89	0.76
HM	0.78^{b}	0.98	1.13	1.09^{a}	1.10^{a}	1.06^{a}	1.11^{a}	0.93	1.04
Strain × Met									
FG+LM	1.42^{ab}	0.74	0.70	0.60^{b}	0.50^{b}	0.69^{bc}	0.78^{bc}	0.73	0.70^{ab}
FG+AM	1.00^{ab}	1.00	1.00	1.00^{ab}	1.00^{ab}	1.00^{ab}	1.00^{ab}	1.00	1.00^{ab}
FG+HM	0.68^{b}	1.19	1.06	1.43^{a}	1.39^{a}	1.08^{a}	1.31^{a}	1.04	1.40^{a}
SG+LM	1.71^{a}	0.69	0.91	0.57^{b}	0.55^{b}	0.58^{c}	0.51^{c}	0.50	0.42^{b}
SG+AM	1.05^{ab}	0.70	1.03	0.63^{b}	0.55^{b}	0.87^{abc}	0.50^{c}	0.77	0.52^{b}
SG+HM	0.88^{b}	0.78	1.20	0.75^{b}	0.81^{b}	1.04^{a}	0.90^{bc}	0.81	0.69^{ab}
SEM	0.10	0.09	0.06	0.07	0.06	0.04	0.05	0.07	0.10
P value									
Strain	0.365	0.186	0.325	0.014	0.020	0.941	0.001	0.105	0.035
Met	0.016	0.547	0.111	0.015	0.008	0.001	0.003	0.148	0.203
Strain × Met	0.854	0.741	0.809	0.137	0.153	0.236	0.689	0.998	0.727

2.4 Protein expression

There was no strain difference in phosphorylation of ERK, mTOR or FoxO4 in breast muscle, although FoxO4 phosphorylation tended ($P = 0.066$) to be higher in the FG broilers (Fig. 1, full-length blots are shown in Supplementary Figure S1). Compared with the LM diets, the HM diets increased ($P < 0.05$) ERK phosphorylation of breast muscle in both strains of broilers, and the AM and HM diets increased ($P < 0.05$) mTOR phosphorylation of breast muscle in the FG broilers. The effect of Met on FoxO4 phosphorylation of breast muscle was not significant within each strain.

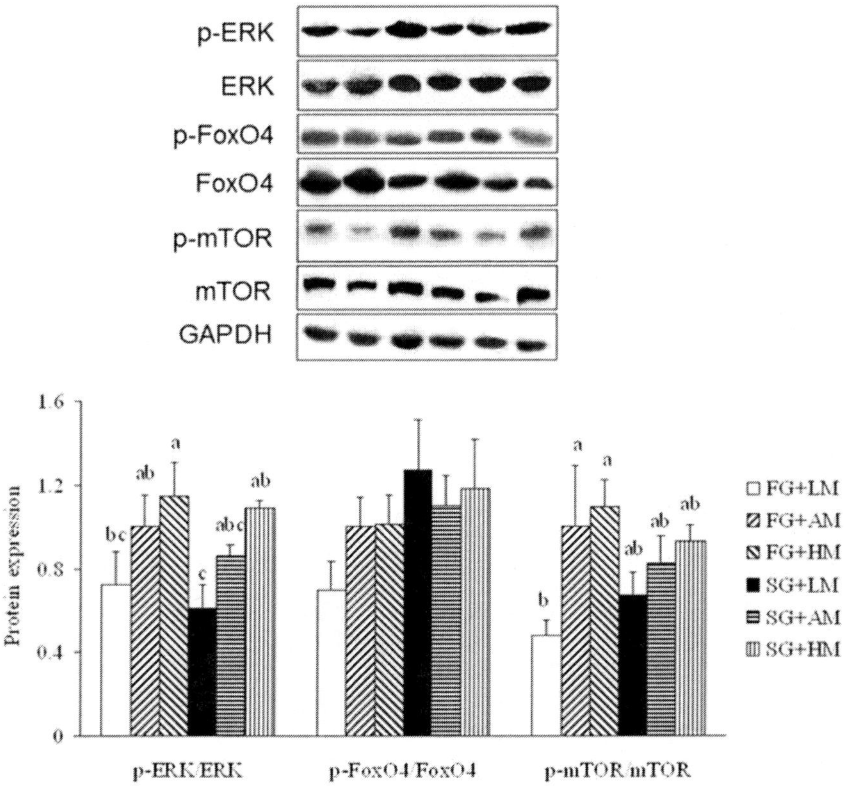

Figure 1 Effects of dietary methionine on protein expression in breast muscle of broilers at 42 d of age. Results are presented as means ± SEM (n = 6). Bars marked with different letters are significantly different at $P < 0.05$. Full-length blots are shown in Supplementary Figure S1.

3. DISCUSSION

As expected, the FG broilers had superior growth performance and breast muscle yield to the SG broilers. However, there was no significant difference in the IGF-I concentration of breast muscle between the strains, supporting the previous study that reported similar IGF-I mRNA expression in breast muscle between FG and SG chickens when they were fed the same diet, although significant differences were observed in live weight and breast muscle weight[14]. It was reported that body weight and circulating IGF-I levels were higher in FG chickens than in SG chickens from 1 to 6 weeks[15]. Thus, the present results implied that the strain differences in growth performance and breast muscle weight might be associated with circulating but not autocrine/paracrine IGF-I levels.

The FG broilers had higher mRNA expression of MRF4, Myf5, MEF2B and MEF2D in breast muscle than the SG broilers, indicating that these genes might be involved in the faster breast muscle development for the FG broilers. It has been demonstrated that MyoD and Myf5 are required for myogenic determination whereas MyoG and MRF4 are important for terminal differentiation[16,] and MEF2 family also play an important role in regulating muscle differentiation[17]. Thus, our findings suggest that the FG broilers may enhance myogenic determination and differentiation by upregulating the expression of MRF4, Myf5, MEF2B and MEF2D. Yin *et al.*[18] also found greater mRNA abundance of Myf5 and MRF4 in lines of chickens selected for high body weight than in those selected for low body weight at 28 d of age.

There was no strain difference in phosphorylation of ERK, FoxO4 or mTOR, although FoxO4 phosphorylation tended to be higher in the FG broilers, suggesting that the faster breast muscle development of the FG broilers might be regulated by some other signaling pathways, which deserves further investigation. Similarly, Wang *et al.*[7] found no difference in mTOR mRNA expression in the pectoralis major muscle between the FG and SG broilers at 63 or 105 d of age.

This study demonstrated that Met supplementation improved growth performance and breast muscle weight of the FG broilers as previously reported[19], which might be due to simultaneously increased IGF-I concentration. Dietary Met has been reported to affect IGF-I concentration in breast muscle of broilers[20]. However, the growth performance and breast muscle weight did not differ between the broilers fed the AM and HM diets, which did not agree with the data of Ahmed and Abbas[21], who reported that dietary Met higher than NRC (1994)5 increased body weight gain and breast muscle weight of broilers compared with the control. The discrepancy may be due to higher finisher Met level (0.44%) in the AM diets in our study than that (0.35%) in the control diet in Ahmed and Abbas[21]. Dietary treatments did not affect growth performance or breast muscle weight in the SG broilers, although increased IGF-I concentration was observed. This implies that the SG broilers may be less sensitive to Met supplementation probably due to lower Met requirement for growth and muscle protein accretion[22].

In this study, the HM diets resulted in strain-dependent changes in myogenic gene expression in breast muscle compared with the LM diets, with MRF4, Myf5, MEF2A and MEF2B upregulated in the FG broilers but MEF2A upregulated and MSTN downregulated in the SG broilers. This implied that dietary Met affected myogenic determination and differentiation in a strain-dependent way, particularly in the FG broilers.

We previously found that increasing dietary Met levels increased Myf5 and MEF2B mRNA expression and decreased MSTN mRNA expression in broilers[23]. It was reported that methylation status of MSTN gene was negatively correlated with the gene expression in skeletal muscle of broilers fed various levels of Met[24]. The effect of dietary Met on methylation of DNA has been reported[25]. Thus it can be inferred that the changes in myogenic gene expression in the FG and SG broilers may be attributed to altered DNA methylation.

The ERK1/2 signaling pathway is required for myoblast terminal differentiation[26], and mTOR and FoxO pathways play an important role in muscle protein synthesis and degradation, respectively[27]. In this study, the HM diets increased ERK phosphorylation of breast muscle in both strains of broilers compared with the LM diets, but increased mTOR phosphorylation only in the FG broilers, implying that Met may affect myoblast differentiation and protein synthesis in breast muscle in a strain-related manner. The changes of myogenic gene expression induced by the HM diets described above may be attributed to increased ERK phosphorylation as shown by Li *et al.*[28], who found that the expression of muscle regulatory factors and ERK activation were suppressed during the inhibition of myogenic differentiation by hypoxia, and increasing ERK activity by forced expression of mitogen-activated protein kinase kinase 1 could partly reverse the inhibition of myogenic differentiation by hypoxia. Increased mTOR phosphorylation may also play a role in regulating myogenic gene expression as reviewed recently[29]. We previously reported that Met supplementation increased mTOR phosphorylation in breast muscle of broilers with lower hatching weight12. No difference in FoxO4 phosphorylation within each strain suggested that muscle proteolysis was not affected by dietary Met.

In conclusion, enhanced myogenic gene expression may be involved in the superior growth performance and breast muscle yield of the FG broilers. The HM diets improved growth performance and breast muscle yield of the FG broilers probably by regulating myogenic gene expression, IGF-I synthesis and phosphorylation of ERK and mTOR, but the SG broilers were less sensitive to Met supplementation.

4. METHODS

All animal handling procedures were performed in strict accordance with guide for the Care and Use of Laboratory Animals central of the Nanjing Agricultural University, and the protocol was approved by the Institutional Animal Care and Use Committee of the Nanjing Agricultural University.

4.1 Bird husbandry, diets, and experimental design

The Arbor Acres broiler was selected as the FG strain, and Partridge Shank chicken, a typical indigenous meat-type strain in China, was selected as the SG strain. A total of 180 broilers from each strain were obtained from a local hatchery and raised from 1 to 42 d of age. The broilers were divided into a 2 × 3 factorial arrangement of treatments with six replicate cages of ten broilers per treatment Diets were formulated according to the NRC (1994)5 to contain low (LM. 0.38 and 0.28 g/100 g), adequate (AM, 0.51 and 0.42 g/100 g) and high (HM, 0.65 and 0.52 g/100 g) Met during 1–21 and 22–42 d, respectively. The LM diets contained no supplemental DL-Met, and the AM and HM diets were formulated by supplementing DL-Met (99%; Adisseo Inc.) in the LM diets at the expense of maize gluten meal (Table 4). Dietary crude protein content was determined according to the procedures of AOAC (2000)30, and amino acid composition was determined using an L-8900 Amino Acid Analyzer (Hitachi) as previously described31. Methionine and cystine were analyzed as Met sulfoxide and cysteic acid after performic acid oxidation. Chicks were allowed free access to mash feed and water in 3-layer cages in a temperature-controlled room with a 23 L:1D lighting program. The temperature of the room was maintained at 32 to 34 °C for the first 3 d and then reduced by 2 to 3 °C per week to a final temperature of 20 °C. Feed intake was recorded by replicate cage at 21 and 42 d of age. At 42 d of age, birds were weighed after feed deprivation for 12 h to calculate body weight gain and feed conversion ratio (feed intake:weight gain).

4.2 Sample collection

At 42 d of age, one male broiler per replicate was selected and weighed after feed deprivation for 12 h. Birds were killed by cervical dislocation. After dissection, the whole breast muscle (pectoralis major and pectoralis minor without bones) was weighed. Then samples were immediately collected from pectoralis major muscles and stored in liquid nitrogen until analysis.

4.3 Measurement of insulin-like growth factor-I in breast muscle

After thawing at room temperature, the breast muscle samples were homogenised with ice-cold physiological saline solution, and then centrifuged at 5000 × g for 10 min at 4 °C to collect the supernatant. Total protein of the supernatant was determined by the Bradford method[32]. The concentration of IGF-I was measured by a commercial chicken-specific

ELISA kit (Nanjing Jiancheng Bioengineering Institute) and expressed as ng/mg protein.

Table 4: Ingredients and nutrient composition of experimental diets (as-fed basis).

	1–21 d			22–42 d		
Item	**LM**	**AM**	**HM**	**LM**	**AM**	**HM**
Ingredient (g/100 g)						
Maize	57.09	57.09	57.09	61.08	61.08	61.08
Soybean meal	31.37	31.37	31.37	28.7	28.7	28.7
Maize gluten meal	3.71	3.56	3.41	1.70	1.57	1.44
Soybean oil	3.00	3.00	3.00	4.01	4.01	4.01
Dicalcium phosphate	2.00	2.00	2.00	1.60	1.60	1.60
Limestone	1.20	1.20	1.20	1.30	1.30	1.30
L-Lysine, HCl	0.33	0.33	0.33	0.31	0.31	0.31
DL-Methionine	0	0.15	0.30	0	0.13	0.26
NaCl	0.30	0.30	0.30	0.30	0.30	0.30
Vitamin and mineral mix[1]	1.00	1.00	1.00	1.00	1.00	1.00
Analysed composition (g/100 g)[2]						
Crude protein	21.16	21.32	21.27	19.04	19.07	19.12
Lysine	1.14	1.18	1.14	1.03	1.06	1.08
Methionine	0.38	0.51	0.65	0.28	0.42	0.52
Total sulphur amino acids	0.73	0.89	1.00	0.61	0.75	0.88

4.4 mRNA quantification

The mRNA expression was determined as previously described[33]. Briefly, total RNA of samples was isolated using RNAiso reagent (TaKaRa Biotechnology) and diluted in diethyl pyrocarbonate treated water to appropriate concentration. Then the diluted RNA was immediately reverse transcribed into cDNA with PrimeScript RT reagent Kit (TaKaRa), and the cDNA was quantified using SYBR Premix Ex Taq II (TaKaRa) on ABI 7300 Real-Time PCR System (Applied Biosystems). Optimised cycling conditions were 95 °C for 30 s followed by 40 cycles of 95 °C for 5 s, 60 °C for 31 s, and final dissociation stage of 95 °C for 15 s, 60 °C for 1 min, 95 °C for 15 s and 60 °C for 15 s. The geometric means of glyceraldehyde 3-phosphate dehydrogenase (GAPDH) and β-actin were used to normalise the genes of interest as recommended[34]. The primers for GAPDH, β-actin, MSTN, MyoD, MyoG, and MRF4, Myf5, MEF2A, MEF2B, MEF2C and MEF2D were synthesised as previously reported[23]. Relative mRNA levels (arbitrary units) were calculated on the basis of PCR efficiency and threshold cycle

(Ct) values[35]. The mRNA level of each target gene in the FG broilers fed the AM diets was assigned a value of one.

4.5 Western blot

The samples were homogenised in ice-cold RIPA lysis buffer (Beyotime Institute of Biotechnology) and then centrifuged at 12000 × g for 10 min at 4 °C to collect the supernatant. Protein concentration in the supernatant was determined using a bicinchoninic acid protein assay kit (Bioworld Technology, Inc.). The proteins were separated by SDS-PAGE and transferred to polyvinylidene fluoride membrane. Then the membrane was blocked and incubated with appropriate antibodies: mTOR, FoxO4, p-mTOR (Ser2448), p-FoxO4 (Ser193) and p-ERK (Thr202/Tyr204) were purchased from Cell Signaling Technology, Inc., and ERK from Bioworld Technology, Inc. The GAPDH antibody (Bioworld Technology, Inc.) was used as a loading control. After washing, membranes were incubated with a secondary antibody (Rockland Immunochemicals). The bands were visualised by infrared fluorescence using the Odyssey Imaging System (LI-COR) and quantified by Odyssey infrared imaging system software.

4.6 Statistical analysis

Two-way ANOVA was performed to determine the main effects of strain and Met and their interaction using the general linear model procedure of SPSS software (version 16.0; SPSS Inc.). Differences among treatments were examined by Duncan's multiple range test, which were considered significant at $P < 0.05$. Data are presented as means and standard error of means (SEM).

ACKNOWLEDGEMENTS

This study was supported by Natural Science Foundation of Jiangsu Province (BK20140710) and the Fundamental Research Funds for the Central Universities (KYZ201640).

REFERENCES

1. Niculescu, M. D. & Zeisel, S. H. Diet, methyl donors and DNA methylation: interactions between dietary folate, methionine and choline. *J. Nutr.* 132, 2333S–2335S (2002).
2. Hickling, D., Guenter, W. & Jackson, M. E. The effects of dietary methionine and lysine on broiler chicken performance and breast meat yield. *Can. J. Anim. Sci.* 70, 673–678 (1990).

3. Chamruspollert, M., Pesti, G. M. & Bakalli, R. I. Influence of temperature on the arginine and methionine requirements of young broiler chicks. *J. Appl. Poult. Res* 13, 628–638 (2004).
4. Lumpkins, B. S., Batal, A. B. & Baker, D. H. Variations in the digestible sulfur amino acid requirement of broiler chickens due to sex, growth criteria, rearing environment, and processing yield characteristics. *Poult. Sci.* 86, 325–330 (2007).
5. NRC. *Nutrient Requirements of Poultry*. 9 edn, (National Academy Press, 1994).
6. Fanatico, A. C. *et al.* Performance, livability, and carcass yield of slow- and fast-growing chicken genotypes fed low-nutrient or standard diets and raised indoors or with outdoor access. *Poult. Sci.* 87, 1012–1021 (2008).
7. Wang, X. Q. *et al.* Effects of breed and dietary nutrient density on the growth performance, blood metabolite, and genes expression of target of rapamycin (TOR) signalling pathway of female broiler chickens. *J. Anim. Physiol. Anim. Nutr* 97, 797–806 (2013).
8. Naya, F. J. & Olson, E. MEF2: a transcriptional target for signaling pathways controlling skeletal muscle growth and differentiation. *Curr. Opin. Cell Biol.* 11, 683–688 (1999).
9. Zanou, N. & Gailly, P. Skeletal muscle hypertrophy and regeneration: interplay between the myogenic regulatory factors (MRFs) and insulin-like growth factors (IGFs) pathways. *Cell. Mol. Life Sci.* 70, 4117–4130 (2013).
10. Adams, G. R. Invited Review: Autocrine/paracrine IGF-I and skeletal muscle adaptation. *J. Appl. Physiol.* 93, 1159–1167 (2002).
11. Sandri, M. *et al.* Foxo transcription factors induce the atrophy-related ubiquitin ligase atrogin-1 and cause skeletal muscle atrophy. *Cell* 117, 399–412 (2004).
12. Wen, C., Chen, Y., Wu, P., Wang, T. & Zhou, Y. MSTN, mTOR and FoxO4 are involved in the enhancement of breast muscle growth by methionine in broilers with lower hatching weight. *PLoS One* 9, e114236 (2014).
13. Dozier, W. A., Kidd, M. T. & Corzo, A. Dietary amino acid responses of broiler chickens. *J. Appl. Poult. Res* 17, 157–167 (2008).
14. Lu, F. Z., Chen, J., Wang, X. X. & Liu, H. L. Investigation of the insulin-like growth factor system in breast muscle during embryonic and postnatal development in Langshan and Arbor Acres chickens subjected to different feeding regimens. *Asian-Australasian Journal of Animal Sciences* 22, 471–482 (2009).
15. Beccavin, C., Chevalier, B., Cogburn, L., Simon, J. & Duclos, M. Insulin-like growth factors and body growth in chickens divergently selected for high or low growth rate. *J. Endocrinol.*168, 297–236 (2001).
16. Megeney, L. A. & Rudnicki, M. A. Determination versus differentiation and the MyoD family of transcription factors. *Biochem. Cell Biol.* 73, 723–732 (1995).
17. Olson, E. N., Perry, M. & Schulz, R. A. Regulation of muscle differentiation by the MEF2 family of MADS box transcription factors. *Dev. Biol.* 172, 2–14 (1995).
18. Yin, H. *et al.* Expression profiles of muscle genes in postnatal skeletal muscle in lines of chickens divergently selected for high and low body weight. *Poult. Sci.* 93, 147–154 (2014).
19. Zhai, W. *et al.* Protein expression in pectoral skeletal muscle of chickens as influenced by dietary methionine. *Poult. Sci.* 91, 2548–2555 (2012).
20. Wen, C., Wu, P., Chen, Y. P., Wang, T. & Zhou, Y. M. Methionine improves the performance and breast muscle growth of broilers with lower hatching weight by altering the expression of genes associated with the insulin-like growth factor-I signaling pathway. *Br. J. Nutr.* 111, 201–206 (2014).
21. Ahmed, M. E. & Abbas, T. E. Effects of dietary levels of methionine on broiler performance and carcass characteristics. *Int. J. Poult. Sci.* 10, 147–151 (2011).
22. Burley, H. K., Patterson, P. H. & Anderson, K. E. Alternative feeding strategies and genetics for providing adequate methionine in organic poultry diets with limited use of synthetic amino acids. *Worlds Poult. Sci. J* 72, 168–177 (2016).
23. Wen, C. *et al.* Methionine improves breast muscle growth and alters myogenic gene expression in broilers. *J. Anim. Sci.* 92, 1068–1073 (2014).

24. Liu, G., Zong, K., Zhang, L. & Cao, S. Dietary methionine affect meat quality and myostatin gene exon 1 region methylation in skeletal muscle tissues of broilers. *Agric. Sci. China* 9, 1338–1346 (2010).
25. Waterland, R. A. Assessing the effects of high methionine intake on DNA methylation. *J. Nutr.* 136, 1706S–1710S (2006).
26. Jones, N. C., Fedorov, Y. V., Rosenthal, R. S. & Olwin, B. B. ERK1/2 is required for myoblast proliferation but is dispensable for muscle gene expression and cell fusion. *J. Cell. Physiol.* 186, 104–115 (2001).
27. Métayer, S. *et al.* Mechanisms through which sulfur amino acids control protein metabolism and oxidative status. *J. Nutr. Biochem.*19, 207–215 (2008).
28. Li, X. *et al.* Extracellular signal-regulated kinase 1/2 mitogen-activated protein kinase pathway is involved in inhibition of myogenic differentiation of myoblasts by hypoxia. *Exp. Physiol* 97, 257–264 (2012).
29. Ge, Y. & Chen, J. Mammalian target of rapamycin (mTOR) signaling network in skeletal myogenesis. *J. Biol. Chem.* 287, 43928–43935 (2012).
30. AOAC. *Official Methods of Analysis*. 17 edn, (Association of Official Analytical Chemists, 2000).
31. Wang, Y. *et al.* Energy and ileal digestible amino Acid concentrations for growing pigs and performance of weanling pigs fed fermented or conventional soybean meal. *Asian Australas. J. Anim. Sci.* 27, 706–716 (2014).
32. Bradford, M. M. A rapid and sensitive method for the quantitation of microgram quantities of protein utilizing the principle of protein-dye binding. *Anal. Biochem.* 72, 248–254 (1976).
33. Wen, C., Wang, L. C., Zhou, Y. M., Jiang, Z. Y. & Wang, T. Effect of enzyme preparation on egg production, nutrient retention, digestive enzyme activities and pancreatic enzyme messenger RNA expression of late-phase laying hens. *Anim. Feed Sci. Technol.*172, 180–186 (2012).
34. Vandesompele, J. *et al.* Accurate normalization of real-time quantitative RT-PCR data by geometric averaging of multiple internal control genes. *Genome Biol.* 3, 0034.0031–0034.0011 (2002).
35. Pfaffl, M. W. A new mathematical model for relative quantification in real-time RT–PCR. *Nucleic Acids Res* 29, e45 (2001).

CHAPTER 15

An endoparasitoid wasp influences host DNA methylation

Sunil Kumar & Yonggyun Kim

Department of Bioresource Sciences, Andong National University, Andong 36729, Korea

ABSTRACT

Parasitism by endoparasitoid wasps changes the expression of various host genes, and alters host immune and developmental processes. However, it is not clearly understood how parasitism changes host gene expression in a whole genome scale. This study focused on an epigenetic control of Cotesia plutellae, an endoparasitoid wasp, against its host, Plutella xylostella. Two DNA methyltransferases (DNMT-1 and DNMT-2) are encoded in the genome of P. xylostella. In addition, methyl-binding domain proteins (MBDs) and DNA demethylation factor, ten-eleven translation protein (TET) are encoded. DNA methylation of P. xylostella genomic DNA was confirmed by restriction digestion with Gla I specific to 5-methylcytosine. DNA methylation intensity in parasitized (P) larvae was decreased compared to that in nonparasitized (NP) larvae, especially at late parasitic stage, at which expression levels of both DNMT-1 and DNMT-2 were also decreased. DNA demethylation of P. xylostella was confirmed in both NP and P larvae by restriction digestion with PvuRts1I recognizing 5-hydroxymethyl cytosine. Parasitism also suppressed expression levels of TET and MBDs. Treatment of 5-aza-2′-deoxycytidine (AZA) reduced DNA methylation intensity of NP larvae, causing suppression of hemocyte-spreading behavior and delay of immature development. RNA interference of DNMT-1 or DNMT-2 mimicked the adverse effects of AZA.

1. INTRODUCTION

Parasitism is a non-mutualistic symbiosis. Its success requires host regulation that is beneficial to parasites. Endoparasitoids including some braconid and ichneumonid wasps exhibit koinobiotic life, in which wasp larvae grow inside developing hosts[1]. These wasps can regulate hosts through significant immunosuppression to defend the attack of the host's immune system and induce the delay of host development to allow endoparasitoid larvae to mature first[2]. To achieve various ranges of host regulation, massive changes in host gene expression levels occur as seen in different parasitic patterns induced by braconid or ichneumonid wasps[3,4]. Indeed, some parasitic factors derived from polydnaviruses symbiotic to endoparasitoids are transcriptional regulators, such as viral ankyrins (vankyrins) and viral histone H4 (vH4). Vankyrins are truncated variants of host inhibitor κB that plays a crucial role in inhibiting the activity of nuclear factor κB, a transcriptional factor[5,6]. Some vankyrins have been demonstrated to be able to inhibit host gene expression[7]. A vH4 has been identified in Cotesia plutellae bracovirus (CpBV). It has an extended N-terminal tail (38 amino acids containing 9 lysine residues) compared to host histone H4[8]. vH4 joins host nucleosomes and alters host gene expression[9]. Thus, some parasitic factors can modulate host gene expression. However, it was not clearly understood how the massive changes in host gene expression occurred during parasitism.

Epigenetics deals with heritable changes in gene expression without change in DNA sequence. DNA methylation, histone modification, and microRNA expression are examples of epigenetic control of gene expression[10]. Environmental changes such as nutrients[11], exposure to pesticides[12], pathogens[13], and climate change[14] can influence epigenetic control. Especially, DNA methylation on a specific genome in response to environmental change is heritable by its *de novo* maintenance devices[15]. DNA methyltransferases (DNMTs) are classified into three types (DNMT-1, DNMT-2, and DNMT-3), in which DNMT-2 is no longer considered as a DNA methylation agent due to its specificity to tRNA in vertebrates[16,17]. De novo DNA methylation is performed by DNMT-3, while the methylation state is maintained by catalytic activity of DNMT-1. In insects, DNMT-1 and DNMT-2 are well conserved whereas DNMT-3 is uncommon except some hymenopteran and hemipteran insects[18]. As reported in Drosophila, insect DNA methylation levels are much lower compared to those of vertebrates[19]. Unlike vertebrate genomic DNAs that are globally methylated, insect genomes are locally methylated in gene body probably to regulate the transcription and mRNA splicing of target genes[20].

Endoparasitoid wasp C. plutellae parasitizes young larvae of the diamondback moth (DBM), Plutella xylostella[21]. Parasitized larvae exhibit significant alterations in immune response and development[22]. A proteomic analysis has indicated that more than 25% of host genes annotated into various categories of physiological functions are altered in gene expression after being parasitized3. In this study, we hypothesized that C. plutellae parasitism could alter host gene expression in an epigenetic mode by changing DNA methylation level in addition to manipulating the activities of host transcriptional factors. To test this hypothesis, DNA methylation was monitored in P. xylostella and the DNA methylation levels in parasitized (P) larvae were compared to those in nonparasitized (NP) larvae. Based on the presence of DNA methylation, DNA methylation/demethylation-associated genes were identified from P. xylostella genome and their expressions in both NP and P larvae were assessed. Finally, this study demonstrated the effect of down-regulation of DNA methylation on immune response and immature development of P. xylostella.

2. RESULTS

2.1 DNA methylation on P. xylostella genome

To test any presence of 5-methylcytosine (5-mC) on P. xylostella genome, its genomic DNA (gDNA) was digested with Gla I restriction enzyme to specifically cleave at 5-mC (Fig. 1A). Gla I cut gDNAs of both NP and P larvae of P. xylostella. A monoclonal antibody specific to 5-mC was used to measure the relative amounts of 5-mC on P. xylostella genomes (Fig. 1B). As expected, a vertebrate gDNA (a positive control) obtained from calf thymus possessed high amount of 5-mC, while yeast gDNA (a negative control) did not possess any 5-mC. The presence of 5-mC was found in different developmental stages of P. xylostella. The 5-mC levels were expressed as relative amounts compared to 5-mC level of the amount of calf thymus DNA (250 ng = 10 RPx, Fig. S1). There were significant differences in 5-mC levels among developmental stages. Egg had significantly ($P < 0.05$) more 5-mC level than other developmental stages. Among different body parts of larvae, the head region had higher ($P < 0.05$) levels of 5-mC than other body parts (Fig. 1C). The DNA methylation amounts of P. xylostella larvae were also compared among different insect species. The amount of 5-mC in P. xylostella gDNA was similar to that in other insects except honey bee which had the highest ($P < 0.05$) amount of DNA methylation (Fig. 1D).

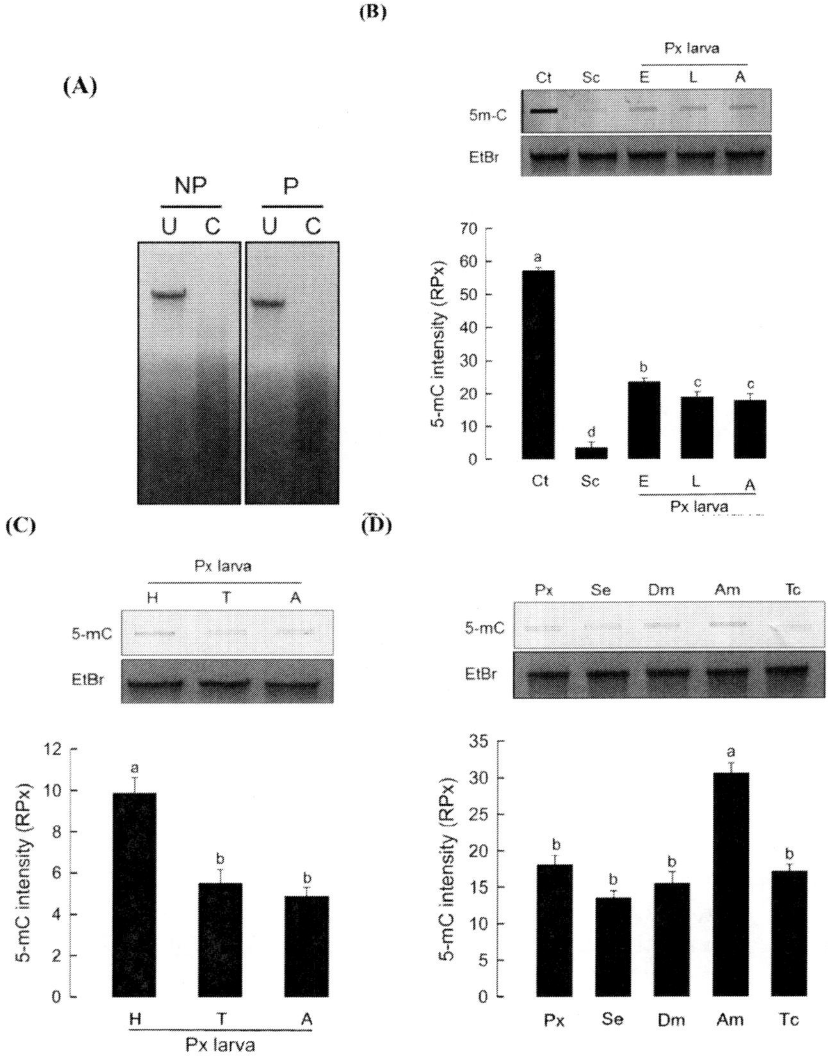

Figure 1: DNA methylation in P. xylostella ('Px') genome. (A) Restriction enzyme analysis for the presence of DNA methylation in P. xylostella. Gla I (4 units) digestion of P. xylostella genomic DNA (2.0 µg) from non-parasitized ('NP') or parasitized ('P') larvae. 'U' indicates undigested and 'C' indicates digested genomic DNAs (gDNAs) separated on 1.5% agarose gel. (B) Quantitative analysis of 5-methyl cytosine (5-mC) of different developmental stages of P. xylostella: egg ('E'), larvae ('L'), adult ('A') along with calf thymus gDNA ('Ct') as positive control and Saccharomyces cerevisiae gDNA ('Sc') as negative control. (C) Variation in DNA methylation of different tagmata of Px larvae: head ('H'), thorax ('T'), and abdomen ('A').

(D) Comparison of Px gDNA methylation with those of different species: Spodoptera exigua ('Se'), Drosophila melanogaster ('Dm'), Apis mellifera ('Am'), and Tribolium castaneum ('Tc'). DNA methylation was detected in slot blot experiment using specific monoclonal antibody against 5-mC. Each slot contained 2.0 μg of gDNA. The experiment was repeated 3 times using different genomic DNA samples. Methylation was quantified by ImageJ analysis. 'RPx' represents relative pixel units on the basis of 250 ng of calf thymus gDNA (=10 RPx). EtBr represents ethidium bromide staining. Different letters above standard deviations indicate significant difference among means at Type I error = 0.05 (LSD test).

2.2 Down-regulation of 5-mC level in *P. xylostella* gDNA by C. plutellae parasitism

Change in 5-mC levels was monitored during the development of P larvae to determine whether parasitism could influence host DNA methylation (Fig. 2). In P larvae, 5-mC level was transiently increased at 2 days after parasitization, but significantly decreased thereafter compared to the levels in NP larvae (Fig. 2A). In contrast, NP larvae did not show any difference during the larval stages corresponding to P larvae. 5-mC was observed in the nuclei of hemocytes of NP P. xylostella larvae (Fig. 2B). However, its intensity was clearly decreased in the hemocytes of P larvae. P larvae contained endoparasitoid eggs or larvae within the body. Parasitoid wasp eggs and larvae also possessed 5-mC in their genomes (Fig. S2). Parasitoid wasps also showed significant variations in 5-mC levels among developmental stages, in which the egg stage had the highest ($P < 0.05$) level. Thus, 5-mC amounts assessed in P larvae represented methylation amounts in both host and parasite.

2.3 Two different DNA methyltransferases (*DNMT-1 and DNMT-2*) of *P. xylostella*

From the genome database of P. xylostella, two DNMT genes (Px008220 and Px011175) were annotated. When these two DNMT genes were aligned with known DNMT genes, Px008220 and Px011175 were clustered with DNMT-1 and DNMT-2, respectively (Fig. 3A). Domain analysis indicated that Px008220, similar to human DNMT-1, possessed DNA methyltransferase 1-associated protein (DMAP), microtubule-interacting protein associated with TRAF3 (MIP-T3), replication foci targeting sequence (RFTS), zinc finger domain (CXXC), bromo-associated homology (BAH), and 5-mC-specific DNA methylase (DNA methylase) domain (Fig. 3B, Fig. S3).

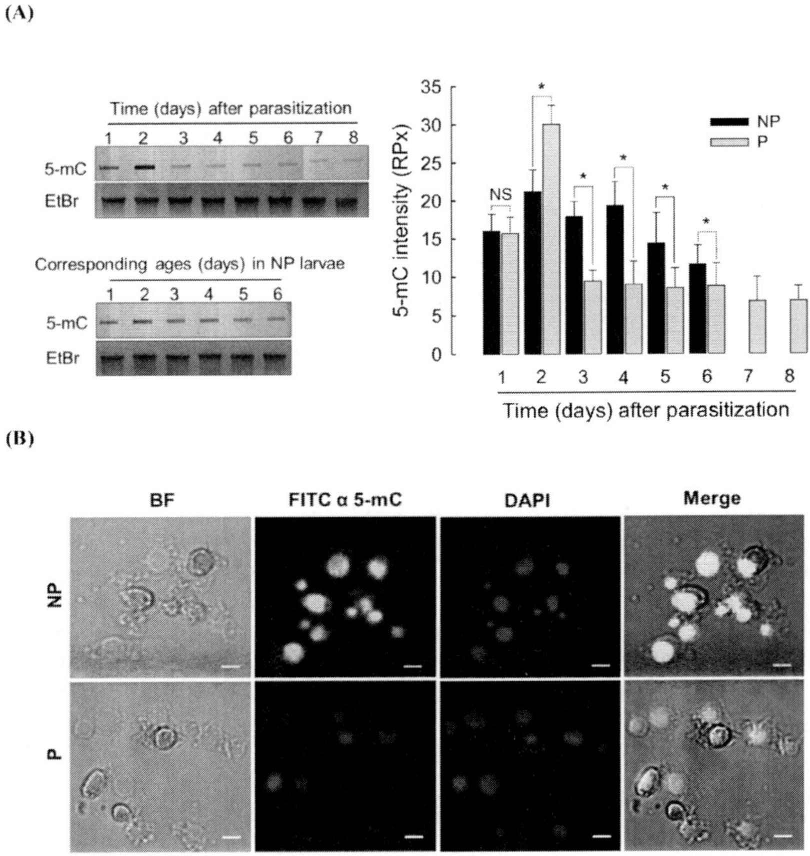

Figure 2: Decrease in DNA methylation of P. xylostella larvae parasitized by C. plutellae. (A) Decrease of 5-methyl cytosine (5-mC) in parasitized ('P') larvae compared to non-parasitized ('NP') larvae. Slot blot analysis of 5-mC (left). Each slot contains 2 µg of gDNA. Ethidium bromide ('EtBr') staining indicates that equal amounts of gDNA are used in different samples. Quantification of 5-mC amounts on the slot blot (right). The amounts are expressed in relative pixel ('RPx') units on the basis of 250 ng of calf thymus gDNA (=10 RPx). 'NS' and '*' indicate 'not significant difference' or 'significant difference' between NP and P larvae at Type I error = 0.05 (LSD test). Each measurement was replicated three times with different biological samples. (B) Immunofluorescence assay of 5-mC in hemocytes of P. xylostella larvae. Total hemocytes were visualized at bright field ('BF') mode. 5-mC was detected with a monoclonal antibody and visualized with a FITC-conjugate secondary antibody. Nuclei were visualized by DAPI. Merged view combined BF, FITC, and DAPI pictures. Scale bar indicates 10 µM.

However, Px011175 contained only 5-mC-specific DNA methylase (DNA methylase) domain, similar to human DNMT-2. These indicate that P. xylostella possesses two DNMTs: PxDNMT1 and PxDNMT2.

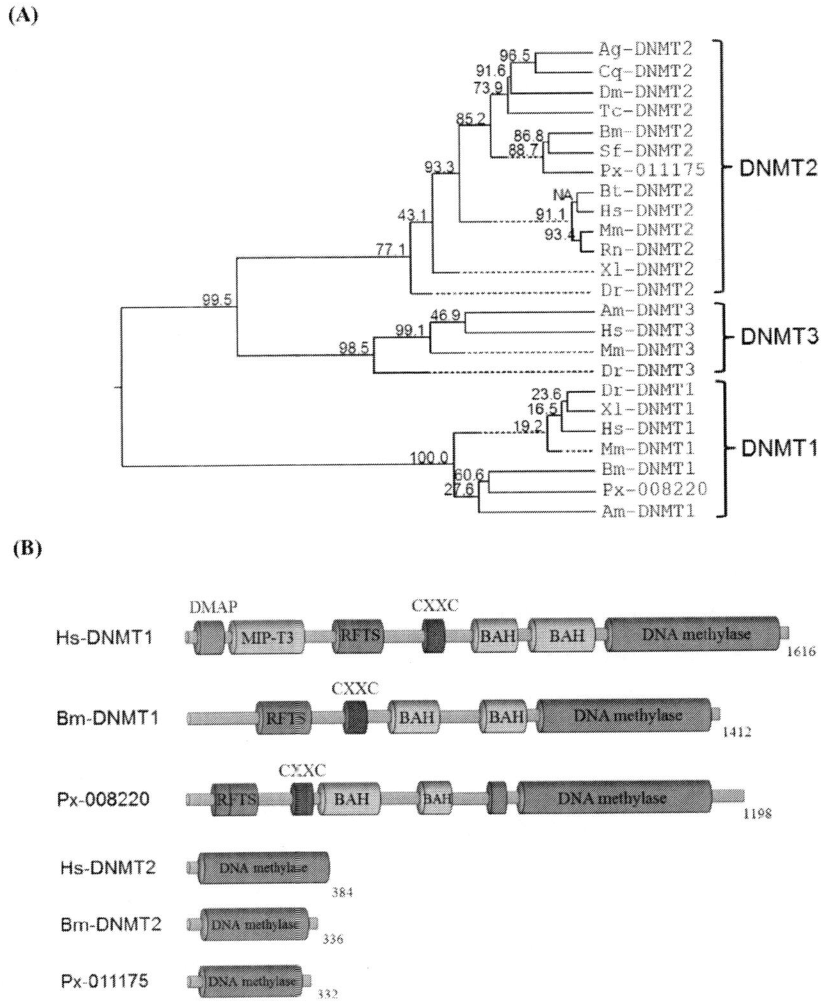

Figure 3: Two DNA methyltransferases PxDNMT1 and PxDNMT2 of P. xylostella. (A) Phylogenetic analysis of PxDNMT1 (Px008220) and PxDNMT2 (Px011175) with known DNMT genes using MEGA6[52] and ClustalW programs of DNASTAR (Version 5.01). Bootstrap values were obtained after 500 repetitions. Amino acid sequences of DNMTs were retrieved from GenBank with accession numbers list in Table S1. (B) Schematic diagram of domain structure of PxDNMT1 and PxDNMT2 with

reference to Homo sapiens (Hs) and Bombyx mori (Bm) DNMTs. Conserved domains include DNA methyltransferase 1 associated protein (DMAP), microtubule-interacting protein associated with TRAF3 (MIP-T3), replication foci targeting sequence (RFTS), Zn-finger-like motif (CXXC), bromo-associated homology (BAH), and C-5 cytosine specific DNA methylase (DNA methylase) domain predicted with Pfam domain search engine

Expressions of PxDNMT1 and PxDNMT2 were monitored in different tissues of late larvae of P. xylostella (Fig. 4A). All tested tissues expressed these two DNMTs. However, the expression patterns were different between the two DNMTs. PxDNMT1 in the gut or epidermis was expressed more than that in the hemocyte or fat body. In contrast, fat body exhibited the highest expression level of PxDNMT2. The expression levels of two DNMTs in NP and P larvae were compared (Fig. 4B). In NP larvae, the expression levels of the two DNMTs were not changed as larvae grew. However, in P larvae, the expression levels of both DNMTs were significantly ($P < 0.05$) decreased as larvae grew. The expression level of PxDNMT1 was significantly ($P < 0.05$) decreased at 4 days post parasitization, while that of PxDNMT2 was significantly ($P < 0.05$) decreased at 3 days post parasitization (Fig. 4C).

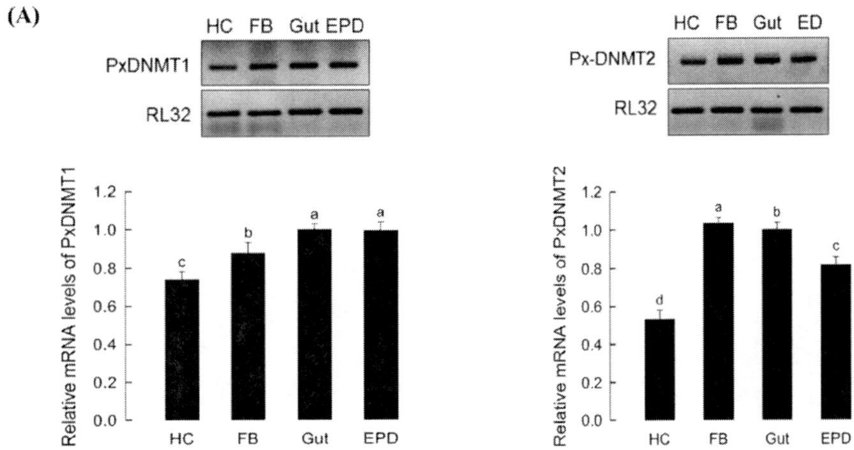

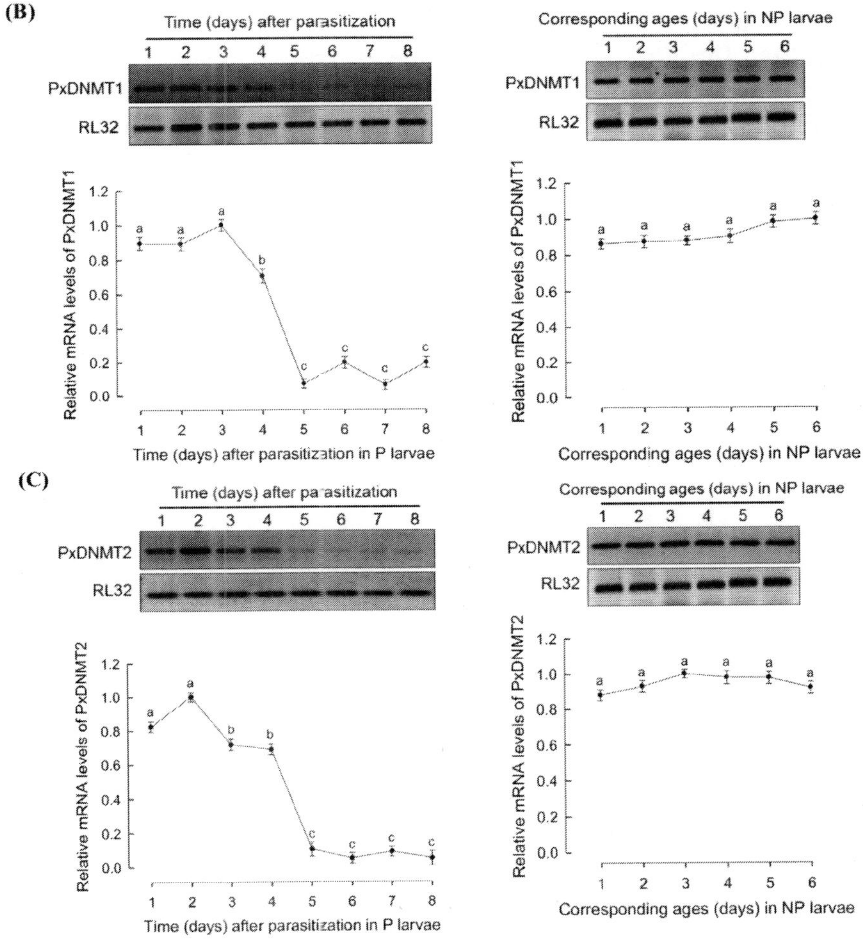

Figure 4: Down-regulation of PxDNMT1 and PxDNMT2 expressions in P. xylostella larvae parasitized by C. plutellae. (A) Expression profile of PxDNMT1 and PxDNMT2 in different tissues (hemocyte ('HC'), fat body ('FB'), midgut ('gut'), and epidermis ('EPD')) of fourth instar larvae of P. xylostella determined by RT-PCR (above) and RT-qPCR (below). (B) Decrease of PxDNMT1 expression levels in P. xylostella larvae parasitized by C. plutellae. PxDNMT1 mRNA levels in parasitized ('P') and nonparasitized ('NP') larvae determined by RT-PCR (above) and RT-qPCR (below). (C) Decrease of PxDNMT2 expression levels in P. xylostella larvae parasitized by C. plutellae determined by RT-PCR (above) and RT-qPCR (below). Expression of ribosomal protein RL32 was used to confirm cDNA integrity. Each treatment was replicated three times. Different letters above standard deviations indicate significant difference among means at Type I error = 0.05 (LSD test).

2.4 Influence of parasitism on expression of *MBD genes* of *P. xylostella*

5-mC can be recognized by specific recognition protein MBD that recruits and mediates chromatin remodeling to control gene expression23. Three MBD genes (PxMBD3, PxMBD4, and PxMBD5) were annotated in the DBM genome database (Fig. 5). These three PxMBDs share high sequence homologies with human MBDs (Fig. 5A). Especially, PxMBD4 was co-clustered with a canonical MBD protein, MeCP2, of human. In domain analysis, all three PxMBDs possess the characteristic MBD domain (Fig. 5B, Fig. S4).

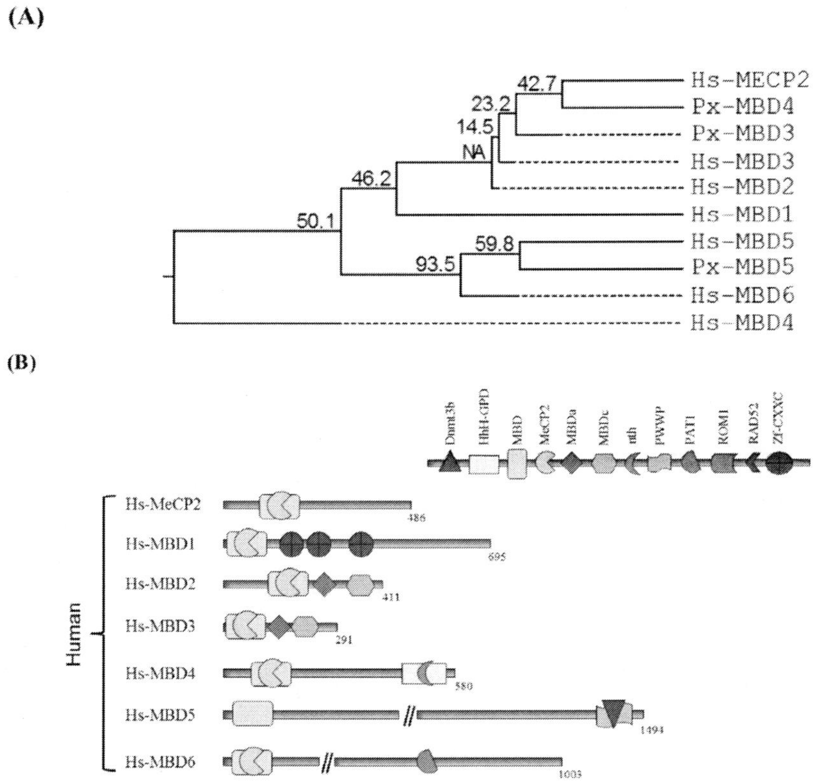

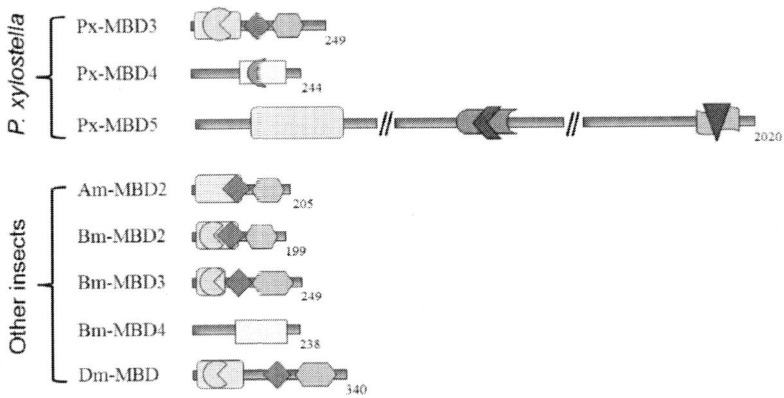

Figure 5: Three methylation-binding domain (MBD) proteins of P. xylostella. (A) Phylogenetic analysis of PxMBD3 (Px003962), PxMBD4 (XP_011549067.1), and PxMBD5 (Px004348) compared to other types of human MBDs using MEGA6[52] and ClustalW programs of DNASTAR (Version 5.01). Bootstrap values were obtained after 500 repetitions. Amino acid sequences of human MBDs were retrieved from GenBank with accession numbers listed in Table S1 (B) Schematic diagram of domain structure of PxMBDs with reference to human and other insect species. Conserved domains include DNA methyltransferase 3b (Dnmt3b), hallmark helix-hairpin-helix and Gly/Pro rich loop domain (HhH-GPD), methyl-CpG binding domain (MBD), methyl-CpG-binding protein 2 (MeCP2), methyl-CpG binding domain at P55 region (MBDa), methyl-CpG binding protein C-terminal (MBDc), endonuclease III (nth), Pro-Trp-Trp-Pro (PWWP), DNA topoisomerase II-associated protein PAT1 (PAT1), retinal outer segment membrane protein 1 (ROM1), DNA recombination/repair protein Rad52 (RAD52), and zinc finger CXXC domain (Zf-CXXC). They were predicted with Pfam domain search engine

Out of the three MBDs of P. xylostella, only PxMBD4 and PxMBD5 were analyzed for their expression levels in NP and P larvae of P. xylostella (Fig. 6) because the expression of PxMBD3 was not detected in our RT-PCR assay. All tested tissues expressed these two MBDs, in which fat body and epidermis exhibited higher expression levels of the two MBDs than hemocyte or gut (Fig. 6A). The expression levels of the two MBDs in NP and P larvae were compared (Fig. 6B). In NP larvae, the expression levels of the two MBDs were not changed during development. However, in P larvae, the expression levels of both MBDs were significantly ($P < 0.05$) decreased as larvae grew. Expression level of PxMBD4 in NP larvae was significantly ($P < 0.05$) decreased at 4 days post parasitization, while that of PxMBD5 was significantly ($P < 0.05$) decreased at 5 days post parasitization (Fig. 6C).

2.5 Influence of parasitism on DNA methylation-degrading enzyme (*TET*) of *P. xylostella*

Methylation levels on cytosine residues were modulated earlier than the down-regulation of DNMT expression, suggesting that there might be an expressional control in methylation-degrading enzyme before down-regulation of DNMT expression after parasitization. PvuRts1I is an enzyme that specifically digests 5-hydroxymethylcytosine. Indeed, this restriction enzyme could digest both NP and P larvae genomic DNAs, supporting the presence of DNA methylation-degrading enzyme (Fig. 7A).

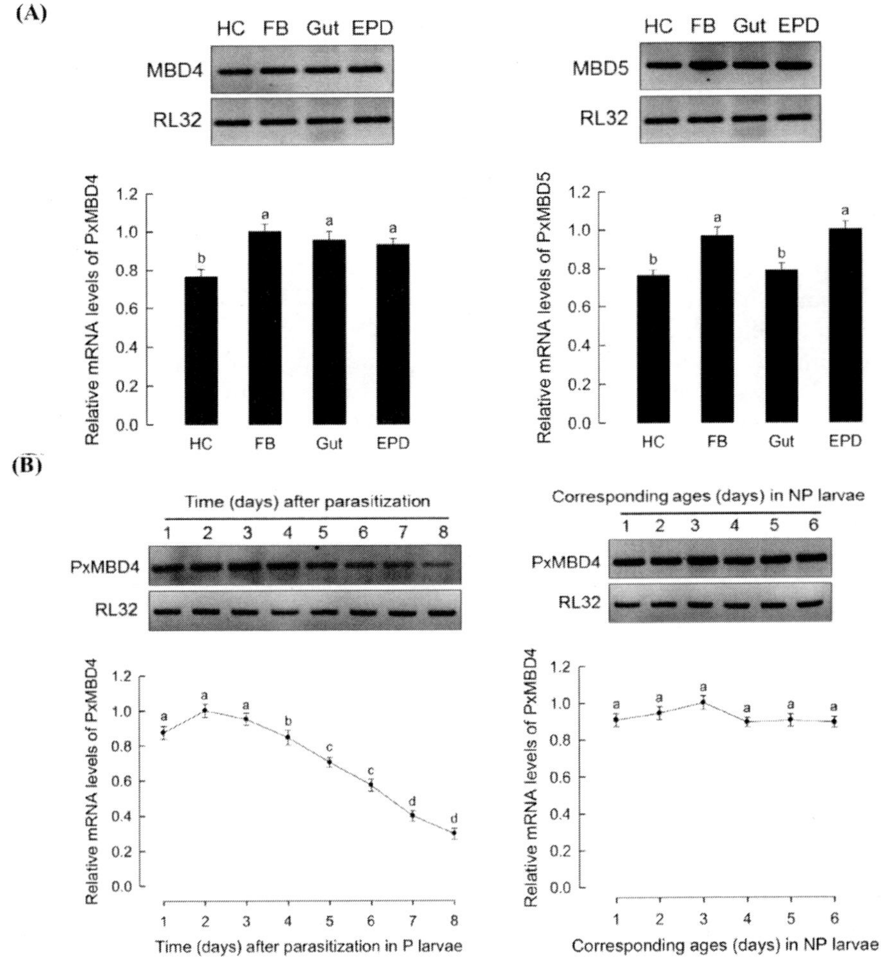

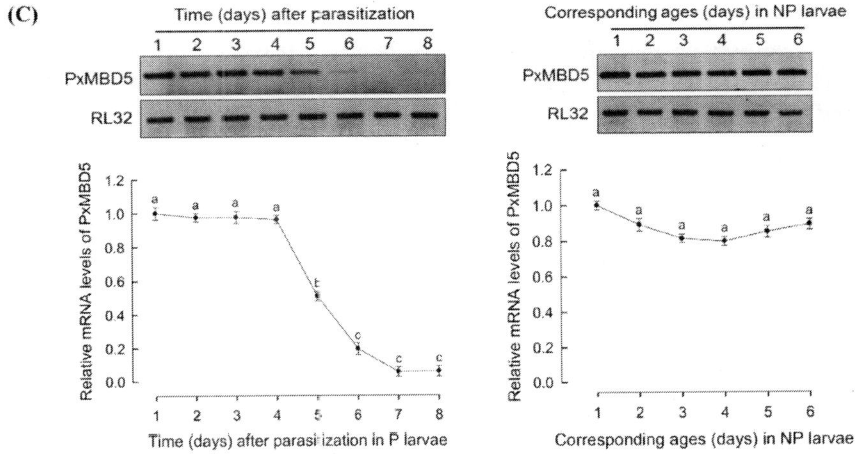

Figure 6: Expression patterns of three MBD genes of P. xylostella. (A) Expression levels of PxMBD4 and PxMBD5 in different tissues (hemocyte ('HC'), fat body ('FB'), midgut ('gut'), and epidermis ('EPD')) of fourth instar larvae of P. xylostella determined by RT-PCR (above) and RT-qPCR (below). (B) Changes in PxMBD4 expression levels in P. xylostella larvae parasitized by C. plutellae. PxMBD4 mRNA levels in parasitized ('P') or nonparasitized ('NP') larvae were determined by RT-PCR (above) and RT-qPCR (below). (C) Decrease of PxMBD5 expression levels in P. xylostella larvae parasitized by C. plutellae based on RT-PCR (above) and RT-qPCR (below). Expression of ribosomal protein RL32 was used to confirm cDNA integrity. Each treatment was replicated three times. Different letters above standard deviations indicate significant difference among means at Type I error = 0.05 (LSD test).

From the P. xylostella genome database, Px000976 (PxTET1) and Px004714 (PxTET2) were annotated to be TET dioxygenase, a DNA methylation-degrading enzyme (Fig. 7B). Based on their predicted amino acid sequences, both PxTET1 and PxTET2 shared similar structures with human TET. They all possessed two characteristic domains of zinc finger CXXC (Zf-CXXC) and Tet_JBP (2-oxoglutarate and iron-dependent dioxygenase) containing HxD, Hxs, and Rx5a motifs (Fig. 7C).

The expression levels of both PxTETs were assessed in different tissues of P. xylostella larvae. Results are shown in Fig. 8A. All tissues exhibited marked expressions of the two TETs, in which hemocytes exhibited the lowest expression levels of both TETs. The expression levels of the two TETs in NP and P larvae were compared (Fig. 8B). In NP larvae, the expression levels of both TETs were not changed during larvae development. However, in P larvae, the expression levels of both TETs were significantly decreased as larvae grew. Expression level of PxTET1 was significantly ($P < 0.05$) decreased at 2 days post parasitization, while that of PxTET2 was significantly ($P < 0.05$) decreased at 5 days post parasitization (Fig. 8C).

2.6 Significance of 5-mC in *P. xylostella* genome in immune response and development

AZA, a specific inhibitor of cytosine methylation, was used to test the significance of 5-mC in immune response and developmental processes of P. xylostella. AZA treatment at 48 h significantly ($P < 0.05$) decreased 5-mC levels of P. xylostella gDNA in a dose-dependent manner (Fig. 9A).

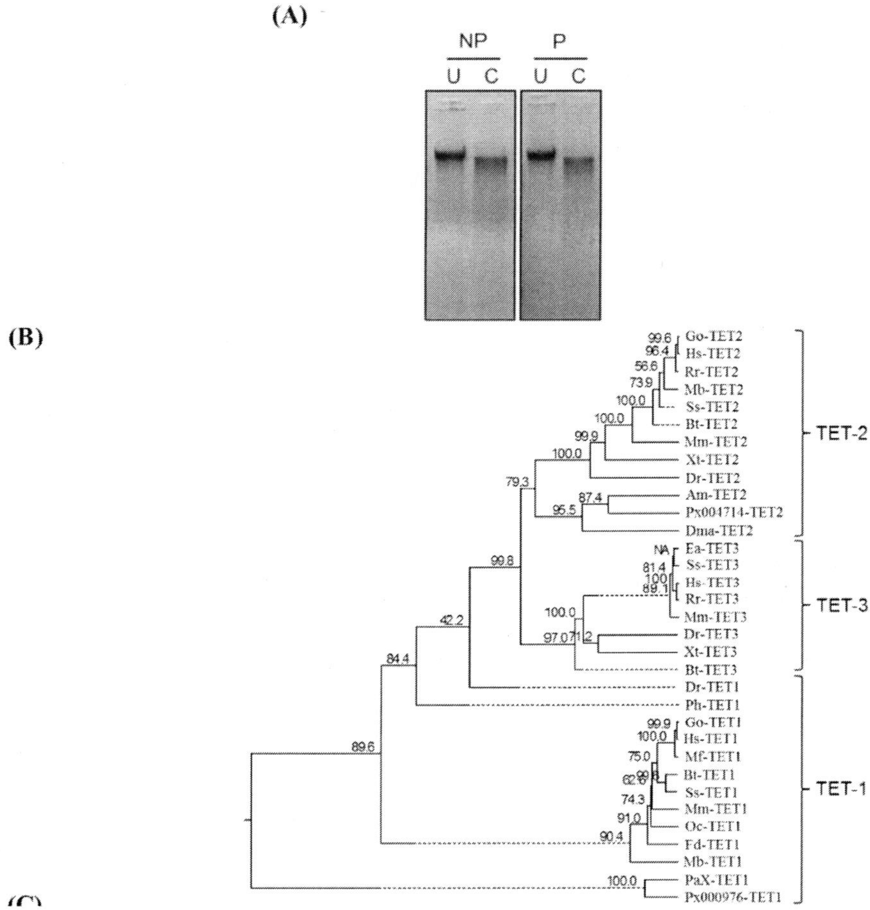

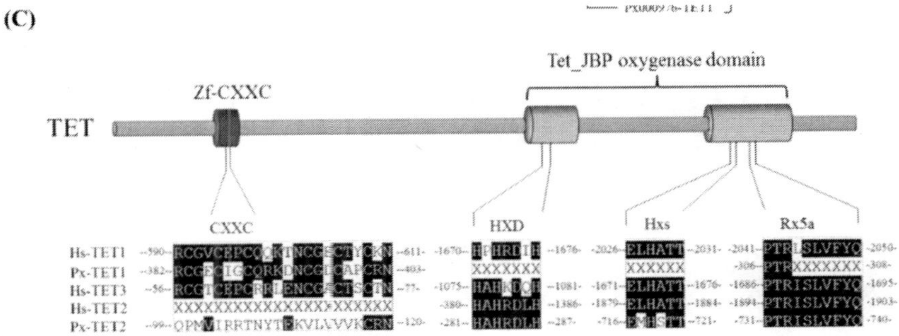

Figure 7: DNA methylation-degradation enzyme (ten-eleven translation (PxTET)) in P. xylostella. (A) Restriction enzyme analysis for the presence of DNA hydroxymethylation in P. xylostella. PvuRts1I (1 unit) digestion of P. xylostella genomic DNA (2.0 µg) from nonparasitized ('NP') and parasitized ('P') larvae. 'U' indicates undigested and 'C' indicates digested genomic DNA (gDNAs) separated on 1.5% agarose gel. (B) Phylogenetic analysis of PxTET1 (Px000976) and PxTET2 (Px004714) with known TET genes using MEGA6[52] and ClustalW programs of DNASTAR (Version 5.01). Bootstrap values were obtained after 500 repetitions. Amino acid sequences of TETs were retrieved from GenBank with accession numbers listed in Table S1. (C) Schematic diagram of domain structure of PxTET1 and PxTET2 with reference to human TETs. Conserved domains include zinc finger CXXC (Zf-CXXC) and 2-oxoglutarate and iron-dependent dioxygenase (Tet_JBP) containing HxD, Hxs, and Rx5a motifs. All domains and motifs were predicted with Pfam domain search engine

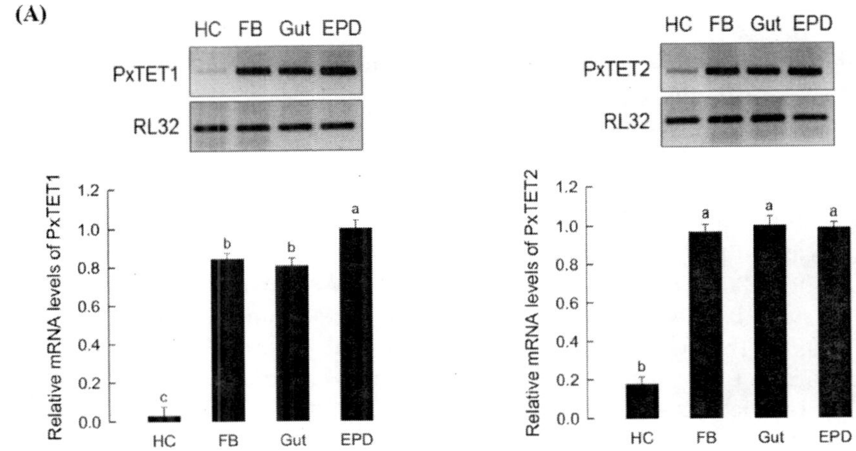

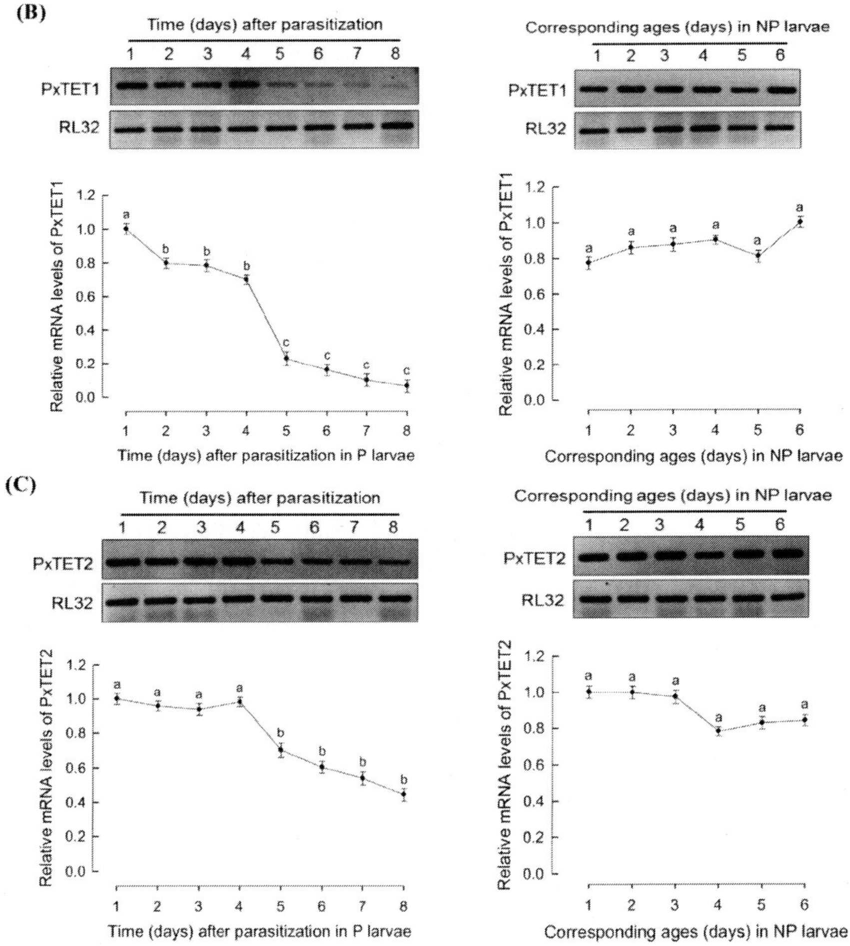

Figure 8: Expression patterns of ten-eleven translation proteins (TETs) in P. xylostella. (A) Expression profile of PxTET1 and PxTET2 in different tissues (hemocyte ('HC'), fat body ('FB'), mid gut ('gut'), and epidermis ('EPD')) of fourth instar larvae of P. xylostella based on RT-PCR (above) and RT-qPCR (below). (B) Down-regulation of PxTET1 in P. xylostella larvae parasitized by C. plutellae. PxTET1 mRNA levels in parasitized ('P') and nonparasitized ('NP') larvae based on RT-PCR (above) and RT-qPCR (below) are shown. (C) Down-regulation of PxTET2 in P. xylostella larvae parasitized by C. plutellae based on RT-PCR (above) and RT-qPCR (below). Expression of a ribosomal protein RL32 was used to confirm cDNA integrity. Each treatment was replicated three times. Different letters above standard deviations indicate significant difference among means at Type I error = 0.05 (LSD test).

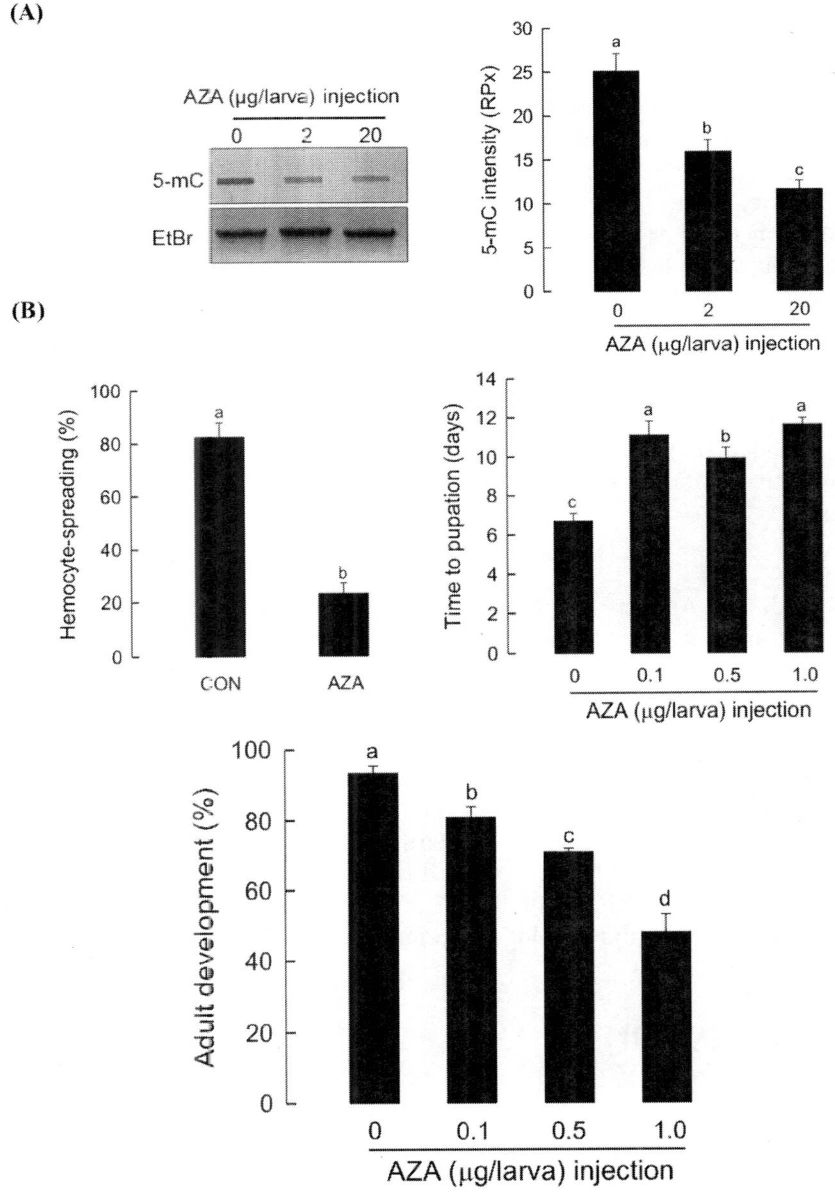

Figure 9: Influence of 5-Aza-2′-deoxycytidine (AZA) on immune response and development of P. xylostella. (A) Dose-dependent inhibitory effect of AZA on 5-methyl cytosine (5-mC) levels. Its inhibitory effect was assessed at different concentrations (0, 2.0, and 20 μg/larva) of AZA at 48 h post-injection to third instar larvae. For slot blot of 5-mC (left), each slot contained 2.0 μg genomic DNA (gDNA). Quantification of 5-mC

amounts of the slot blot bands is shown in the right panel. The amounts are expressed in relative pixel (RPx) units on the basis of 250 ng of calf thymus gDNA (=10 RPx). Ethidium bromide stain indicates that equal amounts of gDNA are used in different samples. As control, solvent DMSO was injected. Each treatment was replicated three times. (B) Effect of AZA on immune response and development of P. xylostella. For measurement of immune response, hemocyte-spreading behavior was checked at 48 h after the injection of 2 μg of AZA. Hemocyte-spreading behavior was assessed by counting cytoplasmic extension along with filopodial or pseudopodial growth. As control, DMSO was injected. Each treatment used 10 larvae. For measurement of developmental alterations, AZA (2 μg) was injected to each first day third instar (L3D1) larvae. For pupation time (days), time to pupation was measured from L3D1 to pupation. Adult emergence (%) was observed after complete escape from pupal cuticle. Thirty individuals were assessed in each treatment. Different letters above standard deviations indicate significant difference among means at Type I error = 0.05 (LSD test).

In addition, AZA treatment significantly ($P < 0.05$) suppressed immune response based on hemocyte-spreading behavior (Fig. 9B). The AZA treatment also delayed the development of P. xylostella from larvae to pupae.

To confirm the results of AZA assay, both PxDNMT1 and DNMT2 expressions were suppressed by RNA interference (Fig. 10). Treatment with dsRNAs specific PxDNMT1 and PxDNMT2 significantly ($P < 0.05$) suppressed the expression of PxDNMT1 and PxDNMT2 expression, respectively, in the NP larvae of P. xylostella (Fig. 10A). Larvae treated with dsRNA specific to PxDNMT1 (Fig. 10B) or PxDNMT2 (Fig. 10C) exhibited immune response and development similar to larvae treated with AZA.

3. DISCUSSION

DNA methylation is a way to transmit epigenetic information to the next generation. It is a covalent modification of DNA by adding methyl group mostly to cytosine at CpG sequence in animals with catalytic activities of DNMTs. Once methylated, the methylated sites can be recognized by 5-methyl-CpG-binding domain proteins (MBDs), which will recruit chromatin remodeling machineries to control gene expression[24]. DNA methylation can be changed by various environmental factors as described in Introduction section.

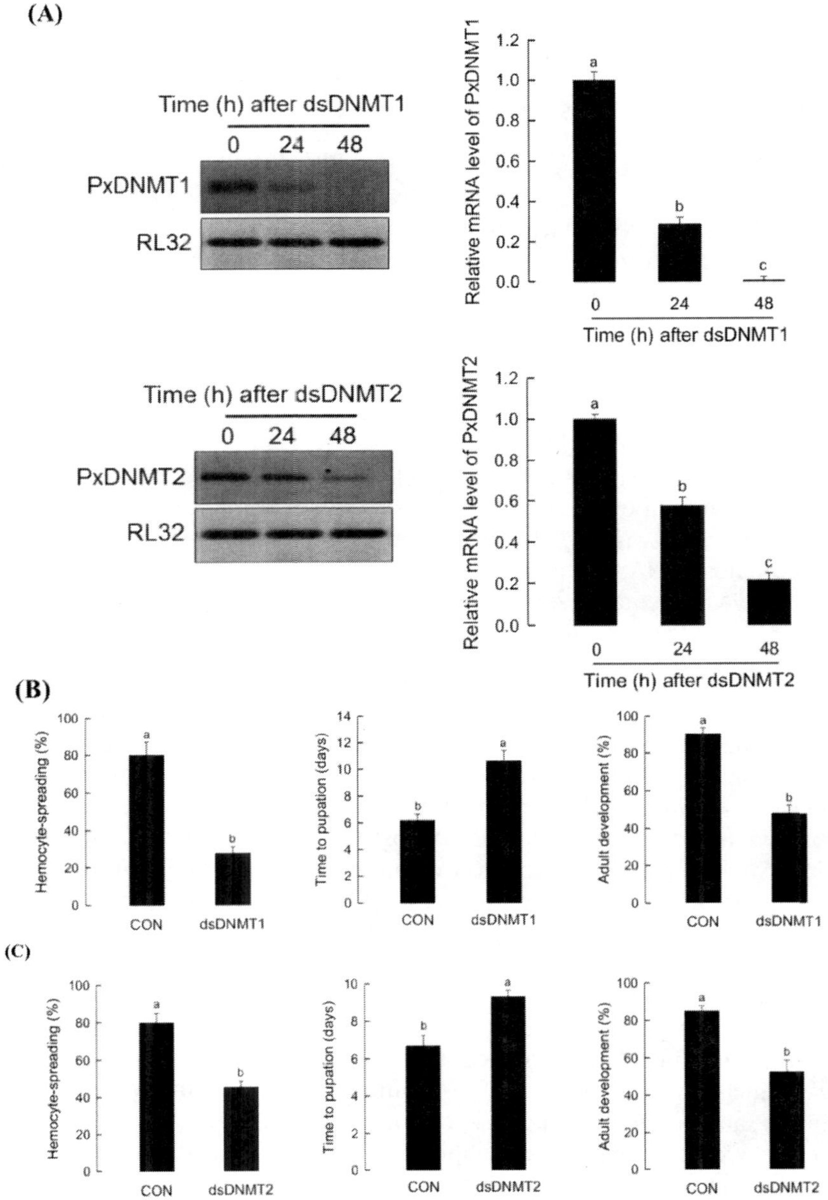

Figure 10: Influence of RNA interference (RNAi) of PxDNMT1 and PxDNMT2 on immune response and development of P. xylostella. (A) RNAi efficacy of PxDNMTs. After injection of dsRNAs specific to PxDNMT1 (dsDNMT1) or PxDNMT2 (dsDNMT2) (250 ng dsRNA per larva), RNAi effect was assessed after 24 and 48 h of dsRNA injection by RT-PCR (left) and RT-qPCR (right). RT-qPCR was replicated three times. (B,C) Effect of RNAi of

PxDNMT1 or PxDNMT2 on immune response and development of P. xylostella. For measurement of immune response, hemocyte-spreading behavior was checked at 48 h post injection of 250 ng of dsDNMT1 or dsDNMT2. Hemocyte-spreading was assessed by counting cytoplasmic extension along with filopodial or pseudopodial growth. As control dsRNA (dsCON), dsRNA specific to a viral gene CpBV-ORF302 was injected. Each treatment used 10 larvae. For measurement of developmental alterations, dsDNMT1 or dsDNMT2 was injected to each first day third instar (L3D1) larvae. For pupation time (days), time to pupation was measured from L3D1 to pupation. Adult emergence (%) was observed after complete escape from pupal cuticle. Thirty individuals were assessed in each treatment. Different letters above standard deviations indicate significant difference among means at Type I error = 0.05 (LSD test).

This study demonstrated that a novel environmental factor could change DNA methylation with respect to parasite and host interaction, in which an endoparasitoid wasp, C. plutellae, suppressed DNMT expressions in order to alter host DNA methylation status to induce immunosuppression and delay developmental rate of host that are necessary for the development of immature endoparasitoid wasp.

Whole-genome methylation analyses in insects such as honeybee (Apis mellifera), silkworm moth (Bombyx mori), and parasitic wasp (Nasonia vitripennis) have shown that 5-mC is by far the most common form of programmed DNA modification[25-27]. This current study showed that P. xylostella genomic DNA was methylated. Two methods can be used to observe the DNA methylation of P. xylostella genome. A restriction enzyme, Gla I, can be used to digests XCGX with methylated cytosine[28]. Both NP and P genomic DNAs were susceptible to Gla I treatment. Slot blot assay using a monoclonal antibody specifically recognizing 5-mC showed positive reactions to both NP and P larval gDNAs. Densitometry analysis after slot blot assay indicated that the relative amounts of 5-mC on P. xylostella gDNAs were similar among different developmental stages, including eggs, larvae, and adults. It also showed that DNA methylation level of P. xylostella was much lower than that of a mammalian species, but was relatively similar among insects such as coleopteran, dipteran, and lepidopteran species except hymenopteran specie. About 4–5% of the cytosines are usually methylated on the entire chromosomes of vertebrates[29], while only 0.5% or less of cytosine residues are methylated in Drosophila spp.[19]. This suggests that P. xylostella genome has methylated cytosines at around 0.5% among total cytosines. Most invertebrates have relatively low DNA methylation amounts in their genome. However, they exhibit the persistence of DNA methylation[30,31]. In contrast to genome-wide DNA methylation in vertebrates, DNA methylation in invertebrates is largely confined to genes and most of them are not in the intergenic regions[32,33]. Especially, DNA methylation on transposable and repetitive elements is almost nonexistent in insects[34,35]. These facts suggest that DNA methylation plays little role in suppressing the proliferation of transposable elements in insects, unlike vertebrates. Indeed, intragenic DNA methylation characteristic of insects is involved in gene expression through mRNA splicing[36]. Clearly, DNA methylation plays crucial roles in insect physiological processes. Gene-specific DNA methylation can mediate insecticide resistance by up-regulating esterase gene expression to increase detoxification in the green peach aphid, Myzus persicae[37]. In the European honey bee, A. mellifera, DNA methylation controls alternative splicing of mRNAs in the fat body[38]. These results suggest that any alteration of DNA methylation may cause significant physiological alterations in P. xylostella.

Parasitism by C. plutellae suppressed DNMT gene expressions of P. xylostella. Two DNMT genes, PxDNMT1 and PxDNMT2, were identified from P. xylostella genome. PxDNMT-1 is likely to play a role in maintaining DNA methylation status because it shares structural similarity with human DNMT-1. In contrast, PxDNMT2 is highly homologous to human DNMT-2. Thus, P. xylostella has two types of DNMT. However, it does not have DNMT-3. In insects, there are variations in retaining DNMT genes in their genome[18]. Hemipteran and hymenopteran species possess all three types of DNMT genes. In the pea aphid, Acyrthosiphon pisum, two isoforms of DNMT1 and DNMT3 are encoded in its genome[39]. In addition, honey bee and other social hymenopteran insects possess both DNMTs that presumably perform maintenance of DNA methylation (DNMT-1) and de novo DNA methylation (DNMT-3)[40,41]. However, coleopteran and lepidopteran species possess only DNMT-1 and DNMT-2, but not DNMT-3[31]. Furthermore, dipteran species do not have DNMT-1 or DNMT-3. They only have DNMT-2[19,42]. Thus, lepidopteran species including P. xylostella analyzed in this study appear to lose DNMT-3, indicating that lepidopteran species may not form de novo DNA methylation during development in their life cycle. However, epigenetic control through DNA methylation plays a significant role in domestication of silkworm from a wild population, in which demethylation is associated with down-regulation of gene expression during domestication[26]. Furthermore, DNMT-1 of B. mori has been confirmed to have DNA methylation activity and its knockdown of gene expression will lead to decreased egg hatch[26,43]. This suggests that there must be de novo DNA methylation in B. mori. A structural prediction of DNMT-2 suggests that it may perform DNA methylation in cytosines of non-CpG sequences[44]. Thus, the two DNMTs encoded in P. xylostella may play roles in maintaining DNA methylation and de novo DNA methylation to epigenetically control gene expression. Thus, suppression of both PxDNMT1 and PxDNMT2 in P larvae would significantly alter the physiological processes of P. xylostella. In fact, down-regulation of DNMT gene expression resulted in decreased 5-mC contents in P. xylostella. Significant decrease of 5-mC levels in P larvae began at 3 days after parasitization. This appears to be well explained by significant decrease of PxDNMT2 expression in P larvae. However, there was a significant increase of 5-mC level in P larvae at 2 days after parasitization. This may be explained by changes in 5-mC levels in parasitoid egg or larvae within host P larvae. Assessment of 5-mC levels in different C. plutellae developmental stages showed that 5-mC levels were decreased with development (Fig. S2). Although P larvae at 2 days after parasitization had egg or larvae of parasitoid wasp, P larvae at 1 day after parasitization also had the parasitoid eggs that had the highest 5-mC level. Thus, the increase of 5-mC level in P larvae at 2 days after parasitization is not likely to be explained by the 5-mC of endoparasoids. Alternatively, such increase might be due to decreased degradation of DNA methylation by host metabolizing enzyme such as TET that can demethylate 5-mC by oxidation processes[45]. The expressional levels of both TET genes were decreased during parasitism. Interestingly, the expression level of PxTET1, but not PxTET2, began to decrease at 2 days after the parasitization. This suggests that the early increase of 5-mC level in the P larvae might be due to decreased 5-mC metabolism by TET under no change in expression levels of both DNMTs. However, any causal relationship between down-regulation of PxTET1 expression and increase of 5-mC level is yet clear until an activity of PxTET1 to catalyze demethylation of 5-mC is confirmed. The transient increase may be significant in terms of host defense because high methylation on host genome is not favored for parasitoid development (see below). Decrease of 5-mC levels is associated with physiological alterations such as immunosuppression and delay of larval development. AZA is a specific inhibitor of 5-mC in mammals[46]. AZA treatment significantly decreased 5-mC level in P. xylostella and altered its physiological processes. Similarly, RNAi of PxDNMT1 or PxDNMT2 resulted in significant physiological alterations. These results suggest that DNA methylation is required for normal physiological processes of P. xylostella. In addition, endoparasitoid wasp C. plutellae can alter host physiological processes through an epigenetic mode by suppressing DNMT expressions. DNA methylation is closely associated with insect development for cell growth and differentiation[47]. The best example is the royal jelly effect on queen development of A. mellifera, in which royal jelly contains a specific inhibitor to histone deacetylase (HDAC) that is linked to DNMT3[48]. The HDAC inhibitor can reduce DNA methylation level to extend larval period. With respect to immunity, epigenetic control plays a crucial role in modulating gene expression of antimicrobial peptides (AMPs). It has been reported that histone acetylation can promote AMP expression in the greater wax moth, Galleria mellonella[49]. In Drosophila, DNMT-2 is required for efficient innate immune responses especially in antiviral defense by directly binding to viral RNAs with subsequent modification through RNA methylation[50]. Considering that C. plutellae depends on its symbiotic virus CpBV for successful parasitism, the parasitoid wasp should suppress PxDNMT2. The question is, which parasitic factor suppresses the expression of PxDNMTs? As explained in the Introduction, several CpBV genes can modulate host gene expression. Among these CpBV genes, vH4 might modulate host gene expressions by chromatin remodeling through direct joining to host nucleosomes[9]. This prediction needs to be explored in subsequent studies. Alternatively, down-regulation of DNA methylation in parasitized larvae of P. xylostella may be indirect due to alteration of host gene expression by various parasitic factors as demonstrated in another parasitoid wasp against P. xylostella[4].

In summary, P. xylostella genome is methylated and its methylation levels can be modulated by its methylation and demethylation machineries. Environmental factor such as parasitism by C. plutellae can influence its host genome by altering DNA methylation through significant reduction in gene expression of both DNMT-1 and DNMT-2. The reduction of host DNA methylation might be accompanied by massive changes in gene expression. This epigenetic alteration of host gene expression would favor successful parasitism of C. plutellae.

5. METHODS

5.1 Insects and parasitization

All procedures are described in Park and Kim[51].

5.2 Bioinformatics for DNA methylation-associated genes

From a genome database of P. xylostella and NCBI-GenBank, the following genes were obtained: two DNA methyltransferase genes (DNMT-1, Px008220; DNMT-2, Px011175), three methyl-binding domain proteins (MBD-3, Px003962; MBD-4, XP_011549067.1; and MBD-5, Px004348), and two ten-eleven translation proteins (TET-1, Px000976; TET-2, Px004714). Predicted amino acid sequences of these P. xylostella genes were then compared to those of other genes obtained from GenBank (Table S1).

Phylogenetic analyses were performed using MEGA6[52] and ClustalW programs of DNASTAR (Version 5.01, Madison, WI, USA). Bootstrapping values to support branching were obtained with 500 repetitions. Protein domains of these genes were predicted using Pfam version 29.0 (http://pfam.xfam.org) provided by European Bioinformatics Institute.

5.3 RNA interference (RNAi)

Double-stranded RNAs (dsRNAs) were prepared using a method described by Park and Kim[51]. For microinjection of dsRNA, each dsRNA was mixed with a cell transfection reagent (Metafectene PRO, Biontex, Planegg, Germany) in a ratio of 1:1 (v/v) and incubated at 25 °C for 20 min. As dsRNA control, dsRNA specific to a viral gene CpBV-ORF30251 was used. A hundred nL of dsRNA (250 ng) complex was injected into each hemocoel of early third instar larvae of P. xylostella and its effect on gene expression was quantified by RT-qPCR as described below.

5.4 Real time-quantitative PCR (RT-qPCR)

Total RNA extraction and cDNA preparation followed a method described in Park and Kim[51]. RT-qPCR was performed in 20 µL reaction volume consisting of 2× SYBR® Green Realtime PCR Master Mix (TOYOBO, Dojima Hama, Osaka, Japan), 5 µM of gene-specific forward and reverse primers (Table S2), and 50 ng of cDNA as template. As an endogenous control, RL32 gene was assessed along with test samples. PCR was performed at 95 °C for 5 min for initial denaturation followed by 40 cycles at 98 °C for 15 s, 54 °C for 30 s, and 72 °C for 45 s with a final extension at 72 °C for 7 min. Melting curves were assessed to confirm unique PCR products. Relative quantification method[53] was used to estimate mRNA expression levels. The experiment was independently replicated three times.

5.5 Analysis of DNA methylation using slot-blot apparatus

gDNAs were isolated from different developmental stages of P. xylostella larvae. DNA concentration and purity were determined with GeneQuant spectrophotometer and electrophoresis on 1.5% agarose gel, respectively. Two µg of gDNA was first denatured by heating at 100 °C for 10 min in a tube containing denaturing solution (0.4 M NaOH, 10 mM EDTA). The reaction was neutralized by adding equal volume of 2 mM ice cold ammonium acetate (pH 7.0). As positive and negative controls of DNA methylation, a calf thymus DNA (Sigma-Aldrich Korea, Seoul, Korea) and Saccharomyces cerevisiae yeast DNA (Invitrogen, Seoul, Korea) were used, respectively. After denatured DNA samples were then passed through slot-blot apparatus (Bio Rad, Hercules, CA, USA), the membranes were air-dried and exposed twice to a UV crosslinker (CL-1000, UVP, LLC, Upland, CA, USA) for each 30 sec. After blocking the membranes with 5% skimmed milk, membranes were incubated with primary antibody raised against 5-mC (Abcam, Seoul, Korea) at 1:1,000 dilution for 2 h. The bound primary antibody was captured with secondary antibody conjugated with alkaline phosphatase (Sigma-Aldrich Korea) and visualized with 5-bromo-4-chloro-3-indolyl phosphate/nitro blue tetrazolium (Sigma-Aldrich Korea). For densitometry quantification of bands, the membranes were air-dried and images were captured with HP Scanjet 5470c scanner (Hewlett-Packard Company, Greeley, CO, USA). Band intensity was estimated using ImageJ 2.1.4.9 software At every run, 250 ng of calf thymus DNA was loaded to normalize band intensity and calculate relative pixel (RPx) amounts, in which 25 ng of calf thymus DNA was set to be 1 RPx.

5.6 Restriction enzyme assay

Purified gDNA was digested with 4 units of *Gla* I (SibEnzyme, West Roxbury, MA, USA) at 30 °C for 16 h or 1 unit of *PvuRts*1I (Diagenode, Seraing, Belgium) at 32 °C for 2 h in a total volume of 50 μL. The resulting DNA was separated on 1.5% agarose gel and stained with ethidium bromide.

5.7 Analysis of DNA methylation using immunofluorescence assay

For fluorescence microscope observation, hemocytes were collected from 50 larvae of P. xylostella in anticoagulant buffer (ACB: 1.32% sodium citrate, 0.48% citric acid, and 1.47% dextrose, pH 4.5). The suspension was incubated on ice for 40 min and replaced with TC-100 insect cell culture medium (Hyclone, Daegu, Korea). The hemocyte suspension was transferred onto glass slide and incubated at 25 °C for 12 h in the dark. After washing once with 100 mM phosphate buffered saline (PBS, pH 7.4), cells were fixed with 4% paraformaldehyde at RT for 10 min and permeabilized with cold methanol (−20 °C) for 2 min. Cells were washed with PBS and incubated with primary antibody against 5-mC (1:5,000 dilution) at RT for 1 h. After washing twice in PBS, secondary antibody-conjugated with fluorescein isothiocyanate (FITC, Sigma-Aldrich Korea) was added in 1:5,000 dilutions and incubated at RT for 1 h. Cells were washed twice in PBS and incubated with 4′,6-diamindino-2-phenylindole (DAPI, Thermo Scientific, Meridian, Rockford, USA) at 1:1,000 dilution in PBS for 2 min. After washing twice with PBS, cells were observed under a fluorescence microscope (DM 2500, Leica, Wetzlar, Germany) at 400× magnification.

5.8 5-Aza-2′-deoxycytidine (AZA) treatment

AZA was purchased from Sigma-Aldrich Korea and dissolved in dimethylsulfoxide (DMSO, Amresco, Solon, Ohio, USA) to prepare 50 mg/mL stock solution. For microinjection, AZA was mixed with Metafectene PRO as described above. DMSO was used as control. At 24–48 h post injection, total gDNA was isolated from larvae to assess methylation levels using slot blot analysis. Other larvae treated with AZA were subjected to assessment of hemocyte-spreading behavior assay or developmental assessment (see below).

5.9 Hemocyte-spreading behavior assay

Hemolymph was collected in ACB from NP larvae of P. xylostella at 48 h after injection of AZA or dsRNA specific to PxDNMT1 or PxDNMT2. After centrifuging at 800× g for 3 min, the pellet was resuspended in equal volume of ice cold ACB solution and incubated on ice for 20 min. The suspension was centrifuged and the pellet was resuspended in 1 mL of TC100 insect cell culture medium. Fifty μL of hemocyte suspension was seeded onto 96 well culture plates and incubated at 25 °C for 40 min in the dark. Hemocyte-spreading behavior was observed under a phase contrast microscope (Olympus S730, Tokyo, Japan) by counting hemocytes exhibiting cytoplasmic extension such as filopodial growth and pseudopodial growth.

5.10 Developmental assay

NP larvae at L3D1 stage were microinjected with 0.1, 0.5, and 1.0 μg of AZA and 250 ng of dsRNA specific to *PxDNMT1* or *PxDNMT2* as described above. Treated larvae were fed with cabbage leaves ad libitum and kept at 25 °C for growth. For pupal development assessment, time to pupation was measured by the elapsed time from L3D1 to pupation. Pupation was confirmed by pupal molting within cocoon. For adult emergence assessment, whole body escape from the pupal cuticle was considered as complete adult emergence. The number of adult emerged was counted. All bioassays used three replicates per treatment. For each replication, 30 larvae were used.

ACKNOWLEDGEMENTS

This study was supported by a general researcher program (2013R1A1A2A10058197) funded by National Research Foundation of Korea.

REFERENCES

1. Beckage, N. E., Thompson, S. A. & Federici, B. A. Parasites and pathogens of insects. Vol. 1. Parasites. Academic Press, NY, USA 364 (1993).
2. Vinson, S. B. & Iwantsch, G. F. Host regulation by insect parasitoids. Q. Rev. Biol. 55, 143–165 (1980).
3. Song, K. H., Jung, M. K., Eum, J. H., Hwang, I. C. & Han, S. S. Proteomic analysis of parasitized Plutella xylostella larvae plasma. J. Insect Physiol. 54, 1270–1280 (2008).
4. Etebari, K. et al. Deep sequencing-based transcriptome analysis of Plutella xylostella larvae parasitized by Diadegma semiclausum. BMC Genomics 12, 446 (2011).
5. Bae, S. & Kim, Y. IkB genes encoded in Cotesia plutellae bracovirus suppress an antiviral response and enhance baculovirus pathogenicity against the diamondback moth. Plutella xylostella. J. Invertebr. Pathol 102, 79–87 (2009).
6. Bitra, K., Suderman, R. J. & Strand, M. R. Polydnavirus Ank proteins bind NF-κB homodimers and inhibit processing of Relish. PLoS Pathog. 8, e1002722 (2012).
7. Magkrioti, C., Iatrou, K. & Labropoulou, V. Differential inhibition of BmRelish1-dependent transcription in lepidopteran cells by bracovirus ankyrin-repeat proteins. Insect Biochem. Mol. Biol. 41, 993–1002 (2011).
8. Gad, W. & Kim, Y. A viral histone H4 encoded by Cotesia plutellae bracovirus inhibits haemocyte-spreading behaviour of the diamondback moth, Plutella xylostella. J. Gen. Virol. 89, 931–938 (2008).
9. Hepat, R., Song, J. J., Lee, D. & Kim, Y. A viral histone h4 joins to eukaryotic nucleosomes and alters host gene expression. J. Virol. 87, 11223–11230 (2013).
10. Bonasio, R., Tu, S. & Reinberg, D. Molecular signals of epigenetic states. Science 330, 612–616 (2010).
11. Mazzio, E. A. & Soliman, K. F. Epigenetics and nutritional environmental signals. Integr. Comp. Biol. 54, 21–30 (2014).
12. Collotta, M., Bertazzi, P. A. & Bollati, V. Epigenetics and pesticides. Toxicology 307, 35–41 (2013).
13. Cheeseman, K. & Weitzman, J. B. Host-parasite interactions: an intimate epigenetic relationship. Cell Microbiol. 17, 1121–1132 (2015).
14. Bonduriansky, R., Crean, A. J. & Day, T. The implications of nongenetic inheritance for evolution in changing environments. Evol. Appl. 5, 192–201 (2012).
15. Bird, A. DNA methylation patterns and epigenetic memory. Genes Dev. 16, 6–21 (2002).
16. Goll, M. G. et al. Methylation of tRNAAsp by the DNA methyltransferase homolog Dnmt2. Science 311, 395–398 (2006).
17. Jurkowska, R. Z., Jurkowski, T. P. & Jeltsch, A. Structure and function of mammalian DNA methyltransferases. Chembiochem. 12, 206–222 (2011).
18. Glastad, K. M., Hunt, B. G., Yi, S. V. & Goodisman, M. DNA methylation in insects: on the brink of the epigenomic era. Insect Mol. Biol. 13, 117–123 (2011).
19. Marhold, J. et al. Conservation of DNA methylation in dipteran insects. Insect Mol. Biol. 13, 117–123 (2004).
20. Xiang, H. et al. Single base-resolution methylome of the silkworm reveals a sparse epigenomic map. Nat. Biotechnol. 28, 516–520 (2010).
21. Bae, S. & Kim, Y. Host physiological changes due to parasitism of a braconid wasp, Cotesia plutellae, on diamondback moth, Plutella xylostella. Comp. Biochem. Physiol. A138, 39–44 (2004).
22. Ibrahim, A. M. & Kim, Y. Parasitism by Cotesia plutellae alters the hemocyte population and immunological function of the diamondback moth, Plutella xylostella. J. Insect Physiol. 52, 943–950 (2006).
23. Du, Q., Luu, P. L., Stirzaker, C. & Clark, S. J. Methyl-CpG-binding domain proteins: readers of the epigenome. Epigenomics 7, 1051–1073 (2015).
24. Bogdanovic, O. & Veenstra, G. J. DNA methylation and methyl-CpG binding proteins: developmental recruitments and function. Chromosoma 118, 549–565 (2009).

25. Cingolani, P. et al. Intronic non-CG DNA hydroxymethylation and alternative mRNA splicing in honey bees. BMC Genomics 14, 666 (2013).
26. Xiang, H. et al. Comparative methylomics between domesticated and wild silkworms implies possible epigenetic influence on silkworm domestication. BMC Genomics 14, 646 (2013).
27. Beeler, S. M. et al. Whole-genome DNA methylation profile of the jewel wasp (Nasonia vitripennis). G34, 383–388 (2014).
28. Tarasova, G. V., Nayakshina, T. N. & Degtyarev, S. K. Substrate specificity of new methyl-directed DNA endonuclease GlaI. BMC Mol. Biol. 9, 7 (2008).
29. Gama-Sosa et al. Tissue-specific differences in DNA methylation in various mammals. Biochim. Biophys. Acta 740, 212–219 (1983).
30. Suzuki, M. M., Kerr, A. R., De-Sousa, D. & Bird, A. CpG methylation is targeted to transcription units in an invertebrate genome. Genome Res. 17, 625–631 (2007).
31. Zemach, A., McDaniel, I. E., Silva, P. & Zilberman, D. Genome-wide evolutionary analysis of eukaryotic DNA methylation. Science 328, 916–919 (2010).
32. Suzuki, M. M. & Bird, A. DNA methylation landscapes: provocative insights from epigenomics. Nat. Rev. Genet. 9, 465–476 (2008).
33. Feng, S., Jacobsen, S. E. & Reik, W. Epigenetic reprogramming in plant and animal development. Science 330, 622–627 (2010).
34. Regev, A., Lamb, M. & Jablonka, E. The role of DNA methylation in invertebrates: developmental regulation of genome defense? Mol. Biol. Evol. 15, 880–891 (1998).
35. Schaefer, M. & Lyko, F. Lack of evidence for DNA methylation of invader4 retroelements in Drosophila and implications for Dnmt2-mediated epigenetic regulation. Nat. Genet. 42, 920–921 (2010).
36. Lyko, F. & Maleszka, R. Insects as innovative models for functional studies of DNA methylation. Trends Genet. 27, 127–131 (2011).
37. Field, L. M. Methylation and expression of amplified esterase genes in the aphid Myzus persicae (Sulzer). Biochem J. 349, 863–868 (2000).
38. Li-Byarlay, H. et al. RNA interference knockdown of DNA methyltransferase 3 affects gene alternative splicing in the honey bee. Proc. Natl. Acad. Sci. USA 110, 12750–12755 (2013).
39. Walsh, T. K. et al. A functional DNA methylation system in the pea aphid. Acyrthosiphon pisum. Insect Mol. Biol. 19, 215–228 (2010).
40. Wang, Y. et al. Functional CpG methylation system in a social insect. Science 314, 645–647 (2006).
41. Bonasio, R. et al. Genomic comparison of the ants Camponotus floridanus and Harpegnathos saltator. Science 329, 1068–1071 (2010).
42. Hung, M. S. et al. Drosophila proteins related to vertebrate DNA (5-cytosine) methyltransferases. Proc. Natl. Acad. Sci. USA 96, 11940–11945 (1999).
43. Mitsudome, T. et al. Biochemical characterization of maintenance DNA methyltransferase DNMT-1 from silkworm, Bombyx mori. Insect Biochem. Mol. Biol. 58, 55–65 (2015).
44. Borsatti, F. & Mandrioli, M. The structure of insect DNA methyltransferase 2 (DNMT2) DNA binding domain is responsible for the non-CpG methylation in insect genomes. Caryologia 57, 305–311 (2004).
45. Bhutani, N., Burns, D. M. & Blau, H. M. DNA demethylation dynamics. Cell 146, 866–872 (2011).
46. Christman, J. K. 5-Azacytidine and 5-aza-2′-deoxycytidine as inhibitors of DNA methylation: mechanistic studies and their implications for cancer therapy. Oncogene 21, 5483–5495 (2002).
47. Mukherjee, K., Tuyman, R. M. & Vilcinskas, A. Insects as models to study the epigenetic basis of diseases. Prog. Biophys. Mol. Biol. 118, 69–78 (2015).
48. Kucharski, R., Maleszka, J., Foret, S. & Maleszka, R. Nutritional control of reproductive status in honeybees via DNA methylation. Science 319, 1827–1830 (2008).
49. Mukherjee, K., Fischer, R. & Vilcinskas, A. Histone acetylation mediates epigenetic regulation of transcriptional reprogramming in insects during metamorphosis, wounding and infection. Front. Zool. 9, 25 (2012).

50. Durdevic, Z. et al. Efficient RNA virus control in Drosophila requires the RNA methyltransferase Dnmt 2. Sci. Rep. 14, 269–275 (2013).
51. Park, B. & Kim, Y. Transient transcription of a putative RNase containing BEN domain encoded in Cotesia plutellae bracovirus induces an immunosuppression of the diamondback moth, Plutella xylostella. J. Invertebr. Pathol. 105, 156–163 (2010).
52. Tamura, K. et al. MEGA5: molecular evolutionary genetic analysis using maximum likelihood, evolutionary distance and maximum parsimony methods. Mol. Biol. Evol. 28, 2731–2739 (2011).
53. Livak, K. J. & Schmittgen, T. D. Analysis of relative gene expression data using real time quantitative PCR and the 2–ΔΔCT method. Methods 25, 402–408 (2001).

CHAPTER 16

Muscular Anatomy of the *Podocoryna carnea* Hydrorhiza

Robert E. Steele

1 Department of Ecology and Evolutionary Biology, Yale University, New Haven, Connecticut, United States of America, 2 Department of Geology and Geophysics, Yale University, New Haven, Connecticut, United States of America

ABSTRACT

The muscular anatomy of the athecate hydroid *Podocoryna carnea* hydrorhiza is elucidated. The polyp-stolon junction is characterized by an opening, here called the chloe, in the otherwise continuous hydrorhizal perisarc. The chloe is elliptical when the polyp first arises, but takes on a more complex outline as multiple stolons anastomose to communicate with that polyp. Surrounding the polyp base are spots, here called anchors, which autofluoresce at the same wavelengths as perisarc and which, like perisarc, contain chitin as assessed by Calcofluor White, Congo Red and wheat germ agglutinin staining. Anchors remain after living tissues are digested using KOH. Collagen IV staining indicates that the mesoglea is pegged to the anchors and rhodamine phallodin staining detects cytoskeletal F-actin fibers of the basal epidermis surrounding the anchors. Longitudinal muscle fibers of the polyp broaden at the polyp base and are inserted into the mesoglea of the underlying stolon, but were neither observed to extend along the stolonal axis nor to attach to the anchors. Circular muscular fibers of the polyp extend into stolons as a dense collection of strands running along the proximal-distal axis of the stolon. These gastrodermal axial muscular fibers extend to the stolon tip.

Epidermal cells at the stolon tip and the polyp bud display a regular apical latticework of F-actin staining. A similar meshwork of F-actin staining was found in the extreme basal epidermis of all stolons. Immunohistochemical staining for tubulin revealed nerves at stolon tips, but at no other hydrorhizal locations. These studies bear on the mechanisms by which the stolon tip and polyp bud pulsate, the manner in which the stolon lumen closes, and on the developmental origin of the basal epidermis of the hydrorhiza.

1. INTRODUCTION

Hydrozoans typically adhere to surfaces by hydrorhizal attachment structures. These may take the form of a basal disc, a system of root-like rhizoids, or an extensive system of stolons. Stolons are diploblastic tubes suspended from an encasing acellular perisarc and bound to the surface by basal epidermal rivets [1]. The stolonal lumen is co-extensive with that of the gastrovascular cavities of the polyps of the colony. In many encrusting forms the stolons elongate, branch, and anastomose to form a complex gastrovascular network. *Podocoryna carnea*, the object of this investigation, is such a hydroid.

Hydrorhizal tissues display coordinated patterns of movement. Fluids within the gastrovascular hydrorhizal network are driven by contractions of the polyps [2]–[5]. The radius of the stolonal lumen undergoes cycles of expansion and contraction, as fluids enter from the polyp and return to it. In addition to the bi-directional gastrovascular flows, the stolon tip and the polyp bud display growth pulsations [6]–[10]. The tip undergoes rhythmic changes in shape as it progresses over the surface [11], [12], as does the bud as it develops from a rudiment to a polyp [12]–[14]. Severed stolons continue to elongate and isolated buds continue to develop so are driven by motility systems distinct from the polyp contractions that drive gastrovascular flows [8], [9], [15], [16].

Curiously, the physiological and anatomical basis for these movement patterns have been little investigated. We have sought to ask whether the polyp-stolon junction is morphologically specialized, whether the cells of the stolons bear muscle fibers and, if so, how they are related to muscle fibers of polyp epitheliomuscular cells, and finally, whether the stolons bear specializations at the growing tip and polyp bud. Our principal approach has been the use of confocal microscopy with phalloidin-linked fluorescent dyes labeling F-actin to visualize muscle fibers and cytoskeletal features, supplemented by immunohistochemical stains to visualize mesoglea and nerves. We find the polyp-stolon junction to be anatomically simple, the stolon to bear extensive muscular fibers co-extensive with the circular muscle fibers of the polyps and the apical regions of tip and bud epidermal cells to support a distinctive F-actin meshwork reminiscent of adherens junction rings.

2. METHODS

Animal care at Yale University is conducted under the supervision of the Yale Animal Care and Use Committee. Hydroids are diploblastic organisms lacking a brain and are not governed by specialized guidelines. Animals sacrificed for confocal microscopy were clonal explants of animals maintained in continuous culture and were narcotized prior to fixation as detailed below.

Colonies of *Podocoryna carnea* (P3 strain) were maintained under standard conditions [17]. Briefly, colonies are grown on glass microscope slides, glass cover slips, or glass-bottomed Petri dishes. Clonal replicates are generated by explanting a small region of the hydrorhiza bearing 1–3 polyps and affixing them to a glass surface with a loop of quilting thread. After 2 days the colonies have attached and the thread is removed. Colonies are maintained in recirculating aquaria with daily exchanges of 25% of the seawater (31 ppm). Colonies are fed to repletion every other day with 3–4 day old *Artemia salina* nauplii.

To prepare specimens for microscopy, colonies growing on cover slips were relaxed in 2% urethane (Sigma) for 90 seconds, then fixed in 4% formaldehyde for 2 hours at room temperature. After fixation, colonies were washed 3×5 minutes in 1× PBS, and permeabilized in 0.5% Triton in 1× PBS for 10 minutes at room temperature. Colonies were washed an additional 3×5 min in 1× PBS, after which they were blocked with 3% dry milk in 1× PBS for 1 hour at 4 degrees C. To stain for F-actin, colonies were washed once in 1× PBS and stained with a 1:40 dilution of rhodamine phalloidin (Biotium #00027, Hayward CA) for 30 minutes at room temperature. After 2×5 min washes in 1× PBS the coverslips bearing colonies were mounted inverted on microscope slides in ProLong Gold (Invitrogen, Eugene OR), supported on two thin strips of double stick tape.

To detect mesoglea, we stained for collagen IV using a mouse monoclonal antibody (M-39) kindly provided by Xiaoming Zhang [18], [19]. Nerves were visualized by detecting tubulin using a 1:1 mixture of mouse monoclonal supernatants 12 g10 and E7 (Developmental Studies Hybridoma Bank, University of Iowa, Iowa City, IA). Colonies were treated as with rhodamine phalloidin staining up to and including the permeabilization step. Thereafter colonies were blocked with 10% goat serum in 1× PBS for 1 hour at room temperature. Blocking solution was replaced with type IV collagen antibody at 1:400 in 10% goat serum/1×PBS or the tubulin antibody mixture at 1:10 and incubated overnight at 4 degrees C. The colonies were then washed 3×5 minutes in 1× PBS, and incubated with secondary antibody (1:80 FITC goat anti mouse, Invitrogen) with or without rhodamine phalloidin (1:100, Biotium) in a 10% goat serum/1× PBS solution for 2 hours at room temperature. After 2×5 min washes in 1× PBS the coverslips were mounted as described above.

In some preparations nuclei were stained prior to mounting using Hoechst 33342 (Molecular Probes, 1:200) or DAPI. Cover slips were immersed in Hoechst stain for 20 minutes followed by 2×5 min 1× PBS washes. For DAPI staining, a ProLong Gold product containing DAPI was used.

Unless otherwise noted images were captured on a Zeiss LSM 500 laser scanning confocal microscope. Confocal images were acquired and processed using either Zeiss Axiovision or Zeiss Efficient Navigation (ZEN) software. Projection and three-dimensional rendering were performed using this same software. Optical sections were made through 152 polyp-stolon junctions, 149 stolons, and 18 polyp buds.

Perisarc was studied in animals from which tissues were removed by KOH digestion. To remove tissue, colonies were relaxed for 90 sec. in 2% urethane (Sigma) and placed in dH_2O for 10 minutes in a Coplin jar. Colonies are then placed horizontally in petri dishes and digested for 10–15 minutes in 10% KOH with intermittent gentle rinsing of surface with the KOH solution using a Pastuer pipette until tissue has digestion was complete (ca. 10–15 min). Slides were then washed 3×2 minutes in dH_2O. Perisarc autofluorescence was documented at blue (Ex., 760–780 in 2-photon configuration, chameleon IR laser, no pinhole, Em. 435–485,), green (Ex., 488 nm, Em. 500–550 nm), and red (Ex., 561 nm, Em. 575 nm) wavelengths.

Perisarc contains chitin [20], [21]. Three conventional chitin stains were employed for chitin staining. Calcofluor White binds to chitin, cellulose and other β-1,4-linked carbohydrates [22], [23]. To stain using Calcofluor White, KOH treated slides were placed, inverted, over ca. 1 ml of 1 g/l solution of Calcofluor White solution (Fluka Analytical, Sigma) in a glass-bottomed petri dish (Warner Instruments). Following a one minute incubation, stolons were immediately imaged using a DAPI filter set (Ex 365 nm, Em 450 nm) on a Zeiss Axiovert microscope. Congo Red stains chitin and collagen [24]–[26]. Perisarc does not contain collagen [27]. KOH treated colonies were stained overnight at room temperature in a Coplin jar containing a filtered 1.5 mg/ml solution of Congo Red (Sigma) in dH_2O. Coverslips were rinsed until no further staining solution could be seen in the wash, and imaged in dH_2O on a Zeiss LSM 500 confocal microscope. Finally, wheat germ agglutinin binds to the N-acetyl-D-glucosamine residues of chitin. To stain with wheat germ agglutinin, slides were incubated for 20 minutes at room temperature in 25 µg/ml WGA-AlexaFluor 555 (Invitrogen) in PBS, washed 3×5 minutes in PBS, imaged in dH_2O using a Texas Red filter set (Ex 560 nm, Em 645 nm) on a Zeiss LSM 500 confocal microscope.

We found it useful to image hyperplastic stolons. These are specialized stolonal tips that are differentiated in either xenogeneic encounters with *Hydractinia symbiolongicarpus* or allogeneic encounters with unrelated *P. carnea* colonies [28]–[30]. The tip region is extended and the encasing periderm reduced in hyperplastic stolons [29], offering the opportunity to image tip-specific features more fully. To study hyperplastic stolons, we established clonal explants of *P. carnea* and *H. symbiolongicarpus* on a common surface, which induced development of hyperplastic stolons for imaging.

3. RESULTS

The Polyp-Stolon Junction

3.1 The chloe

Polyps emerge from a hole in the otherwise continuous perisarc. This opening appears to lack a name and, since we will have occasion to refer to it frequently, we here designate this opening as the chloe. In young polyps the chloe was found to be elliptical in outline (Fig. 1A), with the major axis aligned with the proximal-distal axis of the stolon. In older polyps, where stolon branches have grown into locations near a polyp's base, the chloe appeared as an opening communicating with both the original polyp and lumen(s) of adjacent stolons

3.2 Anchors

A ring of structures, here called anchors, is invariably found underlying polyps (Fig. 1D–F). Anchors are circular in outline (12.0 +/− 2.4 μm^2, n= 145 in area) and extend to a height of 3.5 +/− 0.7 µm (n=207). In regions where the chloe encompasses a polyp as well as adjacent stolons, anchors are found both in the elliptical pattern found in the original polyp, and in a pattern ringing the perimeter of stolonal regions underlying the chloe (Fig. 1E, F). Anchors are attached to the substratum and are not continuous with the hydrorhizal walls. When living tissues are removed by treatment with KOH, the anchors remain. Anchors autofluoresce with the same intensity as perisarc in all three wavelength bands measured (Figure S1A–D).

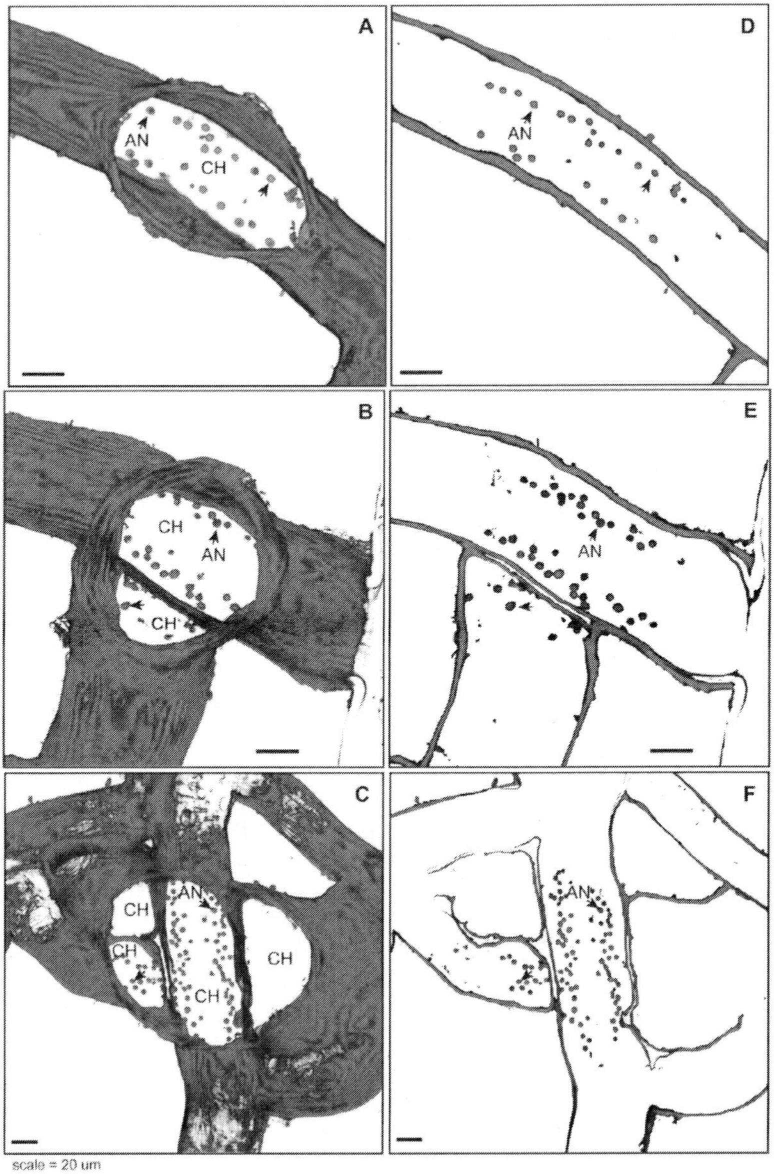

Figure 1. Chloe and anchors. Perisarc stained with WGA, tissues KOH digested. (A–C) Projections of chloe from polyps of differing age and connectivity with surrounding hydrorhiza. Optical depth, number of sections: (A) 32 μm, 17; (B) 34 μm, 18; (C) 38 μm, 20. (D–F) Optical section of same polyp-stolon junctions shown in (A–C) at ca. 3 μm above the substratum, showing anchors (arrowheads). (A–F) AN, anchors; CH, Chloe, Scale: 20 μm.

In KOH digested specimens, Calcafluor White (Figure S1E), Congo Red (Figure S1F), and wheat germ agglutinin (Fig. 1) were found to stain the anchors. Anchors were not found at the base of the youngest polyp buds.

3.3 Muscle fibers at the polyp stolon junction

No specialized valves appear at the polyp-stolon junction (Fig. 2A). Optical sections just above the chloe reveal the circular muscle fibers of the polyp as the expected continuous ring internal to regularly spaced longitudinal muscle strands (Fig. 3A). At the level of the gastrovascular canal of the stolon, the circular muscle fibers of the polyps extend to run as axial strands along the proximal-distal axis of the stolon (Fig. 3B,C). In polyp buds, longitudinal fibers are absent, but circular muscle fibers are distinct and continuous with the axial fibers of the underlying stolon (Fig. 3D). Optical sections of the polyp-stolon junction of a mature gastrozooid and of a polyp bud are available as Movies S1 and S2, respectively.

The longitudinal muscle strands of the polyp broaden into fans at the chloe (Fig. 3E). A 3D rendering of optical sections of longitudinal muscle strands double stained for mesogleal collagen IV shows that the strands terminate in the mesoglea of the stolon (Fig. 2C). In the majority of cases longitudinal muscles strands can only be visualized in apical regions of the stolon. Infrequently muscle fibers were found to extend to the basal side of the stolonal gastroderm, but in no case did they extend through the epidermis to either the anchors or the substratum (Fig. 3F,G). The longitudinal muscle strands of the polyps have not been seen to extend axially along the length of the stolon.

3.4 The Stolon

F-actin staining of the stolon revealed extensive gastrodermal muscle strands (Fig. 2E), terminating at the stolonal tip (Fig. 2F). Thick fibers run along the proximal–distal axis of stolon radially or helically distributed around the stolonal lumen Finer fibers crisscross the axial bands. Muscle fibers are continuous at stolonal branch points (Fig. 2B). In addition to the axial muscle fibers, F-actin staining was observed in a polygonal pattern in the regions of epidermal cells where the cells contact the substratum (Figs. 2E, 4A,D). This polygonal staining was found in all stolons at all locations in the colony.

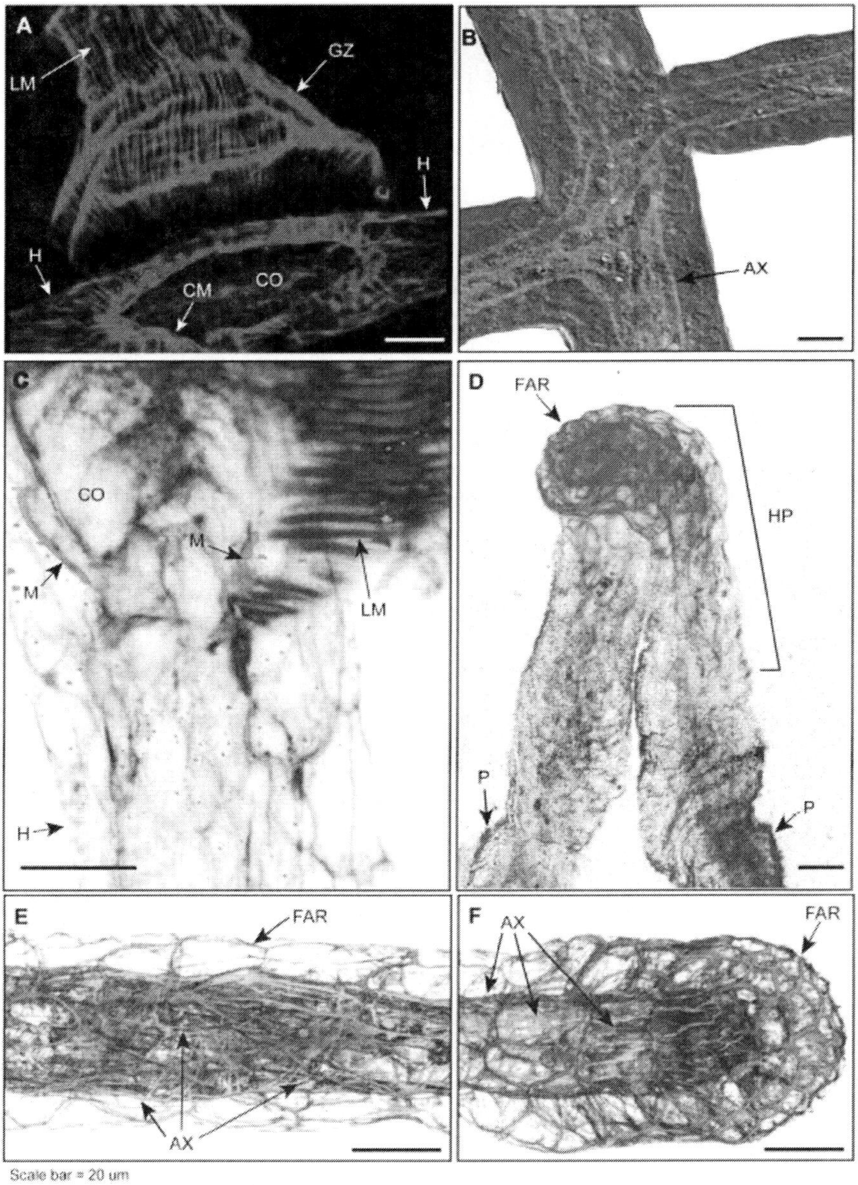

Figure 2. Stolons, hyperplastic stolons and the polyp-stolon junctions. (A) Projection of phalloidin stained gastrozooid, showing longitudinal muscle fibers and circular muscle fibers, the hydrorhiza from which the polyp emanates, and the location of the chloe. Note that polyp-stolon junction is a simple aperture, lacking specialized valves. (B) A portion of the hydrorhiza showing continuity of axial muscle fibers at two stolonal

junctions. Differential interference contrast, grey, axial muscle fibers stained with phalloidin, red, and nuclei, DAPI stained in blue. (C) Projection of polyp stolon junction at the chloe showing longitudinal muscle strands of gastrozooid labeled with phalloidin, red, and mesoglea labeled with antibodies specific for Hydra collagen IV, green. Note that longitudinal muscle strands do not traverse mesoglea. (D) Projection of two hyperplastic stolons, labeled with phalloidin, red. The perisarc, blue, was visualized by autofluorescence. Note extensive meshwork of F-actin rings at tip. (E) Depth coded projection of phalloidin stained stolon at region proximal to the stolon tip, showing F-actin rings of the basal ectodermal cells where they adhere to the substratum (see also Figs. 4A,D) and gastrodermal axial muscle fibers. (F) Depth coded projection of phalloidin stained stolonal tip showing terminus of gastrodermal axial muscle strands and F-actin rings at stolonal tip. (E, F) Color code: red ca. 0–2 μm from substratum; orange 3–6 μm, yellow-green 7–10 μm, blue >10 μm. Optical depth, number of sections: (A) 27 μm, 14; (C) 8 μm, 10; (D) 40 μm, 10; (E) 17 μm, 43; (F) 18 μm, 25. (A–F) Scale: 20 :m. AX, axial muscle fibers; CO, opening of chloe; CM, circular muscle fibers; FAR, F-actin ring; GZ, gastrozooid; H, hydrorhiza; HP, hyperplastic stolons; LM, longitudinal muscle fibers, M, mesoglea, P, perisarc.

The relationship of anchors to F-actin fibers and to the mesoglea was explored. In optical sections nearest the substratum, F-actin fibers are seen in polygonal pattern (Fig. 4A). F-actin cytoskeletal fibers also extend within basal epidermal cells where they are sometimes found to encircle the anchors (Fig. 4B, Movie S3). Optical sections stained for collagen IV indicate that mesoglea lies above the basal epidermal F-actin staining and that anchors are pegged to the mesoglea (Fig. 4C, Movie S4).

3.5 The Stolon Tip and Polyp Bud

The apical surfaces of epidermal cells of both the stolon tip and the polyp bud were found to stain positively for F-actin. Three-dimensional rendering of optical sections revealed a distinctive F-actin meshwork at the tip and bud, comprised of polygons of interconnected actin fibers (Fig. 5A,B, Movies S5 and S6). This meshwork, which is not evident elsewhere, was found to cap the entire polyp bud and to surround the stolon tip from the apical upper epidermis to the lower basal epidermis. Hyperplastic stolons, those induced by allogeneic or xenogeneic encounters, bear a similar F-actin ring meshwork (Fig. 2D). The hyperplastic stolons differ from free growing tips in two respects. The perisarc does not extend to the hyperplastic stolonal tip and the distinctive cytoskeletal epidermal architecture extends proximally to a much larger extent.

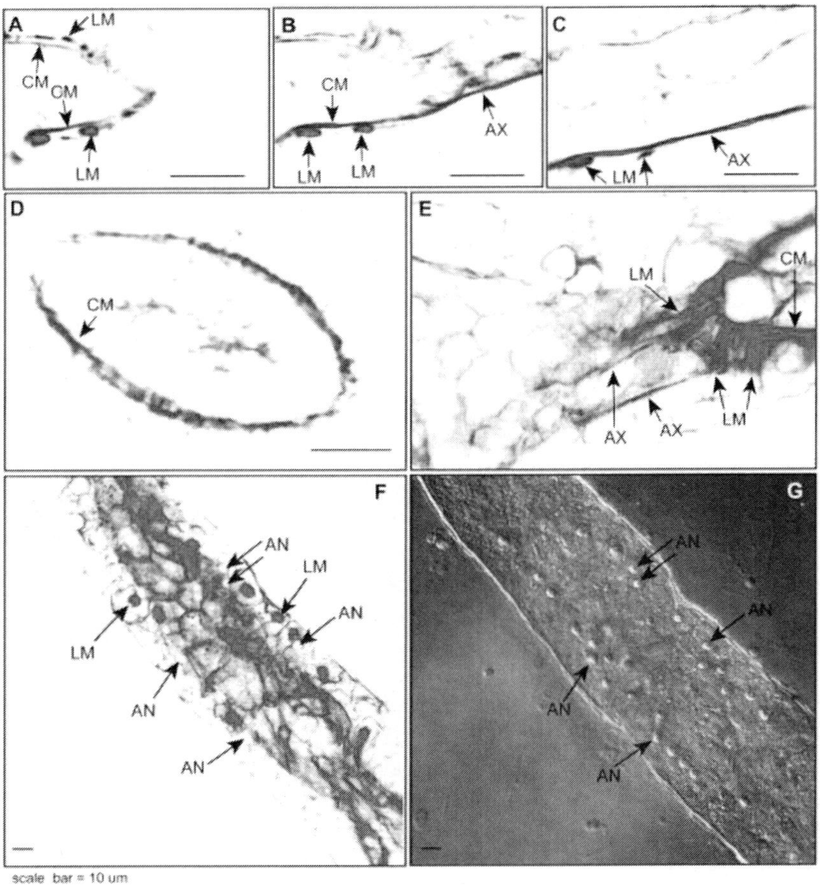

Figure 3. Polyp-stolon junction. (A–C) Phalloidin stain of the same stolon junction shown at (A) 20 μm, (B) 16 μm, (C) 8 μm above substratum, showing the disappearance of epidermal longitudinal muscle fibers of the polyp and the extension of gastrodermal circular muscle fibers of the polyp into the axial muscle strands of the stolon. (D) Optical section of a phalloidin stained polyp bud showing gastrodermal circular muscle fibers, but no longitudinal muscle fibers. (E) Projection of phalloidin stained muscle fibers at the base of a gastrozooid. Note that longitudinal muscle strands of the polyp project into broad fans at the junction with the proximal-distal axis of the stolon. (F) Projection of unusual phalloidin stained polyp-stolon junction showing longitudinal muscle fibers extending to the base of gastroderm. Position of anchors indicated. (G) Differential interference contrast (DIC) ιμαγε of the same stolon as (F) showing that the muscle strands labeled in (F) do not overly anchors. Optical depth, number of sections: (E) 1.3 μm, 3; (F) 2.7 μm, 3. Scale: 10 μm. AN, anchors; AX, axial muscle fibers; CM, circular muscle fibers; LM, longitudinal muscle fibers.

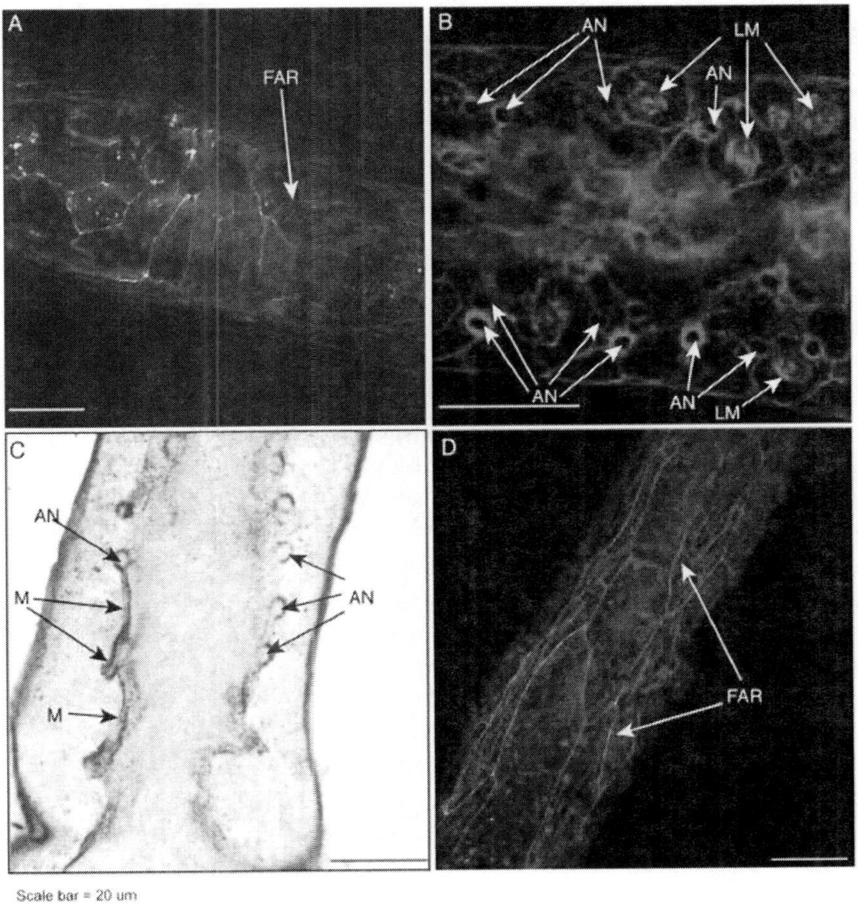

Figure 4. Basal epidermal actin, anchors, and mesoglea. (A) Phalloidin stained F-actin rings in the basal epidermis beneath a polyp-stolon junction. (B) Phalloidin stained F-actin fibers surrounding some, but not all, anchors. Diffuse staining in the center of stolon is gastrodermal. (C) Immunohistochemical staining of collagen IV showing mesoglea (M) overlying and surrounding anchors. Optical depth, number of sections: 3 μm, 4. (D) Phalloidin stained F-actin rings in basal epidermis of stolon at positions proximal to the stolon top. Scale: 20 μm. AN, anchors; FAR, F-actin ring; LM, longitudinal muscle strands; M, mesoglea.

3.6 Nerves

Tubulin staining revealed the stolonal tip is enervated (Fig 5C,D). Nerves were found in the gastroderm in a region just proximal to the tip when the tip is an expansion phase (Fig. 5C) and abutting the tip at the depth of the retraction phase. Nerves are found to extend proximally for a variable distance (mean= 76.3 +/− 45.1 μm, n=10) from the tip. In cases of new stolon branches, nerves have been observed to extend from the tip into the parental stolon (Fig. 5D). Nerves were not observed in other sections of the hydrorhiza under the conditions used to visualize nerves at the stolon tips.

4. DISCUSSION

4.1 Anatomical Observations

Phalloidin staining of the hydrorhiza of *Podocoryna carnea* has revealed an extensive network of hydrorhizal muscular strands. With one exception [31], all earlier attempts have failed to detect muscle fibers in stolons [6], [15], [32]–[35]. The failure to detect muscle fibers, including in a species closely allied to the species we have observed [32], likely speaks to the far greater resolution provided by the use of confocal microscopy.

The stolonal musculature was found to be continuous with the circular muscle strands of the polyp and to extend in a series of gastrodermal axial fibers terminating in blind ends at stolonal tips. Our observation that the polyp bud has circular muscle fibers suggests that the circular muscle strands of newly budded polyps derive from the stolonal musculature. We were unable to unambiguously detect epidermal muscle fibers in the stolon.

We find that that the longitudinal muscle strands of the polyp terminate in the mesoglea of the underlying stolon. These muscle fibers widen at their base. If circular muscle strands were to contract at this location, this arrangement would serve to close off the gastrovascular cavity of the polyp from that of the stolon at the chloe. That such closure can be accomplished is known from the fact that rapid polyp contractions made by hydroids when perturbed are volume-preserving [3], [36]. Prior authors have assumed that the epidermal longitudinal muscle fibers of the polyps were pegged to the substratum via the anchors [37], presumably reasoning that the muscular contraction must be resisted in some fashion and the surface provides the most obvious solution. Our finding that the longitudinal fibers of the polyp terminate in the mesoglea would at first glance seem mechanically implausible. Yet stolons are fluid-filled and when the longitudinal muscle strands of the polyp are contracting, gastrovascular fluid is being ejected from the polyp into the stolon, so the stolon is fully expanded [38]. As water is essentially incompressible, a hydrostatic skeleton opposes the longitudinal muscle fibers.

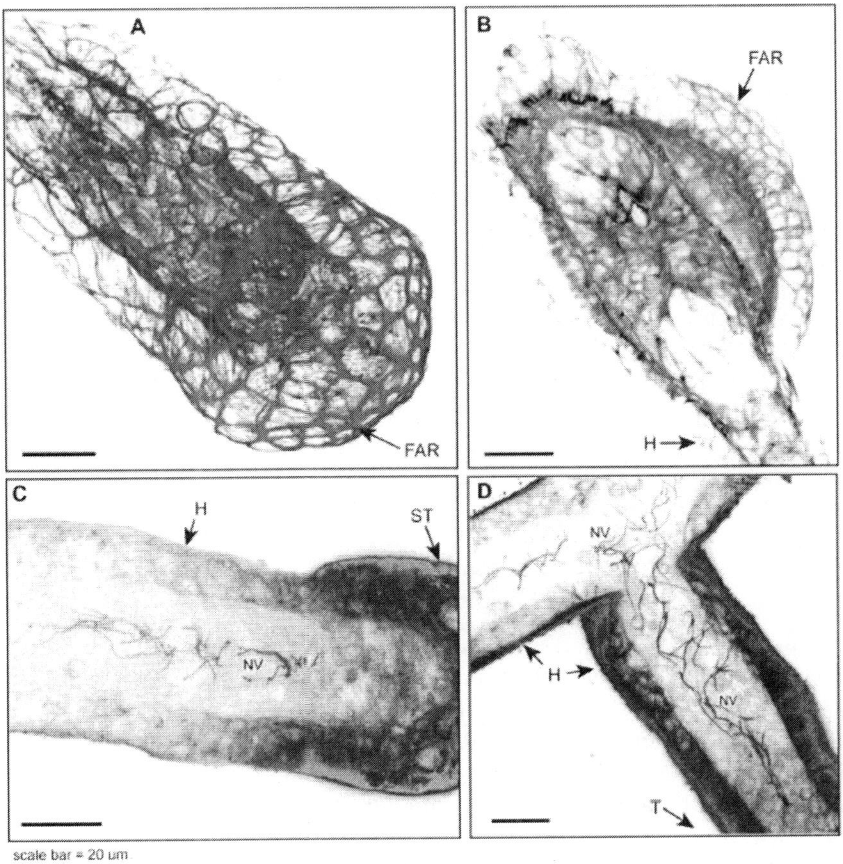

Figure 5. Stolon tip, polyp bud, and nerves. (A) Three-dimensional (3D) reconstruction of phalloidin stained stolonal tip showing F-actin ring structures in apical region of tip epidermal cells. (B) 3D reconstruction of polyp bud showing epidermal F-actin rings. Color code for panel B: red ca. 0–2 μm from substratum; orange 3–6 μm, yellow-green 7–10 μm, blue >10 μm. Projection of gastrodermal nerves, detected immunohistochemically, in locations (C) just proximal to the stolon tip and (D) extending into mother stolon from which a stolon tip had recently developed. Optical depth, number of sections: (A) 22 μm, 52; (B) 20 μm, 11; (C) 8 μm, 11; (D) 6 μm, 9. Scale: 20 μm. FAR, F-actin ring; H, hydrorhiza; NV, nerves; ST, stolon tip.

Our observations indicate that the junction between the polyp and the stolon lacks valves. Rather the junction is simply a blind hole at the base of the polyp. Observations of polyps of different sizes show that the chloe is remodeled as stolons grow into proximity with the polyp. This remodeling requires that the apical perisarc of adjacent stolons is digested to provide a conduit from the stolon to the polyp, which accords with the finding in *Hydractinia* that the basal epidermis of polyps express chitinase [39].The appearance of additional anchors circumscribing regions of adjacent stolons that have come to empty into a single chloe indicates that the polyp base is also remodeled as the polyp enlarges.

The anchors observed at the polyp stolon junction have been observed by others [37], but ours is the first set of anatomical observations. The anchors are not living tissue, as indicated by their presence following KOH digestion. The fact the anchors autofluoresce with the same intensity as does perisarc and that, like perisarc, they stain positively for chitin using Calcofluor White, Congo Red, and wheat germ agglutinin suggest the anchors are localized perisarc secretions. Optical sectioning of a large number of polyps (n=152) did not reveal a single instance of either the axial muscle strands of the stolonal gastroderm or the longitudinal muscle strands of the polyp attaching to the anchors. Indeed, the only F-actin staining found in association with the anchors were cytoskeletal fibers of the basal epidermis that were sometimes found encircling anchors. Our observations of collagen IV staining at and around the apical surface of the anchors indicate that these locations serve to peg the mesoglea to the surface.

The epidermal cells of the hydroid stolonal tip and polyp bud undergo regular oscillations in orientation and volume as the stolon or bud elongates [11]–[14]. The oscillations are a property of the tip and bud tissues, as isolated stolons or buds continue to display essentially similar movements [8]–[11], [40]. We have found that the tip and bud regions are characterized by an extensive intracellular meshwork of F-actin fibers that border and encase the apical margin of epidermal cells. The fact that the meshwork is bound into a coherent multicellular structure is suggestive of adherens junctions. Early electron micrographs indicate cytoplasmic extensions to junctions at the apical margin of stolon tips in *Cordylophora* (Figure 3 in Overton, 1963). Further support for this interpretation is found in the fact that cnidarian genomes include coding sequences for genes similar to the β-catenin, adherens junction protein p120, and cadherins necessary to support adherens junctional rings [41], [42].\

4.2 Functional Implications

Our anatomical observations bear on three distinct functional issues.

4.3 Tip and bud pulsations

The motive forces governing the regular oscillations in tip and bud cell orientation remain unknown. Beloussov and colleagues have showed that the apical surface of tip epidermal cells move as a coherent unit and that changes in cell orientation are largely determined by changes on the angle subtended between the mesoglea and the lateral margins of tip and bud epidermal cells [12]–[14]. Campbell [43], working with the holdfasts of *Corymorpha*, first noted that tip locomotion requires that apical surfaces of the tip cells have a resisting surface and suggested that the perisarc, secreted at the tip, served this purpose. Yet perisarc at the growing tip is not yet hardened and we suggest that the adherens junctional ring may serve this role. The interpretation is supported by the observation that hyperplastic stolons possess the extensive F-actin meshwork, yet appear to secrete minimal perisarc at their tips. Moreover, in athecate hydroids such as the *Podocoryna carnea* colonies we study here, polyps are free of perisarc, and yet polyp buds undergo pulsations.

Early studies showed that cytochalasin B eliminates tip pulsations [10], suggesting a role for F-actin mediated movement. Adherens junction rings are known to be contractile in a number of systems [44]–[47]. If the tip and bud adherens junctional rings are contractile they may provide the motive force responsible for pulsations. This suggestion can be directly tested by treating growing tips and polyp buds with the myosin II specific inhibitor blebbostatin.

4.4 Lumen closure

The stolonal lumen changes in radius as fluids are pumped into and out of polyps [2]–[5]. In the youngest peripheral stolons, the lumen may close completely while in older, more central parts of a colony the lumen oscillations are of smaller amplitude. We are aware of only one attempt to explain how the lumen closes. Schierwater, et al [32] observed that the apical surface of gastrodermal digestive cells are densely populated by structures they called 'zipper-like organelles' (ZLO). They proposed that ZLO's acted as contractile vacuoles that open to admit gastrovascular fluid, thereby expanding the volume of gastrodermal cells and closing the lumen [32]. However, the ZLO's of Schierwater et al [32] are identical in ultrastructural detail with that of discoidal coated vesicles, a feature previously described in hydroid digestive cells [48]. Discoidal coated vesicles act to increase the surface area for absorption. When these vesicles are internalized within the cell they do so without an accompanying change in cell volume beyond that of the nutrients adhering to the coated vesicle surface [48]. Another fact mitigating against the interpretation of Schierwater, et al [32] is the simple observation that fluid is readily observed moving axially as the stolon lumen closes. If the fluid were absorbed into the gastrodermal cells, lumens would not flush axially.

The opening and closing of the lumen need not be an active process. The gastroderm and mesoglea could be behaving as passive elastic materials, expanding when lumen pressure, generated by polyp pumping, exceeds the some resting radius and closing to this resting radius when lumen pressure declines. However, anecdotal observations have been made of stolon branches with closed lumens communicating directly with parent stolons within which gastrovascular fluid is actively flowing. This can only be accounted for if some active process is keeping the stolon branch closed.

Our observations suggest such an active mechanism for lumen closure. If the axial gastrodermal muscle fibers were to contract, gastrodermal cells would be expected to change shape from rhomboidal to columnar with the consequence that the lumen itself would close to the extent that hydroplasmic pressures in the lumen permitted. When a stolon is filling, one or more polyps are pumping fluid into the hydrorhizal network. The pumping of polyps requires the contraction of longitudinal muscle strands and presumably relaxation of some circular muscle fibers. If the state of the circular muscle fibers at the base of a pumping polyp were transmitted to the adjacent stolon, then the axial muscle strands of the stolon would be expected to relax as the polyp empties and the stolon fills. When gastrovascular flow reaches the tip, further expansion of the stolon is no longer possible and the axial fibers of the stolon contract, returning fluid to the polyp in response to which longitudinal muscle strands of the polyp relax to accommodate the incoming fluid. Thus the pattern of polyp driven volume oscillations in hydroplasmic flows is consistent with the suggestion that lumen closure involves muscular contraction. Finally, if muscular contraction is responsible for the shape changes in gastrodermal cells, the entire gastrodermal epithelia must shorten when the stolon lumen empties. Studies of tip pulsations show just such a retraction phase, with the maximal extent of retraction occurring when the stolon lumen has emptied [16].

Our results have localized nerves to stolonal tips. Severing of a tip results in the immediate closure of the stolonal lumen and its disuse until the tip is fully reconstituted and active (unpublished observations). If adherens junctional ring contractions do drive tip pulsations, the nerves are not likely involved in these tip movements, since contraction of adherens junctional rings is driven by Ca^{2+}–calmodulin dependent myosin light chain kinase phosphorylation of myosin II [49]–[52]. Rather, we suggest that the nerves found at stolon tips trigger the stolon axial muscle fiber contraction, initiating return flow to polyps.

An additional role of stolonal muscle fibers may be to increase the rigidity of the mesoglea in the axial direction and, hence reduce any tendency of the stolon lumen to flutter or locally collapse as the water returns from the stolons to the polyps.

4.5 Developmental origin of the basal epidermis and of the anchors

The developmental origin of the basal epidermis of the hydroid hydrorhiza seems not to have been previously addressed. The hydrorhiza is anchored to the substratum by the differentiation of desmocytes which secrete rivets that attach the underlying surface to the overlying mesoglea, after which the desmocyte dies and the rivet lies between adjacent basal epidermal cells [1], [53], [54]. Rivets form on the basal epidermis in regions proximal to the tip. This anchoring arrangement would seem to preclude mitotic activity in the basal epidermis in the immediate regions proximal to the tip. Cell marking studies, conducted on the time scale of seconds up to 20 hours, show that epidermal cells at the tip remain stationary, while apical and lateral cells proximal to the tip region but distal to the youngest polyp move toward the tip [9], [11], [55]. If the basal epidermis is not growing by mitosis in regions proximal to the tip and if epidermal cells on the flanks and apical surfaces of the stolon are moving distalwards, a plausible origin of the basal epidermis is that upper and lateral epidermis is transformed, over a time scale of days, into basal epidermis with the same motion as a track on a treadmill. This model is reminiscent of the well-known tissue dynamics of the polyp epidermis. The polyp is comprised of a two radially symmetric, continuously moving and differentiating cell sheets. Classical studies in *Hydra* have shown that mitosis is concentrated at mid-gastric regions and new cells move toward the hypostome and tentacles above this zone and towards the foot below it, with cells being shed at the termini [56]–[58]. Since stolon tips have not been observed to lose cells, we suggest that epidermal cells on stolon flanks and apical regions become basal epidermal cells as they round the tip. Our suggestion that the basal epidermis is derived from apical and lateral epidermis is further supported by the observation that the substratum-facing surface of basal epidermal cells displays the same putative adherens junctional rings characteristic of the tip. The 'treadmill' hypothesis can be directly tested by marking cells and following their fate over the time course of days.

The anchors observed here suggest another role for the basal stolon ectoderm. We speculate that anchors arise as do rivets [1], [53], [54]. Rivets are secretions of specialized basal ectodermal cells called desmocytes, which, like anchors, serve to link the mesoglea to the substratum. Any correspondence likely ends there, as anchors are peridermal secretions and rivets are not. Our observation that the youngest polyp buds lack both anchors and longitudinal muscles tempt us to suggest the possibility that longitudinal muscle maturation may generate mechanical forces that induce the basal ectodermal epithelium to secrete anchors.

ACKNOWLEDGMENTS

We thank J. Rimas Vaisnys for extensive discussions that motivated much of this work, Evan Buss for animal care, Matthew Nicotra, Steve Dudgeon, and Mark Mooseker for discussion or comments, and Xiaoming Zhang for providing antibodies to *Hydra* collagen IV.

REFERENCES

1. Chapman DM (1969) The nature of cnidarian desmocytes. Tissue Cell 1: 619–632 Available: http://www.ncbi.nlm.nih.gov/pubmed/18631489.
2. Hammett FS, Padis N (1935) Correlation between developmental status of *Obelia* hydranths and direction of hydroplasmic streaming within their pedicels. Protoplasma 23: 81–87.
3. Dudgeon S, Wagner A, Vaisnys JR, Buss LW (1999) Dynamics of gastrovascular circulation in the hydrozoan *Podocoryne carnea*: The 1-polyp case. Biological Bulletin 196: 1–17.
4. Blackstone NW (1996) Gastrovascular flow and colony development in two colonial hydroids. Biological Bulletin 190: 56–68.
5. Van Winkle DD, Blackstone NW (1997) Video microscopic measures of gastrovascular flow in colonial hydroids. Invertebrate Biology 116: 6–16.
6. Berrill NJ (1949) The polymorphic transformations of *Obelia*. Quarterly Journal of Microscopical Science 90: : 235–264. Available: <Go to ISI>://A1949XR00700002.
7. Berrill NJ (1949) Growth and form in gymnoblastic hydroids .1. Polymorphic development in *Bougainvillia* and *Aselomaris*. Journal of Morphology 84: : 1–30. Available: <Go to ISI>://A1949UV17900001.
8. Wyttenbach CR (1968) The dynamics of stolon elongation in the hydroid, *Campanularia flexuosa*. Journal of Experimental Zoology 167: 333–351.
9. Hale LJ (1964) Cell movements, cell division and growth in the hydroid *Clytia johnstoni*. Journal of Embryology and Experimental Morphology 12: 517–538.
10. Donaldson S (1974) Terminal motility in elongating stolons of *Proboscidactyla flavicirrata*. American Zoologist 14: : 735–744. Available: <Go to ISI>:// A1974 T529000026.
11. Wyttenbach CR (1974) Cell movements associated with terminal growth in colonial hydroids. American Zoologist 14: 699–717.
12. Beloussov LV, Labas JA, Kazakova NI, Zaraisky AG (1989) Cytophysiology of growth pulsations in hydroid polyps. Journal of Experimental Zoology 249: 258–270.
13. Beloussov LV (1973) Growth and morphogenesis of some marine Hydrozoa according to histological data and time-lapse studies. Publications of the Seto Marine Laboratory 20: 315–336.
14. Beloussov LV, Badenko LA, Kurilo LF, Katchuri.Al (1972) Cell movements in morphogenesis of hydroid polypes. Journal of Embryology and Experimental Morphology 27: : 317–337. Available: <Go to ISI>://A1972M196500004.
15. Hale LJ (1963) Contractility and hydroplasmic movements in the hydroid *Clytia johnstoni*. Quarterly Journal of Microscopical Science 101: : 39–51. Available: <Go to ISI>://A1960WV88900010.
16. Wyttenbach CR (1973) The role of hydroplasmic pressure in stolonic growth movements in the hydroid, *Bougainvillia*. Journal of Experimental Zoology 186: 79–90.
17. Blackstone NW, Buss LW (1991) Shape variation in hydractiniid hydroids. Biological Bulletin 180: : 394–405. Available: <Go to ISI>://A1991FQ68500007.

18. Zhang X, Sarras MP (1994) Cell-extracellular matrix interactions under in vivo conditions during interstitial cell migration in Hydra vulgaris. Development 120: 425–432.
19. Shimizu H, Aufschnaiter R, Li L, Sarras MP, Borza D-B, et al. (2008) The extracellular matrix of Hydra is a porous sheet and contains type IV collagen. Zoology 111: 410–418.
20. Brown CH (1975) Structural Materials in Animals: Coelenterata. New York: John Wiley & Sons.
21. Tidball JG (1984) Cnidaria: Secreted surface. In: Bereiter-Hahn J, Matolsky AG, Richards KS, editors. Biology of the Integument. Berlin: Springer-Verlag, Vol. 1 , Inverte. 69–78.
22. Hughes J, McCulley ME (1975) The use of an optical brightner in the study of plant structure. Stain Technology 50: 319–329.
23. Maeda H, Ishida N (1967) Specificity of binding of hexopyranosyl polysaccharides with fluorescent brightner. Journal of Biochemistry 62: 276–278.
24. Bely M, Makovitzky J (2006) Sensitivity and specificity of Congo red staining according to Romhanyi. Comparison with Puchtler's or Bennhold's methods. Acta Histochemica 108: 175–180.
25. Elghetany MT, Saleem A, Barr K (1989) The Congo Red stain revisited. Annals of Clinical Laboratory Science 19: 190–195.
26. Michels J, Buntzow M (2009) Assessment of Congo red as a fluorescence marker for the exoskeleton of small crustaceans and the cuticle of polychaetes. Journal of Microscopy 238: 95–101.
27. Chapman G (1973) A note on the composition of some coelenterate exoskeletal materials. Comparative Biochemistry and Physiology (B) 45: 279–282.
28. McFadden CS (1986) Laboratory evidence for a size refuge in competitive interactions between the hydroids *Hydractinia echinata* (Flemming) and *Podocoryne carnea* (Sars). Biological Bulletin 171: : 161–174. Available: <Go to ISI>://A1986E117100006.
29. Buss LW, McFadden CS, Keene DR (1984) Biology of hydractiniid hydroids. 2. Histocompatibility effector system competitive mechanism mediated by nematocyst discharge. Biological Bulletin 167: : 139–158. Available: <Go to ISI>://A1984 TF55400004.
30. Lange R, Plickert G, Müller WA (1989) Histoincompatibility in a low invertebrate, *Hydractinia echinata*: analysis of the mechanism of rejection. Journal of Experimental Zoology 249: : 284–292. Available: <Go to ISI>://A1989T671300006.
31. Hamann O (1882) Der Organismus der Hydroidpolypen. Jena Z Naturiss 15: 473–544.
32. Schierwater B, Piekos B, Buss LW (1992) Hydroid stolonal contractions mediated by contractile vacuoles. Journal of Experimental Biology 162: : 1–21. Available: <Go to ISI>://A1992HD81100001.
33. Allman GJ (1853) On the anatomy and physiology of *Corylophora*. Philosophical Transactions 143: 367–384.
34. Citron E (1902) Beitrage zur Kenntnis des feineren Baus von *Syncoryne sarsii*. Arch Naturgesch 68: 1–26.
35. Schulze FE (1873) Uber den Bau von *Syncoryne sarsii*, Loven und der zughorigen Meduse *Sarsia tubulosa*. Leipzig: Wilhelm Engelmann.
36. Taddei-Ferretti C, Cordella L (1976) Modulation of the *Hydra attenuata* rhythmic activity: phase response curve. Journal of Experimental Biology 65: 737–751.
37. Blackstone NW (1998) Morphological, physiological and metabolic comparisons between runner-like and sheet-like inbred lines of a colonial hydroid. Journal of Experimental Biology 201: : 2821–2831. Available: <Go to ISI>://000077039700004.
38. Dudgeon SR, Kübler JE (2011) Hydrozoans and the shape of things to come. Advances in Marine Biology 59: 107–144.
39. Mali B, Mohrlen F, Frohme M, Frank U (2004) A putative double role of a chitinase in a cnidarian: pattern formation and immunity. Developmental and Comparative Immunology 28: : 973–981. Available: <Go to ISI>://000222815600002.
40. Fulton C (1963) Rhythmic movements in *Cordylophora*. Journal of Cellular and Comparative Physiology 61: : 339–350. Available: <Go to ISI>://A19634098B00002.
41. Magie CR, Martindale MQ (2008) Cell-cell adhesion in the Cnidaria: insights into tissue morphogenesis. Biological Bulletin 214: 218–232.

42. Schneider SQ, Finnerty JR, Martindale MQ (2003) Protein evolution: Structure-function relationships of the oncogene beta-catenin in the evolution of multicellular animals. Journal of Experimental Zoology Part B-Molecular and Developmental Evolution 295B: : 25–44. Available: <Go to ISI>://000182421300003.
43. Campbell RD (1968) Holdfast movement in hydroid *Corymorpha palma*: Mechanism of elongation. Biological Bulletin 134: : 26–34. Available: <Go to ISI>://A1968A701000003.
44. Homem CCF, Peifer M (2008) Diaphanous regulates myosin and adherens junctions to control cell contractility and protrusive behavior during morphogenesis. Development 135: : 1005–1018. Available: <Go to ISI>://000253426300002.
45. Azevedo D, Antunes M, Prag S, Ma XY, Hacker U, et al.. (2011) DRhoGEF2 regulates cellular tension and cell pulsations in the amnioserosa during *Drosophila* dorsal closure. Plos One 6: : e23964. Available: <Go to ISI>://000295173800010.
46. Sawyer JK, Choi W, Jung KC, He L, Harris NJ, et al.. (2011) A contractile actomyosin network linked to adherens junctions by Canoe/afadin helps drive convergent extension. Molecular Biology of the Cell 22: : 2491–2508. Available: <Go to ISI>://00029 2687800007.
47. Carthew RW (2005) Adhesion proteins and the control of cell shape. Curr Opin Genet Dev 15: 358–363 Available: http://www.ncbi.nlm.nih.gov/pubmed/15963712.
48. Slautterback DB (1967) Coated vesicles in absorptive cells of Hydra. Journal of Cell Science 2: : 563–572. Available: <Go to ISI>://A1967A406600010.
49. Monier B, Pélissier-Monier A, Brand AH, Sanson B (2010) An actomyosin-based barrier inhibits cell mixing at compartmental boundaries in Drosophila embryos. Nature Cell Biology 12: : 60–65; sup 1–9. Available: http://www.ncbi.nlm.nih.gov/pubmed/ 19966783.
50. Shewan AM, Maddugoda M, Kraemer A, Stehbens SJ, Verma S, et al.. (2005) Myosin 2 is a key Rho kinase target necessary for the local concentration of E-cadherin at cell-cell contacts. Molecular Biology of the Cell 16: : 4531–4542. Available: <Go to ISI>://WOS:000232257700007.
51. Smutny M, Cox HL, Leerberg JM, Kovacs EM, Conti MA, et al. (2010) Myosin II isoforms identify distinct functional modules that support integrity of the epithelial zonula adherens. Nature Cell Biology 12: 696–702 Available: http://www.ncbi.nlm. nih.gov/pubmed/20543839.
52. Fernandez-Gonzalez R, Simoes SDM, Röper J-C, Eaton S, Zallen JA (2009) Myosin II dynamics are regulated by tension in intercalating cells. Developmental Cell 17: 736–743 Available: http://www.pubmedcentral.nih.gov/articlerender.fcgi?artid= 2854079&tool= pmcentrez&rendertype=abstract.
53. Bouillon J, Levi C (1971) Structure et ultrastructure des attaches hydranthes-hudrotheques chez les polyps Thecata. Z Zellforsch mickrosk Anat 121: 218–231.
54. Marcum BA, Diehl FA (1978) Anchoring cells (desmocytes) in hydrozoan polyp Cordylophora. Tissue Cell 10: : 113–124. Available: <Go to ISI>://A1978ET44500009.
55. Overton J (1963) Intercellular connections in outgrowing stolon of Cordylophora. Journal of Cell Biology 17: : 661–671. Available: <Go to ISI>://A19634066B00011.
56. Campbell RD (1967) Tissue dynamics of steady state growth in Hydra littoralis.I. Patterns of cell division. Developmental Biology 15: : 487–502. Available: <Go to ISI>:// A19679344900006.
57. Campbell RD (1967) Tissue dynamics of steady state growth in *Hydra littoralis*.2. Patterns of tissue movement. Journal of Morphology 121: : 19–28. Available: <Go to ISI>://A19678823600002.
58. Campbell RD (1967) Tissue dynamics of steady state growth in *Hydra littoralis*. 3. Behavior of specific cell types during tissue movements. Journal of Experimental Zoology 164: : 379–391. Available: <Go to ISI>://A19679289400007.

CHAPTER 17

Genome-scale investigation of olfactory system spatial heterogeneity

Torben Noto[1], Derrick Barnagian[2], Jason B. Castro[3*]

1 Department of Cognitive Science, University of California San Diego, La Jolla, California, United States ofAmerica,

2 Department of Brain and Cognitive Sciences, MIT, Cambridge, Massachusetts, United States ofAmerica,

3 Neuroscience Program, Bates College, Lewiston, Maine, United States of America

ABSTRACT

The early olfactory system is organized in parallel, with numerous, specialized subsystems established by the modular and topographic projections of sensory inputs. While these anatomical sub-systems are in many cases demarcated by well-known marker genes, we stand to learn considerably more about their possible functional specializations from comprehensive, genome-scale descriptions of their molecular anatomy. Here, we leverage the resources of the Allen Brain Atlas (ABA)—a spatially registered compendium of gene expression for the mouse brain—to investigate the early olfactory system's genomic anatomy. We cluster thousands of genes across thousands of voxels in the ABA to derive several novel parcellations of the olfactory system, and concomitantly discover novel sets of enriched, subregion-specific genes that can serve as a starting point for future inquiry.

1. INTRODUCTION

A major challenge in neuroanatomy is to partition the brain "at its joints" [1], identifying its minimal sub-systems and their potential heirarchical relationships [2]. While classical histology is essential for this endeavor, it may also be limited in some cases, owing to its unavoidable use of ad-hoc differentiating criteria. Even when discrete subregions can be demarcated by differences in the abundance and/or co-expression of specific molecules, these typically represent only a small and idiosyncratic subset of the molecules that collectively support excitability, synaptic communication, plasticity, neuromodulation, and the developmental specification of neural circuits.

With the advent of accessible, comprehensive data resources such as the Allen Brain Atlas (ABA) [3], these approaches can be complemented by data-driven frameworks for studying the brain's molecular architecture. Because the gene-expression maps of the ABA are panoramic, dense (cataloging the entire mammalian genome), and quantitative, one can develop genomic definitions of brain areas that capture spatial correlations across many genes [3,4]. Indeed, such approaches have been successfully used to recapitulate the brain's major phylogenetic subdivisions, and to map the more granular "genomic anatomy" of the hippocampus[5–8] and neocortex [4,9].

Here, we extend similar methods to investigate the genomic anatomy of the early olfactory system, including the olfactory bulb (OB)—the first central structure in the ascending olfactory system—and its more central cortical targets. As a model anatomical system, the OB has several virtues, the principal of which is the well-described molecular topography and modularity of its sensory inputs. Briefly, the bulb can be partitioned into several non-overlapping subregions, defined by both the projection patterns of molecularly labeled inputs [10–14], and in some cases, the contribution of these regions to specific behaviors or sets of behaviors (e.g. conditioned vs. innate odor responses [14]). At the same time, many basic questions on the olfactory system's organization and substructure persist, highlighting a potentially important role for unbiased genomic tools in understanding this system's functional anatomy.

We describe and implement a workflow for clustering thousands of genes across thousands of olfactory voxels in the ABA, in turn discovering sparse and novel "genomic signatures" for the early olfactory system's constituent structures. Our analysis both recapitulates well-known subdivisions and sub-systems within olfaction, and also identifies novel, candidate subregions and corresponding enriched gene-sets that merit further experimental investigation. The tools described here may be an important complement to ongoing efforts to map and characterize olfactory subsystems.

2. MATERIALS AND METHODS

2.1 Data

All data were obtained from the Allen Brain Institute's (ABI) mouse brain expression project (www.brain-map.org), which comprises a set of in-situ hybridization (ISH) experiments cataloging expression of the entire mammalian genome across the C57BL/6 (P56) mouse brain, at cellular resolution. The data resources of the ABI are documented amply in other publications [3,4] as well as in ABI whitepapers (http://help.brain-map.org/display/mousebrain/Documentation). Briefly, ABI data are available as raw ISH image files, and also as a registered 3D atlas (voxel size: 200 μm x 200 μm × 200 μm) in which each atlas voxel is a vector quantifying expression energy across the genome, at a particular location in the brain. The olfactory bulb comprises 990 such voxels in this data set. In the present study, we used only the registered data, and investigated a quality-controlled subset of genes (3,041 genes out of ~20,000) that has been used in several previous publications [15,16]. This set represents genes showing strong correlation across the coronal and saggital image series of the Allen Brain Expression Atlas.

The complete set of genes we used (3,041 of them) was culled over several publications. The first draft coarse atlas was obtained on 25 micron thick sagittal brain sections taken every 200 microns, for a total of ~20 sections for a single hemisphere [3] From these data, a subset of 4,376 genes of 'neurobiological interest' (described in [17]) were chosen for a more comprehensive and anatomically more granular analysis. While this is evidently a 4-to-5 fold undersampling of the complete genome, it is worth noting that a large fraction of the genome (~10%) is not expressed in the brain, and an additional ~20% is only negligibly expressed [3]. Of the ~ 80% of genes expressed above background, a sizeable fraction (~37%) showed "near ubiquitous" expression, meaning that these genes showed little difference in expression across cell types. Indeed, gene ontological (GO) analyses of the set of near ubiquitous genes was consistent with their involvement in basal cellular function (cellular metabolism, protein metabolism, basal/homeostatic gene function, etc).

For each of the ~4,000 'genes of interest' from the sagittal series described above, additional ISH experiments were done, taking coronal sections, also every 200 microns, for a total of ~56 sections (per gene). From these

~4000 genes, a "higher consistency" [15] subset was obtained that exhibited high expression concordance between coronal and sagittal data sets. The Pearson correlation coefficient was calculated between voxels of the sagittal and coronal data sets for each gene, and those genes in the lowest quartile of the spatial correlation coefficient distribution were discarded. The remaining 75% of genes are those that make up the set of 3,041 genes used in the present study. This set of genes is thus a representative and relatively unbiased subset of "neurobiologically interesting" genes whose expression is consistent across experiments. According to the Allen Brain Institute's release and publication [18] of the Anatomic Gene Expression Atlas (AGEA) data, these genes are strongly enriched in "[GO] categories related to neuronal cell process and function, GABA receptor activity, ion and potassium channel binding, neuron differentiation, axon guidance, synaptic transmission and long-term potentiation." This specific gene set has been used in other publications [15,16,19]. In some analyses, we extracted smaller subsets of 'physiological genes', which consisted of those genes listed in the IUPHAR database () of ion channel genes [20] (146/3041 genes).

2.2 Data handling, dimensionality reduction and clustering

All data analysis was performed in Matlab (Mathworks). Several routines from the 'Brain Gene Expression Analysis' toolbox, written by Grange et al [21], were modified for low-level data handling (extracting matrix columns corresponding to named brain regions, etc). All other analysis functions (dimensionality reduction, clustering, etc) were custom-written, and are available at https://github.com/CastroLab.

Our clustering framework employed non-negative matrix factorization [22,23], in which a low (s) dimensional approximation of a matrix, A, is sought taking the form:$A=WH$(1)Where A is an m x n data matrix, and W and H are m x s and s x n matrices containing feature vectors and their weights, respectively. We performed the factorization using the standard Matlab implementation (nnmf.m), using the alternating least-squares algorithm initially proposed by [22]. Briefly, this algorithm iterates over the following steps:

1) assume W is known and solve the least squares problem for H using$(W^TW)H=W^TA$(2)
2) set negative elements of $\mathbf{H} \rightarrow 0$
3) assume H is known and solve the least squares problem for W using$(HH^T)W^T=HA^T$(3)
4) set negative elements of $\mathbf{W} \rightarrow \mathbf{0}$

We used 1000 iterations as our stopping criterion for the maximum number of iterations.

3. RESULTS

3.1 Summary of workflow

Our overall goal was to identify the olfactory system's molecular subdivisions, as specified by regional differences in gene correlations across thousands of genes. To do this, we developed a gene-clustering and dimensionality reduction workflow based on non-negative matrix factorization (NMF) [23](Fig 1). NMF-based methods have enjoyed frequent use in gene-clustering applications owing to the intrinsic non-negativity of expression data, and the ready interpretability of NMF-derived expression profiles. Intuitively, NMF discovers from data a handful of highly informative 'signature' expression profiles (formally termed a basis set) that summarize a large fraction of total variability in gene expression. Brain areas can then be readily clustered by their resemblance to a given one of these expression profiles (formally, their 'weights' in the new basis set). Notably, this decomposition into signature expression profiles can be done at arbitrary levels of granularity. Recent investigations of genomic anatomy in the hippocampus [5] and cortex [4] have used methods similar to ours.

The essential components of our workflow are illustrated in Fig 1. Briefly, the full data set (the matrix labeled 'ABA' in Fig 1) was a 3,041 x 49,742 matrix of genes x voxels cataloging gene expression across the entire mouse brain. Each voxel (i.e. each column of the matrix) catalogs the expression of thousands of genes, and was obtained by analyzing thousands of in-situ-hybridization (ISH) experiments, and spatially registering these data to a common 3D grid (the Atlas proper) with 200 cubic-micron voxels. This matrix catalogs ~ 15% of the entire genome (~20,000 genes), and is a previously published, quality controlled subset of the ABA for genes exhibiting highly correlated expression between both coronal and saggital ISH experiments [3,21]. Although contiguous columns of this matrix tend to represent spatially contiguous voxels, coordinates in anatomical space are not formally represented in column ordering. Indeed, it is important to emphasize that all matrix operations are performed in 'gene space', and blind to relative spatial positions. Visualization of voxels in their correct spatial context was achieved through separate lookup tables that assign voxels to grid coordinates.

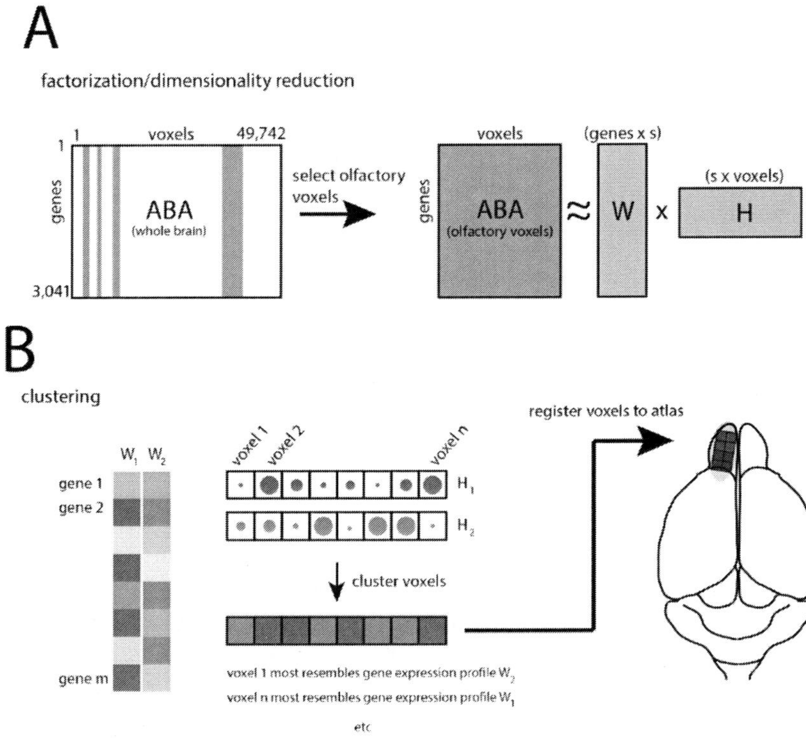

Figure 1. Workflow **A)** The principal data set was a genes x voxels matrix [3,21] cataloging brain-wide expression. Columns corresponding to known olfactory areas were selected, and the dimensionality of the resultant sub-matrix was reduced using non-negative matrix factorization (NMF). The cartoon shows an example of reducing the original data from m-dimensional to s-dimensional, where m>>s. **B)** NMF yields a new basis set (column vectors) as well as weights of voxels in the new basis (row vectors). Voxels are readily clustered by selecting the largest weight, and registration of clustered voxels to the anatomical atlas reveals the spatial patterning of clusters. Note that the clustering itself is blind to spatial relationships among voxels. The cartoon shows an example where s = 2.

We first extracted a sub-matrix of 2,310 columns corresponding to all olfactory areas, and performed NMF (see Methods) to approximate this as the product of a matrix of basis vectors (W) and matrix of corresponding weights (H) (Fig 1A). This was done for subspace sizes (s) ranging from 2 to 7. Owing to the inherent sparseness of NMF decompositions [23], clustering (Fig 1B) was readily performed by selecting the largest resultant weight for each voxel. Clustered voxels were then registered back to the Atlas to visualize their spatial relationships.

3.2 Genomic definitions of gross olfactory system subdivisions

Applying the framework described above, we decomposed the olfactory system into its genomically defined subdivisions, at several levels of granularity. Fig 2A and 2B show the olfactory system's major, classically defined *anatomical* subdivisions, for the ABA's most granular available taxonomy. Fig 2C shows several *genomic* parcellations of the olfactory system, for NMF decompositions at several levels of granularity (subspace size = s). For a choice of s = 2, the olfactory system was clearly partitioned into its major sensory recipient (bulbar) and higher cortical subdivisions. At choices of s = 3 and s = 5, the decomposition began to reveal substructure within the anterior and posterior subdivisions of the olfactory bulb, as well as clearly demarcating the limbic (amygdala-associated) divisions of olfactory cortex. For the most granular decompositions we performed (s = 5 and s = 7), the olfactory system was partitioned into subdivisions that recapitulated many well-known anatomical subdivisions, including the accessory olfactory bulb (AOB), the anterior olfactory nucleus (AON), the amygdalar portions of olfactory cortex, and the anterior and posterior subdivisions of piriform cortex.

The brain's 'components' can be defined by a variety of molecular criteria, including the patterned expression of developmental markers, transcription factors, ion channel genes, genes for the synthesis and loading of neurotransmitter, as well as many others. To develop a 'functional' parcellation of the olfactory system—that is, one that is likely to reflect differences in biophysical and synaptic properties—we performed an additional analysis using a more constrained set of physiologically relevant genes. Specifically, we performed the NMF decomposition for the subset of genes (143 of the 3,041) listed in the IUPHAR database [20] corresponding to voltage or ligand-gated ion channels, and G protein coupled receptors (GPCRs). The results of this decomposition are shown (for a choice of s = 5) in Fig 3. Fig 3A shows the resultant matrix of weights, H (see workflow in Fig 1), and its block-diagonal structure highlights the 'categorical' nature of NMF-derived representations. Briefly, a given voxel tends to be well-characterized by a single basis vector, to the relative exclusion of others.

The spatial organization of these 'physiological' clusters is illustrated in Fig 3B, and these map on well to several of the olfactory system's known anatomical subdivisions (compare to Fig 2). While these clusters and their spatial relationships are in themselves interesting, their greater potential utility comes from the fact that each cluster is defined by a unique and characteristic signature expression profile, discovered through dimensionality reduction. Fig 4A shows the five signature profiles for the decomposition described above. Note that these profiles tend to be sparse: there are only strong 'hits' for a small subset of genes, with the great majority of genes contributing little to nothing to the definition of a given cluster. The sparseness of the expression profiles is also illustrated and quantified in the histograms (Fig 4B).

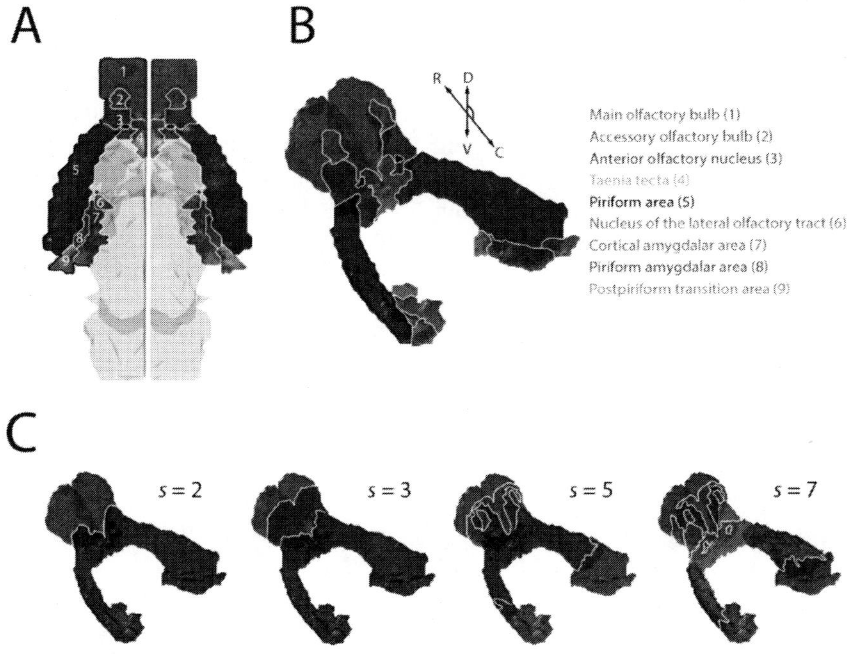

Figure 2. Genomically-defined divisions of the olfactory system. **A)** Horizontal-plane projection of the mouse olfactory system as demarcated in the Allen Brain Atlas (named subregions shown to the right). Diencephalic, midbrain, and brainstem structures are shown in gray for spatial context, though voxels comprising these structures were not used in the analysis. Abbreviations: R-rostral, C-caudal, D-dorsal, V-ventral) **B)** 3D rendering of the olfactory system, showing its various subdivisions, as in (A). The main (1) and accessory (2) olfactory bulbs define the rostral-most pole of the mouse brain and receive direct input from the olfactory periphery; the remaining structures receive both direct and indirect bulbar input, and lie along the brain's ventral surface. **C)** Non-Negative Matrix Factorization (NMF)-based clustering of the olfactory system, for various choices of subspace size, s (see text for details).

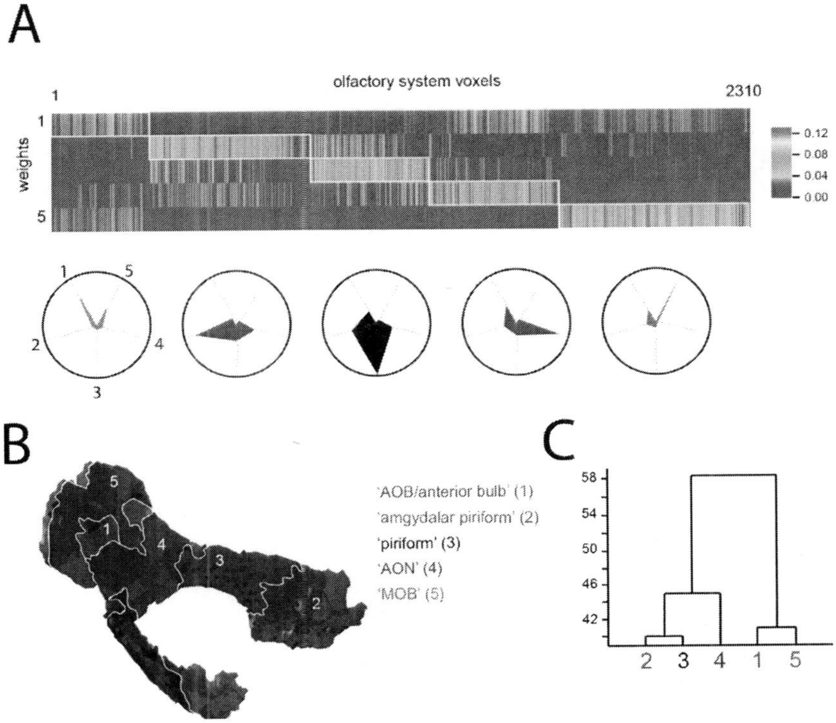

Figure 3. Olfactory system subdivisions defined by 'physiological' genes. **A)** Top: matrix of weights (H matrix) for all olfactory system voxels, for a 5 dimensional NFM decomposition. Columns have been sorted by peak value (preserving the order and values of row-contents) to reveal the block-diagonal structure of the matrix. Bottom: Polar plots of the average column-vector for each diagonal block, showing categorical assignment of voxels to one the five weight dimensions, to the relative exclusion of others. **B)** Spatial arrangement of clustered voxels shown in (A), in Brain-Atlas coordinates. Note the strong spatial contiguity of most of the clusters (with region 1 being a notable exception). Colors correspond to the polar plot colors in (A). Region names in quotes are given as a heuristic summary—they do not necessarily align with ABA or other atlas definitions. **C)** Dendrogram showing inter-cluster distances (Euclidian distance between basis vectors).

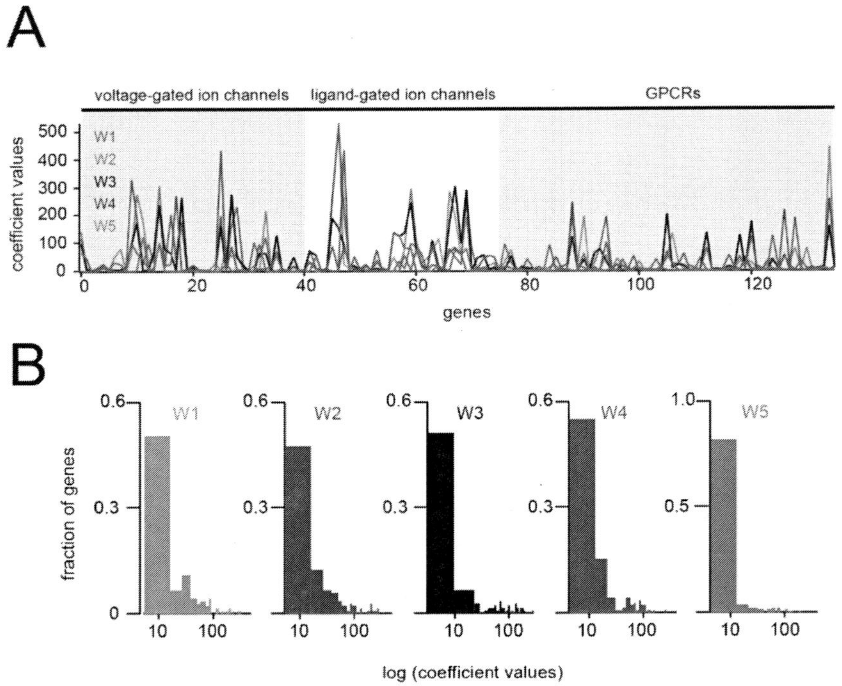

Figure 4. Gene expression profiles derived from NMF are sparse.

A) Expression profiles (the basis vectors W1-W5) for the three families of IUPHAR genes investigated (see text). **B)** Histograms of each expression profile. The x-axis is on a log-scale to reveal structure (gene 'hits') in the distribution tails that is not otherwise evident.

3.3 Validating olfactory subdivisions, and discovering candidate subdivisions using the ABA

The parcellations described above recapitulate many of the major known olfactory structures and subregions. While these are important for identifying sets of key differentiating molecules for major structures, it is also potentially interesting to investigate more spatially granular parcellations of the olfactory bulb (OB) in an exploratory context. Understanding the degree to which the OB is molecularly heterogeneous and modular vs homogeneous and 'equipotential' could help constrain models of early olfactory function. Specifically, evidence of genetic modularity and heterogeneity across the bulb would be consistent with current proposals that the OB is a collection of specialized, parallel modules [10,24], and not a single 'multipurpose' columnar circuit simply iterated many times in parallel.

To ask whether our methods are in principle capable of making these more granular distinctions, we first investigated olfactory bulb subsystem organization in a case where we expected to find it: between the anterior and posterior divisions of the accessory olfactory bulb (AOB)—a small, well delineated olfactory subsystem that processes non-volatile compounds, and makes critical contributions to many innate behaviors [25,26].

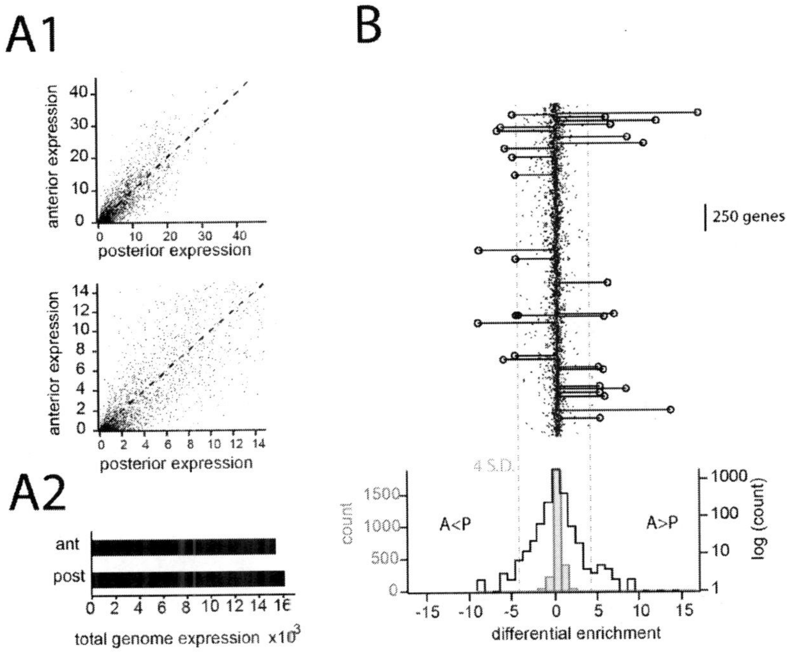

Figure 5. Gene expression and differential enrichment in the anterior vs posterior AOB. A) 1: Scatter plots of gene expression for all 3,041 genes in the anterior vs. posterior AOB. Each point is an average across anterior voxels and posterior voxels (see Methods). Dotted line indicates equal expression in the anterior and posterior. Lower graph is the same data on an expanded scale. A2) total expression of all genes in the Anterior AOB and posterior AOB, illustrating no notable difference between the two. B) top. plot of differential enrichment (anterior AOB enrichment—posterior AOB enrichment; see text) for all genes. The majority of genes were symmetrically or near-symmetrically expressed in the anterior and posterior AOB. Genes with asymmetry scores exceeding 4 standard deviations are shown in lines and markers. The remainder of genes are shown as dots. Bottom: Histogram differential enrichment, shown in both linear (red) and log axes.

The AOB's two divisions are dissociable on the basis of their unique inputs from different receptor classes [27–29], their distinct central projections, and their differential expression of lectins and developmental control molecules [30–32].

Inspecting and quantifying expression between the anterior and posterior halves of the AOB (Fig 5A and 5B), it was evident that, on average, genes were symmetrically distributed across this structure's midline axis. For each of the 3,041 ABA genes, we also calculated a differential enrichment score (see Methods), which quantified expression asymmetry across the A-P axis. A histogram of these scores across genes was sharply peaked at 0 (symmetric expression), but a population of highly differentially enriched genes could be observed in the tails of this distribution.

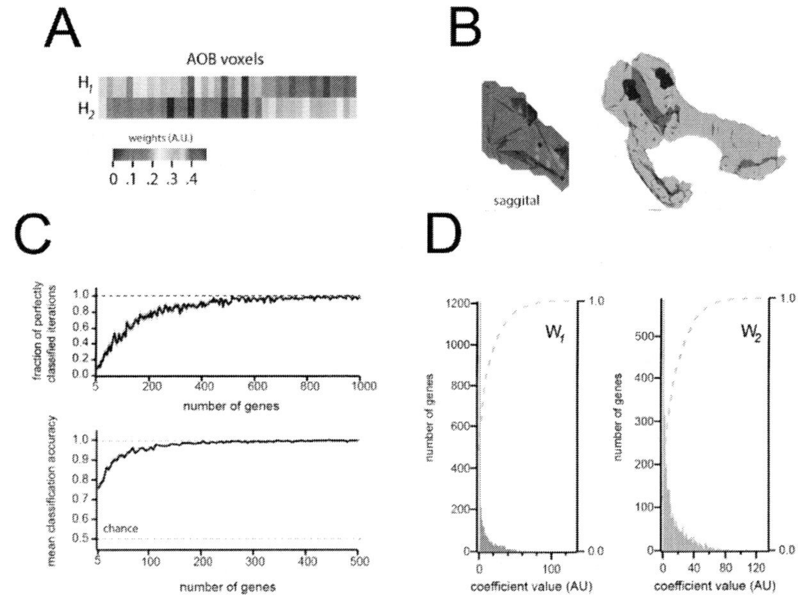

Figure 6. Clustering (s = 2) returns the known anterior and posterior sub-divisions of the AOB. **A)** Weights matrix (See Fig 1) showing dichotomous segregation of AOB voxels. **B)** NMF reveals spatially contiguous subregions of the AOB clearly corresponding to the structure's (known) anterior and posterior sub-divisions. **C)** classification accuracy of NMF vs number of genes included. *Top*: fraction of 'perfectly classified voxels', as a function of number of genes (see Methods). *Bottom*: Mean classification accuracy vs. number of genes used for the factorization. **D)** Histogram of basis-vector values, illustraing sparseness of the basis vectors (i.e. a small number of genes defines membership in anterior vs. posterior).

Interestingly, when we clustered AOB voxels using a 2 dimensional factorization (see Methods), the two resultant clusters clearly corresponded to the structure's anterior and posterior halves (Fig 6). As with the factorizations described above, the expression profiles for these clusters were sparse, with relatively small numbers of genes contributing to the definitions of the AOB's anterior and posterior halves.

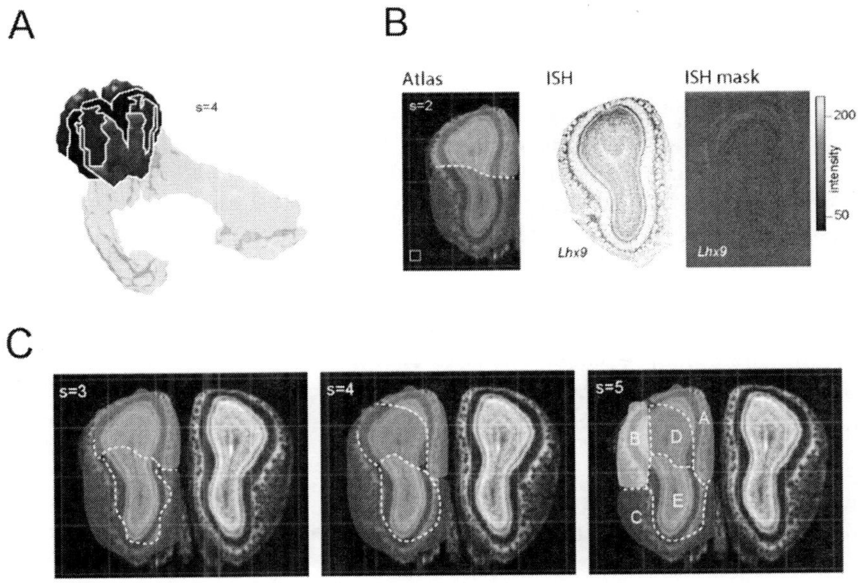

Figure 7. Clustering the main olfactory bulb reveals candidate genomic subdivisions. **A)** Location of NMF-derived clusters (s = 2) in ABA coordinates (see S1 File for genes comprising the clusters). **B)** left: NMF-decomposition (s = 2) of the fifth coronal section of the expression atlas reveals clear dorsal v. ventral domains (white box indicates ABA voxel-size); middle: ISH image from the ABA of *Lhx9* –the highest ranked gene in the dorsal cluster; right: segmented expression mask showing strong dorsal pattering of Lhx9 (contrast unadjusted from the raw, downloaded mask). **C)** NMF decompositions of the OB (5th coronal section) at progressively greater granularity, revealing the ventral glomeruli as a clear and contiguous genomic territory as well as two distinct domains within the dorsal glomeruli. The lettering of clusters in the s = 5 case corresponds to the lettering in Fig 8. See supplementary info (S1 File) for genes comprising each cluster.

Given our methods' successful identification of granular substructure within the AOB (a 'validatory' case), we next derived a candidate parcellation of the main olfactory bulb (an 'exploratory' case). Recent work supports the idea that bulb is comprised of molecularly and behaviorally specialized subsystems [10,14,33–35], and understanding these at the scale of the genome could provide important clues about their local computations and potential functional roles. Fig 7A shows a 4-dimensional decomposition of the olfactory bulb (with the AOB excluded), illustrating that the bulb can be parceled into molecularly distinct territories on the basis of spatial correlations between genes. The rank-ordered genes comprising each of these territories can be found in the supplemental information (S1 File).

While this analysis revealed large, spatially contiguous clusters along the rostro-caudal extent of the bulb, it was generally difficult to map these clusters on to distinct lamina and known bulbar domains. To get a clearer portrait of how and whether clusters corresponded to histological layers, we applied our workflow on a single (2D) coronal section of the bulb (specifically, the 5th coronal section of the reference atlas) (Fig 7B and 7C). For a choice of s = 2, the bulb was clearly segregated into broad dorsal and ventral domains, and inspection of the ISH data from the leading hit from the dorsal domain revealed a marked enrichment of the LIM homeodomain transcription factor *Lhx9* in the bulb's dorsal aspect (Fig 7B). This is, to our knowledge, the first report of strong dorsal patterning of this gene in the bulb.

As the number of clusters was increased from s = 3 to s = 5, the bulb was partitioned into subregions that revealed distinct glomerular domains. For the choice of s = 5, we observed a cluster corresponding to the ventral glomeruli, and two separate clusters demarcating the dorsomedial vs. dorsolateral glomeruli. The remaining 2 clusters evidently demarcated the dorsal vs. ventral granule cell and mitral cell layers. The separation of the dorsal glomeruli was especially interesting, as these may map onto the two dorsal domains defined by the projection patterns of Class I and Class II OSNs [36]. Fig 8 shows several examples of raw in-situ data and expression masks for leading genes in each of the five identified clusters (these are also described more in the discussion section). The full list of rank-ordered genes in each of these clusters is available in the supplemental information (S1 File).

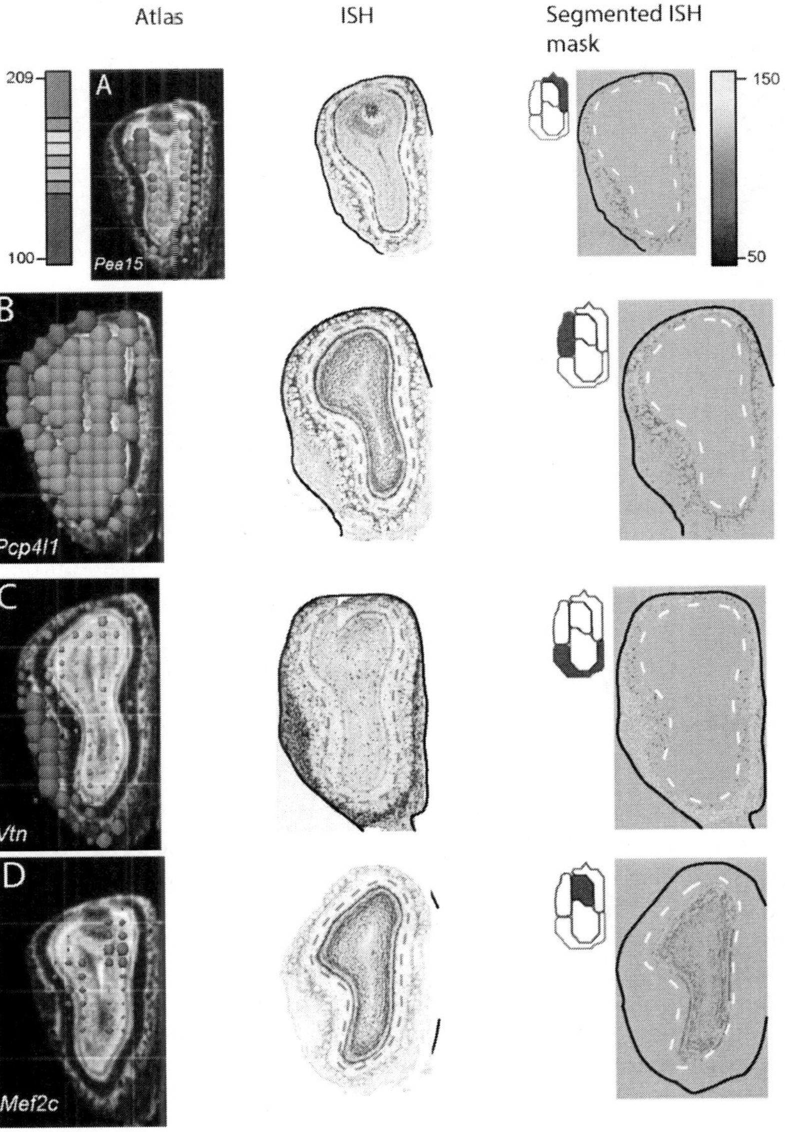
Atlas
ISH
Segmented ISH mask
209
100
A
Pea15
150
50
B
Pcp4l1
C
Vtn
D
Mef2c

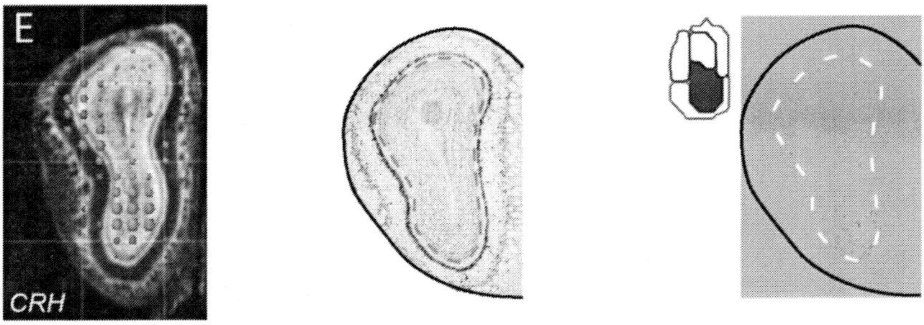

Figure 8. Expression patterns of leading genes in each of the 5 NMF-derived OB clusters. Letters (A-E) correspond to clusters labeled in Fig 7C. Icons at the top-left of the segmented masks show which cluster is being shown. **A)** Phosphoprotein enriched in astrocytes 15 (Pea15); rank # 2/3,041; enriched in the dorsomedial glomerular cluster. **B)** Purkinje cell protein 4-like (Pcp4l1); rank #2/3,04; enriched in the dorsolateral glomerular cluster. **C)** Vitronectin (Vtn); rank #3/3,041; enriched in the ventral glomerular cluster. **D)** Myocyte enhancer factor 2C (Mef2c); rank #7/3,041 (#3 of protein coding genes); enriched in dorsal mitral and granule cells; **E)** Corticotropin releasing hormone (CRH); rank #24/3,041; enriched in ventral granule cell population.

4. DISCUSSION

Using the resources of the Allen Brain Atlas, we have investigated the molecular anatomy of the olfactory system at genomic scale. We described data-driven parcellations of various olfactory sub-regions that capture spatial correlations across thousands of genes. In doing this, we also derived 'signature' gene expression profiles that are characteristic of different subregions. Our analysis recapitulates many of the fundamental olfactory sub-regions familiar from classical anatomy, and also points to the existence of candidate sub-regions (and their most characteristic genes) within the olfactory bulb. These enriched gene sets can immediately form the basis for targeted, follow-up experiments to investigate candidate modules within the bulb. In future studies it will also be important to explicitly test for functional differences between these modules, as differences in expression do not necessarily manifest as differences in excitability or synaptic function.

Several other studies have used the resources of the Allen Brain Institute to investigate the 'genomic anatomy' of CNS's major phylogenetic subdivisions [3], as well as of specific structures including the hippocampus and neocortex [4,5]. Similar to the results we obtained, these studies derived molecular, data-driven parcellations of brain structures, in some cases recapitulating nuclear and laminar structures originally identified using classical histological methods. Among these model anatomical systems, the OB has several virtues that make it attractive for follow-up genomic/anatomic investigation. The principal of these is the well-known and well-characterized topography and modularity of the OB's sensory inputs. At coarse scales, the OB surface can be partitioned into a handful of non-overlapping territories defined by the stereotyped axonal projections of olfactory sensory neurons with common signal transduction pathways[11,14,37]. Superimposed on this large-scale "zonal" map is a more granular and densely tiled map defined by the bulb's numerous glomeruli—regions of dense neuropil that each receive convergent inputs from idiotypic sensory neurons [12,38]. While a considerable amount of recent work has shown that these parallel systems in some cases respond to different characteristic ligands and support distinct behaviors, it is largely unknown whether there is heterogeneous 'readout' of these inputs by specialized intrinsic circuits of the OB. At one extreme, the OB's principal neurons may comprise a molecularly homogeneous group that processes information similarly across inputs. At the other extreme, these principal neurons may be molecularly diverse and/or organized into several classes to support input-specific processing. Our results preliminarily support the latter view, and suggest that the OB is comprised of at least several genomically distinct territories. Additional work will be needed to determine whether these regions are functionally distinct as well.

One especially intriguing finding was our observation of three genomically distinct, spatially non-overlapping glomerular domains. For a choice of 2 clusters, the bulb was partitioned into dorsal vs. ventral subdivisions, and with increasing granularity of clustering the glomerular layer was partitioned into ventral and dorsomedial domains. For the s = 3 and s = 4 decomposition (Fig 7C), the ventral glomeruli extended more dorsally along their lateral aspect, in a manner consistent with glomerular subdivisions demarcated by OCAM [11,39]. For the s = 5 decomposition, two dorsal glomeruli were partitioned into medial and lateral subdivisions, reminiscent of the two separate dorsal domains targeted by Class I and Class II ORNs [36].

Although our analysis has generated large numbers of candidate genes for future experimental inquiry (the set of all rank-ordered genes, by cluster, is provided in S1 File), several specific cases are worth briefly summarizing. First, we note that our identification of vitronectin—an extracellular matrix component—as a marker of the ventral glomerular cluster (ranked #3, Fig 8C) is consistent with results reported in the microarray study by Lin et al [40] (ranked # 13 in that study). Given the concordance between these two large-scale screens, and the known role of vitronectin in regulating neurite outgrowth and neuronal migration, the role of this gene in specifying glomerular topography may merit further investigation. Second, we found that a leading marker (rank #2) of the dorsolateral glomerular cluster was the gene Pcp4l, which has been shown in other studies to be differentially expressed within the OB in a manner regulated by cAMP concentrations [41]. This same gene is also one of only 8 mRNAs (in a genome-wide study) showing activity dependent regulation in mouse OSNs following naris occlusion. Finally, we observed that the LIM homeodomain transcription factor Lhx9 was the highest-ranked gene defining the bulb's dorsal aspect (Fig 7B), and correspondingly was strongly and selectively enriched in dorsal glomeruli, mitral cells, and granule cells. Other Lhx genes have been shown to be important for OSN development (Lhx2)[42] and the genetic specification of innate olfactory behaviors (Lhx6)[43]; Lhx9 may therefore be an interesting candidate for delineating a topographically defined olfactory sub-system.

Although our study is novel in investigating olfactory circuit organization at the scale of the genome, it is important to also summarize its limitations. First, because the ABA ISH data are compiled across experimental subjects, we do not have information about co-expression of molecules at the single cell level. Our analysis can only report on comparatively large-scale 'zonal' molecular heterogeneity across the olfactory system and olfactory bulb. Similarly, the spatial granularity of our analysis is ultimately limited by the voxel size of the ABA, which is large (200 μm on a side) relative to single neurons and neuronal microcircuits. While it would be interesting to systematically study potential molecular variability within and between cell types, and across glomerular columns, this is likely not possible with the registered ABA data.

In sum, we have developed and implemented methods for studying the large scale genomic anatomy of the early olfactory system. The parcellations we derive using these methods identify both known and novel subregions of the olfactory bulb, and the molecular markers identified from this can serve as the starting point for future inquiry into the subsystem organization of olfaction.

REFERENCES

1. Plato , Helmbold WC, Rabninowitz W. Phaedrus. Indianapolis: Bobbs-Merrill Co.; 1956.
2. Swanson LW. Quest for the basic plan of nervous system circuitry. Brain Res Rev. 2007;55: 356–372. pmid:17267046
3. Lein ES, Hawrylycz MJ, Ao N, Ayres M, Bensinger A, Bernard A, et al. Genome-wide atlas of gene expression in the adult mouse brain. Nature. 2007;445: 168–176. pmid:17151600
4. Hawrylycz M, Bernard A, Lau C, Sunkin SM, Chakravarty MM, Lein ES, et al. Areal and laminar differentiation in the mouse neocortex using large scale gene expression data. Methods San Diego Calif. 2010;50: 113–121. pmid:19800006

5. Thompson CL, Pathak SD, Jeromin A, Ng LL, MacPherson CR, Mortrud MT, et al. Genomic anatomy of the hippocampus. Neuron. 2008;60: 1010–1021. pmid:19109908
6. Cembrowski MS, Bachman JL, Wang L, Sugino K, Shields BC, Spruston N. Spatial Gene-Expression Gradients Underlie Prominent Heterogeneity of CA1 Pyramidal Neurons. Neuron. 2016;89: 351–368. pmid:26777276
7. Ramsden HL, Sürmeli G, McDonagh SG, Nolan MF. Laminar and dorsoventral molecular organization of the medial entorhinal cortex revealed by large-scale anatomical analysis of gene expression. PLoS Comput Biol. 2015;11: e1004032. pmid:25615592
8. Dong H-W, Swanson LW, Chen L, Fanselow MS, Toga AW. Genomic-anatomic evidence for distinct functional domains in hippocampal field CA1. Proc Natl Acad Sci U S A. 2009;106: 11794–11799. pmid:19561297
9. Belgard TG, Marques AC, Oliver PL, Abaan HO, Sirey TM, Hoerder-Suabedissen A, et al. A transcriptomic atlas of mouse neocortical layers. Neuron. 2011;71: 605–616. pmid:21867878
10. Munger SD, Leinders-Zufall T, Zufall F. Subsystem organization of the mammalian sense of smell. Annu Rev Physiol. 2009;71: 115–140. pmid:18808328
11. Yoshihara Y, Kawasaki M, Tamada A, Fujita H, Hayashi H, Kagamiyama H, et al. OCAM: A new member of the neural cell adhesion molecule family related to zone-to-zone projection of olfactory and vomeronasal axons. J Neurosci Off J Soc Neurosci. 1997;17: 5830–5842.
12. Vassar R, Chao SK, Sitcheran R, Nuñez JM, Vosshall LB, Axel R. Topographic organization of sensory projections to the olfactory bulb. Cell. 1994;79: 981–991. pmid:8001145
13. Mombaerts P, Wang F, Dulac C, Chao SK, Nemes A, Mendelsohn M, et al. Visualizing an olfactory sensory map. Cell. 1996;87: 675–686. pmid:8929536
14. Kobayakawa K, Kobayakawa R, Matsumoto H, Oka Y, Imai T, Ikawa M, et al. Innate versus learned odour processing in the mouse olfactory bulb. Nature. 2007;450: 503–508. pmid:17989651
15. Bohland JW, Bokil H, Pathak SD, Lee C-K, Ng L, Lau C, et al. Clustering of spatial gene expression patterns in the mouse brain and comparison with classical neuroanatomy. Methods. 2010;50: 105–112. pmid:19733241
16. Menashe I, Grange P, Larsen EC, Banerjee-Basu S, Mitra PP. Co-expression profiling of autism genes in the mouse brain. PLoS Comput Biol. 2013;9: e1003128. pmid:23935468
17. Ng L, Bernard A, Lau C, Overly CC, Dong H-W, Kuan C, et al. An anatomic gene expression atlas of the adult mouse brain. Nat Neurosci. 2009;12: 356–362. pmid:19219037
18. Ng L, Bernard A, Lau C, Overly CC, Dong H-W, Kuan C, et al. An anatomic gene expression atlas of the adult mouse brain. Nat Neurosci. 2009;12: 356–362. pmid:19219037
19. Grange P, Bohland JW, Okaty BW, Sugino K, Bokil H, Nelson SB, et al. Cell-type-based model explaining coexpression patterns of genes in the brain. Proc Natl Acad Sci U S A. 2014;111: 5397–5402. pmid:24706869
20. Alexander SP, Catterall WA, Kelly E, Marrion N, Peters JA, Benson HE, et al. The Concise Guide to PHARMACOLOGY 2015/16: Voltage-gated ion channels. Br J Pharmacol. 2015;172: 5904–5941. pmid:26650441
21. Grange Pascal B JW. Brain Gene Expression Analysis: a MATLAB toolbox for the analysis of brain-wide gene-expression data. 2012;
22. Paatero P. Least squares formulation of robust non-negative factor analysis. Chemom Intell Lab Syst. 1997;37: 23–35.
23. Lee DD, Seung HS. Learning the parts of objects by non-negative matrix factorization. Nature. 1999;401: 788–791. pmid:10548103
24. Mainen ZF. The main olfactory bulb and innate behavior: different perspectives on an olfactory scene. Nat Neurosci. 2007;10: 1511–1512. pmid:18043586
25. Halpern M, Martínez-Marcos A. Structure and function of the vomeronasal system: an update. Prog Neurobiol. 2003;70: 245–318. pmid:12951145
26. Dulac C, Torello AT. Molecular detection of pheromone signals in mammals: from genes to behaviour. Nat Rev Neurosci. 2003;4: 551–562. pmid:12838330
27. Shinohara H, Asano T, Kato K. Differential localization of G-proteins Gi and Go in the accessory olfactory bulb of the rat. J Neurosci Off J Soc Neurosci. 1992;12: 1275–1279.

28. Jia C, Goldman G, Halpern M. Development of vomeronasal receptor neuron subclasses and establishment of topographic projections to the accessory olfactory bulb. Dev Brain Res. 1997;102: 209–216.
29. Berghard A, Buck LB. Sensory transduction in vomeronasal neurons: evidence for G alpha o, G alpha i2, and adenylyl cyclase II as major components of a pheromone signaling cascade. J Neurosci Off J Soc Neurosci. 1996;16: 909–918.
30. Shapiro LS, Halpern M, Ee P-L. Lectin histochemical identification of carbohydrate moieties in opossum chemosensory systems during development, with special emphasis on VVA-identified subdivisions in the accessory olfactory bulb. J Morphol. 1995;224: 331–349. pmid:7595957
31. Taniguchi K, Nii Y, Ogawa K. Subdivisions of the accessory olfactory bulb, as demonstrated by lectin-histochemistry in the golden hamster. Neurosci Lett. 1993;158: 185–188. pmid:7694206
32. Huilgol D, Udin S, Shimogori T, Saha B, Roy A, Aizawa S, et al. Dual origins of the mammalian accessory olfactory bulb revealed by an evolutionarily conserved migratory stream. Nat Neurosci. 2013;16: 157–165. pmid:23292680
33. Liberles SD, Buck LB. A second class of chemosensory receptors in the olfactory epithelium. Nature. 2006;442: 645–650. pmid:16878137
34. Munger SD, Leinders-Zufall T, McDougall LM, Cockerham RE, Schmid A, Wandernoth P, et al. An olfactory subsystem that detects carbon disulfide and mediates food-related social learning. Curr Biol CB. 2010;20: 1438–1444. pmid:20637621
35. Greer PL, Bear DM, Lassance J-M, Bloom ML, Tsukahara T, Pashkovski SL, et al. A Family of non-GPCR Chemosensors Defines an Alternative Logic for Mammalian Olfaction. Cell. 2016;165: 1734–1748. pmid:27238024
36. Bozza T, Vassalli A, Fuss S, Zhang J-J, Weiland B, Pacifico R, et al. Mapping of Class I and Class II Odorant Receptors to Glomerular Domains by Two Distinct Types of Olfactory Sensory Neurons in the Mouse. Neuron. 2009;61: 220–233. pmid:19186165
37. Miyamichi K, Serizawa S, Kimura HM, Sakano H. Continuous and overlapping expression domains of odorant receptor genes in the olfactory epithelium determine the dorsal/ventral positioning of glomeruli in the olfactory bulb. J Neurosci Off J Soc Neurosci. 2005;25: 3586–3592. pmid:15814789
38. Mombaerts P, Wang F, Dulac C, Chao SK, Nemes A, Mendelsohn M, et al. Visualizing an olfactory sensory map. Cell. 1996;87: 675–686. pmid:8929536
39. Kobayakawa K, Kobayakawa R, Matsumoto H, Oka Y, Imai T, Ikawa M, et al. Innate versus learned odour processing in the mouse olfactory bulb. Nature. 2007;450: 503–508. pmid:17989651
40. Lin DM, Yang YH, Scolnick JA, Brunet LJ, Marsh H, Peng V, et al. Spatial patterns of gene expression in the olfactory bulb. Proc Natl Acad Sci U S A. 2004;101: 12718–12723. pmid:15304640
41. Imai T, Suzuki M, Sakano H. Odorant Receptor-Derived cAMP Signals Direct Axonal Targeting. Science. 2006;314: 657–661. pmid:16990513
42. Hirota J, Mombaerts P. The LIM-homeodomain protein Lhx2 is required for complete development of mouse olfactory sensory neurons. Proc Natl Acad Sci U S A. 2004;101: 8751–8755. pmid:15173589
43. Choi GB, Dong H, Murphy AJ, Valenzuela DM, Yancopoulos GD, Swanson LW, et al. Lhx6 Delineates a Pathway Mediating Innate Reproductive Behaviors from the Amygdala to the Hypothalamus. Neuron. 2005;46: 647–660. pmid:15944132

CHAPTER 18

Comparative morphological trade-offs between pre- and post-copulatory sexual selection in Giant hissing cockroaches (Tribe: Gromphadorhini)

Kate L. Durrant[1] Ian M. Skicko[1#] Craig Sturrock[2] & Sophie L. Mowles[2†]

[1]The University of Nottingham, School of Life Sciences, Nottingham, NG7 2RD, United Kingdom.

[2]The Universityof Nottingham, School of Biosciences, Sutton Bonington Campus, LE12 5RD, United Kingdom. #Present address:University of Exeter, Department of Biosciences, Penryn Campus, TR10 9FE, United Kingdom

ABSTRACT

Sperm competition theory predicts that animals face a trade-off between investment in weaponry and investment in ejaculate composition. Within the Madagascan giant hissing cockroaches (Tribe Gromphadorhini) differences in morphology exist that may indicate differing strategies of male-male competition. We compared relative pronotal horn length using high-resolution X-ray CT scanning data, relative testes mass, and male-male agonistic behaviour between two species of hissing cockroaches,

Gromphadorhina oblongonota and *Aeluropoda insignis*. The gross morphology and behaviour of these two species indicated that *G. oblongonota* is selected for pre-copulatory mate acquisition and that *A. insignis* is selected for post-copulatory sperm competition. We found evidence for a trade-off when investing in testes mass vs. horn length between the species. The large, aggressive *G. oblongonota* follows a strategy of greater investment in weapons at the expense of testes mass while the smaller, less-aggressive *A. insignis* invests in relatively greater testes mass and less in pronotal weapon length. We also found evidence of a trade-off within each species, where individuals invest more heavily in weapon length at the expense of testes mass. These findings support the predictions of pre- and post-copulatory competitive investment trade-offs for a relatively understudied Tribe of cockroaches.

1. INTRODUCTION

Pre-copulatory sexual selection is one of the primary forces driving the evolution of extravagant phenotypic traits in males, including body-size and weaponry[1,2]. Sexual selection continues post-copulation if females mate multiply, known as sperm competition[3]. Males subject to both of these selective forces face a potential trade-off between energetically-demanding investment in pre-copulation access to females via weaponry, and competing post-copulation via investment in sperm numbers or quality[4]. Sperm competition models that incorporate the trade-off between expenditure on competitive pre- and post-copulatory male traits predict that as population-level sperm competition intensifies, more investment should be spent on testes and less on weaponry and mate searching[4].

There is evidence for the predicted trade-off between pre- and post-copulation investment in males from organisms as diverse as Arctic charr (*Salevlinus alpinus*)[5]; domestic Fowl (*Gallus gallus domesticus*)6and cetaceans[7]. Among invertebrates there is experimental evidence from dung beetles (*Onthophagus nigriventris*)[8] and the cockroach *Nauphoeta cinerea*[9].

The hissing cockroaches of Madagascar are large, heavily-armoured insects that engage in aggressive male-male contests over females[10]. They use complex acoustic and behavioural displays and have sexually dimorphic pronotal horns of varying pronouncement across species[10,11]. Although data is scarce, and there are no published accounts of the mating systems of wild Gromphadorhini, *Gromphadorhina portentosa* is thought to have a mating system characterised by dominance-based territoriality with female choice of high-ranking males[12,13,14]. Hissing cockroaches are thus predicted to face trade-offs in expenditure on weapons and ejaculates that will vary according to the ecological and social conditions that each species lives under. As with the exoskeleton, the development of testes and production of sperm is complete at adult emergence in *G. portentosa*[15] meaning that investment allocations are fixed upon reaching sexual maturity.

In order to explore the nature of pre-and post-copulatory trade-offs we compared two species, the Wide-horned cockroach *G. oblongonota* (see Fig. 1a; video 1: https://figshare.com/articles/3D_rendered_animation_ of_ Gromphadorhina_oblongonota/3833892), and its monospecific sister genus, the Flat-horned cockroach *Aeluropoda insignis* (see Fig. 1b; video 2: https://figshare.com/articles/3D_rendered_animation_of_Aeluropoda_insignis/3833919)16, which vary in overall body size and the size and shape of pronotal horns as well as observed male-male aggression.

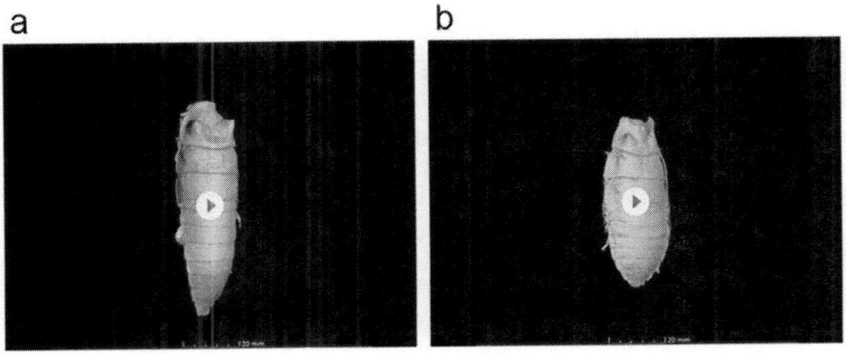

Figure 1 Screenshots from online videos demonstrating the morphological differences in overall body shape and pronotal shield weaponry between (a) *Gromphadorhina oblongonota*https://figshare.com/articles/ 3D_ rendered_animation_of_Gromphadorhina_oblongonota/3833892 and (b) *Aeluropoda insignis*https://figshare.com/articles/3D_rendered_ animation_ of_Aeluropoda_insignis/3833919.

2. MATERIAL AND METHODS

Thirty-three male *Aeluropoda insignis* and 46 male *Gromphadorhina oblongonota* were sourced from breeding colonies at the University of Nottingham. Cockroaches were housed individually as nymphs at 26 ± 1 °C, 12:12 h L:D cycle, fed *ad libitum* a diet of dry dog food and water and raised to maturity in isolation to ensure virginity. They were measured from the distal edge of the pronotal shield to the end of the last abdominal segment using callipers. Pairs of males size-matched by body-length were allowed to interact for up to 20 minutes in a 17 × 10 × 14.5 cm Perspex chamber under red light. Interactions were terminated early if dominance was quickly established to avoid damage to the animals. Interactions were filmed from above in HD video (Sony Handycam HDR-XR160, Japan) and reviewed at half-speed using VLC Media Player. Using the *G. portentosa* aggression ethogram10, aggressive behaviours were scored for each interaction in each species using JWatcher 1.017. The behavioural events scored for each animal in the dyad were hissing, antennate opponent, butt opponent, flip opponent and abdomen thrashing[10].

After data were gathered on male-male agonistic behaviour, cockroaches were euthanised by freezing. Euthanised cockroaches were stored at −80 °C and then defrosted at room temperature prior to CT-scanning. Post-scanning, cockroaches were dissected and combined testes wet weight was measured to an accuracy of 1×10^{-5}g.

We utilised high-resolution X-ray computed tomography to gain precise morphological data. Prior to scanning, individual cockroaches were carefully wrapped in X-ray transparent polyethylene packing foam and placed in 50 mL centrifuge tube. Samples were scanned using a three section multi-scan in 'fast mode' using a GE v|tome|x M 240 kV| X-ray microCT system. The scan consisted of 2638 projection images acquired at 75 kV electron acceleration energy and 200 μA current with a detector timing of 250 ms. The spatial resolution of the scan was 30 μm. Images were reconstructed using Datos Rec v2.3.0 software (GE Inspection and Sensing Technologies, Germany) and VGStudio MAX 2.2 (Volume Graphics, Germany) was used to obtain total body volumes from each specimen.

We used ImageJ[18] to obtain morphological measurements from the CT images from both species. Body length was measured dorsally from the distal leading edge of the pronotum to the tip of the last abdominal segment (see Supplementary Fig. 1a). The 'freehand line tool' was used to measure the length of the crenulated anterior edge of the pronotal shield in *A. insignis* to determine elaboration of this trait (see Supplementary Fig. 1b). For each species horn length was measured laterally from the lower distal corner of the pronotum to the apex of the horn (see Supplementary Fig. 1c,d).

All data were analysed using R version 3.1.0[19]. The R package 'plotrix' was used to produce standard errors of the mean, all means are reported ±1 standard error of the mean. To determine the shape of allometric relationships within and between species, testes mass and horn length were tested for relationships against body volume and body length respectively under three models to determine which was the best-fitting: (1) a simple linear model, (2) a polynomial model and (3) an exponential model; with species as a factor. The most parsimonious final model was selected using a backwards stepwise removal procedure, removing the higher-order interactions first and comparing against the last model calculated using ANOVA to determine whether the inclusion of the interaction term improved the model. The model with the highest Adjusted R^2 when compared to the alternate model and with a significant p-value when tested against the null model for the trait in question was then selected as the best-fitting.

To correct for differing body size between the two species, each morphological trait was divided by a body size indicator (testes mass (g)/total body volume (mm^3) and horn length (mm)/total body length (mm)) and used in a GLM with species as a factor. Thus the final model included relative testes mass as the dependent variable, relative horn length as the explanatory variable and species as a factor to determine if there was a trade-off between pre-copulatory selected traits, horns, and post-copulatory selected traits, testes, that differed in strength or direction between the two species.

3. RESULTS

Forty-two agonistic male-male encounters were recorded for each species, 19 in *Aeluropoda insignis* and 23 in *Gromphadorhina oblongonota*. The two cockroach species differed in the incidence of aggressive behaviours observed during male-male encounters. *G. oblongonota* was observed to engage in antennation, butting and flipping of opponents significantly more frequently than *A. insignis*(Table 1). *A. insignis* hissed more frequently during male-male encounters than *G. oblongonota* (Table 1). There was no significant difference between recorded incidences of abdomen thrashing. We did not score aggressive events per unit of time as there was no significant difference in the duration of male-male encounters between the two species (Wilcoxon rank sum test: $W = 1036$, $n = 84$, $p = 0.139$).

The two cockroach species differed significantly in average body length, horn length and combined testes mass. Mean body length of each species: *A. insignis* = 56.30 mm ± 1.30, *G. oblongonota* = 68.33 mm ± 1.14, Kruskal-Wallis $X^2_1 = 33.82$, $p < 0.001$. Mean horn length of each species: *A. insignis* = 4.04 mm ± 0.13, *G. oblongonota* = 7.22 mm ± 0.23, Kruskal-Wallis $X^2_1 = 55.44$, $p < 0.001$. Mean testes mass of each species: *A. insignis* = 0.04303 g ± 0.00205, *G. oblongonota* = 0.02826 g ± 0.00098, ANOVA, $F_{1,77} = 50.16$, $p < 0.001$.

Horn length in *A. insignis* was related to body length by a linear equation (Adjusted $R^2 = 0.5792$, $F_{1,31} = 45.05$, $p < 0.001$), indicating horn size scales linearly with body size (Fig. 2). No models supported the fitting of any particular equation to the relationship between *A. insignis*testes mass and body volume, the null model was the best fitting (Adjusted $R^2 = 0.0455$, $F_{1,29} = 2.43$, $p = 0.13$) indicating the lack of a strong allometric relationship between overall body size and testes mass (Fig. 3). Total body volume was not available for two specimens. Pronotum crenulation in *A. insignis* was related to body length by a linear equation (Adjusted $R^2 = 0.634$, $F_{1,31} = 56.49$, $p < 0.001$), indicating crenulation or elaboration at the leading edge of the pronotum scales isometrically with body size.

Table 1: Differences in the number of aggressive behaviours per male-male encounter between two species of hissing cockroach including Wilcoxon rank sum test statistics and associated p-values, with significant values in bold.

Behaviour	Species	n	Mean (S.E.M.)	Min.	Max.	W	p
Antennate	*A. insignis*	38	11.00 ± 3.84	0	104		
	G. oblongonota	46	61.37 ± 16.94	0	613	636	**0.0311**
Butt	*A. insignis*	38	0.79 ± 0.29	0	8		
	G. oblongonota	46	5.87 ± 1.45	0	40	560.5	**0.0021**
Flip	*A. insignis*	38	0.05 ± 0.05	0	2		
	G. oblongonota	46	0.80 ± 0.23	0	7	630	**0.0010**
Hiss	*A. insignis*	38	8.87 ± 4.17	0	130		
	G. oblongonota	46	1.04 ± 0.75	0	33	1052	**0.0262**
Abdomen thrashing	*A. insignis*	38	0.58 ± 0.50	0	19		
	G. oblongonota	46	3.76 ± 2.21	0	92	787	0.1476

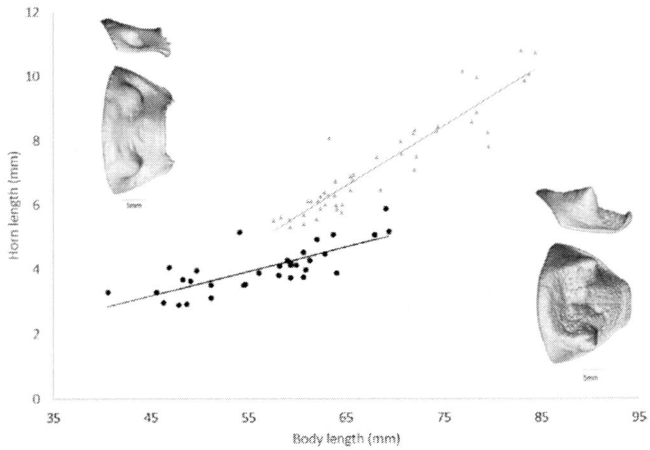

Figure 2: Relationship between total body length and pronotal horn length of *G. oblongonota*, (grey triangles, pronotal shield at bottom right in dorsal and lateral views) and *A. insignis*, (black circles, pronotal shield at top left in dorsal and lateral views).

Trend lines are estimated using the original data and model parameters (Line equations for *A insignis*: y = 0.0765x −0.2665; *G. oblongonota*: y = 0.1865x −5.5199).

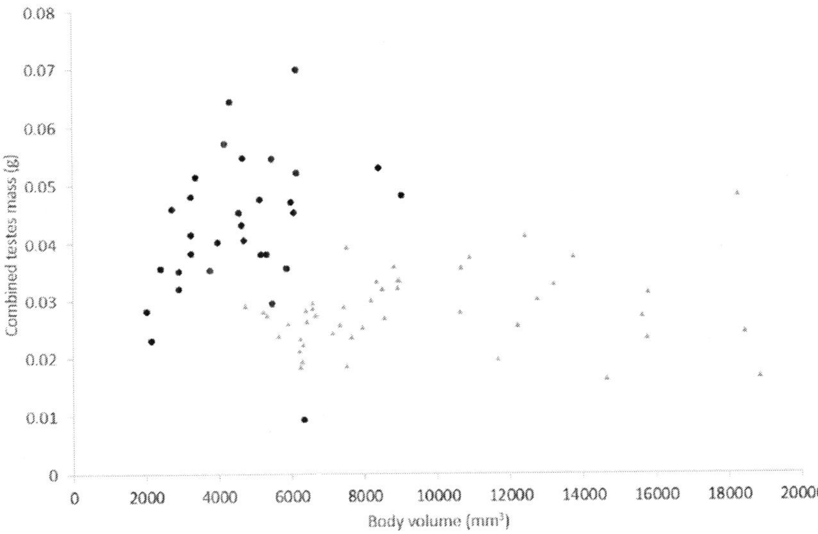

Figure 3: Relationship between total body volume and combined testes mass for *G. oblongonota* (grey triangles, n = 46) and *A. insignis* (black circles, n = 31).

Horn length in *G. oblongonota* was related to body length by a linear equation (Adjusted $R^2 = 0.8519$ $F_{1,44} = 281.8$, $p < 0.001$), indicating horn size scales linearly with body size (Fig. 2). There was a significant difference in the relationship between horn length and body length between the two species, whereby *G. oblongonota* have relatively larger horns at smaller body sizes and the relationship rises at a steeper allometric slope than for *A insignis* (ANOVA Model 1: horn length ~ body length + species + body length:species vs. Model 2: horn length ~ body length + species, F = 43.571, p < 0.001). No models supported the fitting of any particular equation to the relationship between *G. oblongonota*testes mass and body volume, the best fitting model was the null (Adjusted $R^2 = 0.0077$, $F_{1,44} = 1.349$, $p = 0.252$) indicating the lack of a strong allometric relationship between overall body size and testes mass (Fig. 3).

There was a significant negative relationship between relative horn length and relative testes mass (Table 2). Across both species males that invest more in the size of their testes invest less in the length of their horns. Relative testes size differed significantly between the two species, *A. insignis* have relatively larger testes for their size than *G. oblongonota*, and also have greater absolute testes mass despite being the smaller species. There was a significant interaction between relative horn length, relative testes size and species (Table 2). In each species there was an apparent trade-off between testes size and horn length, with animals with relatively larger testes having relatively smaller horns, but the allometric slope of this relationship differs between the two species (Table 2, Fig. 4).

Table 2: GLM results for relative testes size predicted by relative horn length with species as a factor.

Variable	Estimate (S.E.)	df	t value	p
Intercept	2.647e−05 (3.269e−06)	3, 73	8.097	<0.001
Relative horn length	−2.288e−04 (4.572e−05)	3, 73	−5.005	<0.001
Species	−1.635e−05 (4.247e−06)	3, 73	−3.848	<0.001
Relative horn length:Species	1.640e−04 (5.246e−05)	3, 73	3.126	0.0025

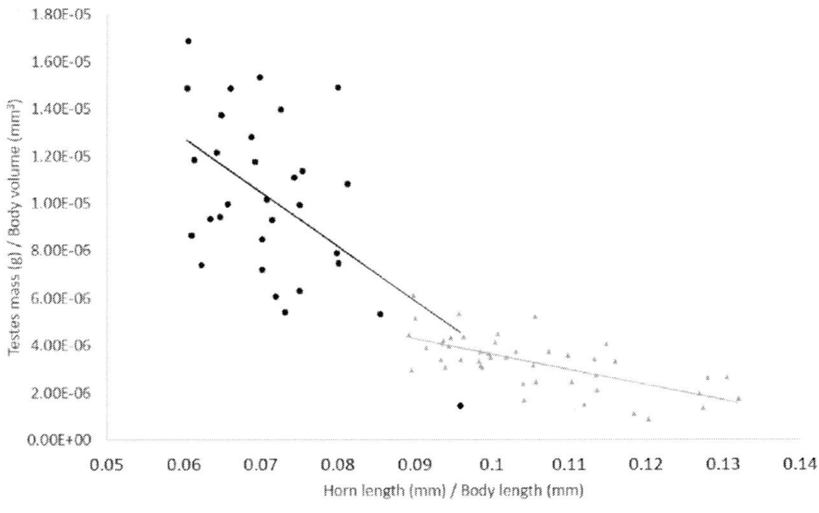

Figure 4: Relative testes size plotted against relative horn length for *A. insignis* (black circles, n = 31) and *G. oblongonota* (grey triangles, n = 46).

Trend lines are estimated using the original data and model parameters (Line equations for *A insignis*: $y = (-2.288e\text{-}04)x + 2.647e\text{-}05$; *G. oblongonota*: $y = (-6.48\ e{-}05)x + 1.012\ e{-}05$).

4. DISCUSSION

Morphometric analyses revealed that the smaller member of the Gromphadorhini tribe, *Aeluropoda insignis*, invested more in pronotal elaboration and horn length at larger body sizes. The linear relationship revealed isometric scaling of horn length with increasing body length. Similarly, isometry was observed in the elaboration of the leading edge of the pronotum. Testes mass was not allometrically related to body size, but absolute testes mass was greater in *A. insignis* than in *Gromphadorhina oblongonota*, and smaller male *A. insignis* invested more in their testes than smaller male *G. oblongonota*, indicating the relative importance of ejaculate-producing testes in competing post-copulation at all body sizes for this species. *A. insignis* males that cannot win in pre-copulation physical contests may have the opportunity to sneak copulations if females remain receptive and populations are aggregated.

G. oblongonota represents a competitive strategy with greater emphasis on weaponry. They are heavily armoured, and weapon development scales isometrically with body size while there is no relationship between testes mass and body volume. The lack of a clear relationship between body size and testes mass indicates that post-copulation competition is not as intense once dominance by male-male contest has been established. Therefore, males of this species do not invest proportionally in their testes as they attain larger body sizes, as would be predicted by isometric or positive allometric models. The small absolute size of the testes when compared to *A. insignis* also indicates that *G. oblongonota* is not likely to engage in intense post-copulatory sperm competition.

When comparing the two species it becomes apparent that while *A. insignis* is a significantly smaller animal with lesser investment in weaponry than *G. oblongonota*, they have significantly larger testes. Sperm competition theory[4] suggests that this is a species that is operating under greater risk of post-copulatory competition and investing relatively more in their spermiogenesis organs than *G. oblongonota*. This may be related to variation in aggregation behaviour of naturally occurring populations of both species. Male Chorusing frogs (*Crinia georgiana*) that aggregate at higher densities invest less heavily in weaponry and more in counter-measures against sperm competition[20]. Natural aggregation patterns in our species are unknown, but in two other members of the Gromphadorhini tribe, *G. portentosa* and *Princisia vanwaerebeki*, males preferred not to aggregate, and female *P. vanwaerebeki* were also not gregarious, although female *G. portentosa* were[21]. This suggests the potential for differing aggregation patterns between species of this tribe.

The linear relationships between body size and weapon size in these cockroaches may appear surprising for traits suspected to be under sexual selection pressure. There is some expectation that if a morphological trait is under sexual selection then it should exhibit positive allometry[22,23], however, positive allometry is not always the signal of a sexually selected trait. A meta-analysis by Bonduriansky[24]demonstrated that isometry or negative allometry are the more commonly observed patterns within a range of sexually selected traits across taxa and suggested that well-known cases of positive allometry resulted from a sampling bias towards a minority of excessively exaggerated traits. Thus our observed isometric scaling is not that unusual when patterns from many taxa are considered. Allometric patterns can differ markedly across species, however. A recent meta-analysis[25] estimated that weakly negative allometry may be the most common way that general traits scale with body size within species, but also found support for positive allometry in sexually selected traits, and interestingly, strong support for negative allometry in genital traits of insects and spiders[26]. Genital traits include the aedeagus or any other organ involved in sperm transfer, and are an important component of sperm competition that remains to be examined in the Gromphadorhini.

A. insignis were less aggressive during male-male interactions than *G. oblongonota*. More physical attacks such as butting and flipping were more frequent in the species with absolutely larger weapons, *G. oblongonota*, while agonistic signalling in the form of hissing was observed more frequently in *A. insignis*. Acoustic signalling is an important element of Gromphadorhini agonistic behaviour, it appears to convey information on body size to opponents[27]. We have shown that the use of acoustic signalling during fights differs between species of differing levels of aggression, with the less aggressive *A. insignis* hissing more frequently during male-male encounters, possibly as part of an assessment strategy that may enable small males to avoid physical violence[28].

Differing levels of aggression and agonistic tactics may indicate different pre-copulation competitive strategies between the two species. Differing levels of aggression between species do not appear random, given the costly nature of agonistic behaviour. The species that displayed the highest level of aggression between three species of the Fence lizard genus *Sceloporus* was also the most dichromatic, and dichromatism was positively correlated with aggression[29]. This is another case of aggression correlating with greater elaboration of a sexually-selected trait across species, as we have demonstrated with aggression and horn length in *G. oblongonota*. Crickets, Orthoptera: Gryllidae, are often used as a model of male-male aggression, and a phylogenetic comparison of aggression across 13 species found that there were marked differences in inter-specific levels of aggression[30]. Bertram *et al.*[30] suggested that high levels of male-male aggression in a cricket species was related to their tendency to use burrows to signal from, and that non-burrowing species were generally less aggressive. Aggression among males of the most aggressive Fence lizard species was attributed to low population densities that reduced the frequency of violent fights and high predation pressure selecting for residents that defended refuges[29]. These studies[29,30] highlight the role of ecological data in aiding the interpretation of behavioural and morphological data, which are currently lacking for the Gromphadorhini.

5. CONCLUSION

Males of the two giant hissing cockroach species appear to face the predicted trade-off between investing in pre- and post-copulatory competition, and each species addresses this trade-off in different ways. *A. insignis* follows a strategy of investing in testes while remaining relatively small-sized with reduced weaponry, and displaying lower levels of male-male aggression while *G. oblongonota* follows a strategy of large body-size, large weaponry and highly aggressive male-male encounters, but no specific investment strategy in testes and sperm production. These observations support sperm competition theoretical predictions but require field observations of natural mating systems in these relatively poorly-known organisms for confirmation. Thus, here we have identified a scenario of both within and between-species differentiation in mating strategies, whereby the two species may gain reproductive success success through differential investment in spermiogenesis or weaponry.

ACKNOWLEDGEMENTS

We would like to thank Sacha Mooney and the University of Nottingham's Hounsfield Facility staff especially Brian Atkinson for technical CT support; Ann Lowe for lab assistance; Tom Reader for statistical advice and Leigh Simmons for critical comments that improved this manuscript. Funding was provided by an Association for the Study of Animal Behaviour grant to SLM and an Anne MacLaren Fellowship to SLM and University of Nottingham postgraduate research support to IMS. The Hounsfield Facility received funding from ERC (FUTUREROOTS; Brussels, Belgium), BBSRC (Swindon, UK), and The Wolfson Foundation (London, UK).

REFERENCES

1. Darwin, C. The Descent of Man and Selection in Relation to Sex. (John Murray, 1871).
2. Andersson, M. Sexual Selection. (Princeton University Press, 1994).
3. Parker, G. A. *Sperm competition and its evolutionary consequences in the insects.* Biol. Rev. 45, 526–567 (1970).
4. Parker, G. A., Lessells, C. M. & Simmons, L. W. *Sperm competition games: A general model for precopulatory male–male competition.* Evolution 67, 95–109 (2013).
5. Rudolfsen, G., Figenschou, L., Folstad, I., Tveiten, H. & Figenschou, M. *Rapid adjustments of sperm characteristics in relation to social status.* Proc. R. Soc. Lond. B 273, 325–332 (2006).
6. Pizzari, T., Cornwallis, C. K. & Froman, D. P. *Social competitiveness associated with rapid fluctuations in sperm quality in male fowl.* Proc. R. Soc. Lond. B 274, 853–860 (2007).
7. Dines, J. P., Mesnick, S. L., Ralls, K., May-Collado, L., Agnarsson, I. & Dean, M. D. *A trade-off between precopulatory and postcopulatory trait investment in male cetaceans.* Evolution 69, 1560–1572 (2015).
8. Simmons, L. W. & Emlen, D. J. *Evolutionary trade-off between weapons and testes.* Proc. Natl Acad. Sci. USA 103, 16346–16351 (2006).
9. Tamara Montrose, V., Edwin Harris, W., Moore, A. J. & Moore, P. J.*Sperm competition within a dominance hierarchy: investment in social status vs. investment in ejaculates.* J. Evol. Biol. 21, 1290–1296 (2008).
10. Clark, D. & Moore, A. *Social interactions and aggression among male Madagascar hissing cockroaches (Gromphadorhina portentosa) in groups (Dictyoptera, Blaberidae).* J. Insect Behav. 7, 199–215 (1994).
11. Bell, W., Roth, L. & Nalepa, C. Cockroaches: Ecology, behavior and natural history. (Johns Hopkins University Press, 2007).
12. Barth, R. H. *The mating behavior of Gromphadorhina portentosa(Schaum) (Blattaria, Blaberoidea, Blaberidae, Oxyhaloinae): An anomalous pattern for a cockroach.* Psyche 75, 124–131 (1968).
13. Leibensperger, L. B., Traniello, J. F. A. & Fraser, J. M. *Olfactory cues used by female Gromphadorhina portentosa (Dictyoptera: Blaberidae) during mate choice.* Ann. Entomol. Soc. Am. 78, 629–634 (1985).
14. Guerra, P. A. & Mason, A. C. *Information on resource quality mediates aggression between male Madagascar hissing cockroaches, Gromphadorhina portentosa (Dictyoptera: Blaberidae).* Ethology 111, 626–637 (2005).
15. Lūsis, O., Sandor, T. & Lehoux, J.-G. *Histological and histochemical observations on the testes of Byrsothria fumigataGuer and Gromphadorhina portentosa Schaum.* Can. J. Zool. 48, 25–30 (1970).
16. Inward, D., Beccaloni, G. & Eggleton, P. *Death of an order: a comprehensive molecular phylogenetic study confirms that termites are eusocial cockroaches.* Biol. Letters 3, 331–335 (2007).
17. Blumstein, D. T. & Daniel, J. C. Quantifying Behavior the JWatcher Way (Sinauer Associates Inc., 2007).
18. Schneider, C. A., Rasband, W. S. & Eliceiri, K. W. *NIH Image to ImageJ: 25 years of image analysis.* Nat. Methods 9, 671–675 (2012).
19. R. Core Team. R: a Language and Environment for Statistical Computing. (R Foundation for Statistical Computing, 2013).
20. Buzatto, B. A., Roberts, J. D. & Simmons, L. W. *Sperm competition and the evolution of precopulatory weapons: Increasing male density promotes sperm competition and reduces selection on arm strength in a chorusing frog.* Evolution 69, 2613–2624 (2015).
21. Varadínová, Z., Stejskal, V. & Frynta, D. *Patterns of aggregation behaviour in six species of cockroach: comparing two experimental approaches.* Entomol. Exp. Appl. 136, 184–190 (2010).

22. Green, A. J. *Positive allometry is likely with mate choice, competitive display and other functions*. Anim. Behav. 43, 170–172 (1992).
23. Petrie, M. *Are all secondary sexual display structures positively allometric and, if so, why?* Anim. Behav. 43, 173–175 (1992).
24. Bonduriansky, R. *Sexual selection and allometry: A critical reappraisal of the evidence and ideas*. Evolution 61, 838–849 (2007).
25. Voje, K. L. *Scaling of morphological characters across trait type, sex, and environment*. Am. Nat. 187, 89–98 (2015).
26. Eberhard, W., Rodriguez, R. L. & Polihronakis, M. *Pitfalls in understanding the functional significance of genital allometry*. J. Evol. Biol. 22, 435–445 (2009).
27. Clark, D. & Moore, A. *Variation and repeatability of male agonistic hiss characteristics and their relationship to social rank in Gromphadorhina portentosa*. Anim. Behav. 50, 719–729 (1995).
28. Arnott, G. & Elwood, R. W. *Assessment of fighting ability in animal contests*. Anim. Behav. 77, 991–1004 (2009).
29. Vinegar, M. B. Comparative aggression in Sceloporus virgatus, S. undulatus consobrinus, and S. u. tristichus (Sauria: Iguanidae). Anim. Behav. 23, 279–286 (1975).
30. Bertram, S. M., Rook, V. L. M., Fitzsimmons, J. M. & Fitzsimmons, L. P. *Fine- and broad-scale approaches to understanding the evolution of aggression in crickets*. Ethology 117, 1067–1080 (2011).

CHAPTER 19

Effects of low-powered RF sweep between 0.01-20 GHz on female *Aedes Aegypti* mosquitoes: A collective behaviour analysis

Abdul Halim Poh[1,2], Mahmoud Moghavvemi[1,2,3*], M. M. Shafiei[1,2], C. S. Leong[4] , Yee Ling Lau[4] , Faisal Rafiq Mahamd Adikan[1] , Majid Bakhtiari[5] , Mahmood Ameen Abdulla Hassan[6]

1 Department of Electrical Engineering, Faculty of Engineering, University of Malaya, Kuala Lumpur, Malaysia

2 Centre of Research in Applied Electronics, Faculty of Engineering, University of Malaya, Kuala Lumpur, Malaysia

3 University of Science and Culture, Tehran, Iran

4 Department of Parasitology, Faculty of Medicine, University of Malaya, Kuala Lumpur, Malaysia

5 Faculty of Computing, University Technology Malaysia, Johor, Malaysia

6 Department of Biomedical Science, Faculty of Medicine, University of Malaya, Kuala Lumpur, Malaysia

ABSTRACT

There are many products claiming to be an electronic solution towards repelling mosquitoes. Several reviews were published in debunking these claims. However, there is a lack of a systematic study on effects of electromagnetic (EM) or more specifically, radio frequency (RF) waves against mosquitoes due to the conclusions made in those years. Therefore, we attempt to establish a fundamental study on female *Aedes Aegypti* (Linnaeus) mosquitoes by quantifying the collective behavior of the mosquitoes against a continuous stream of low-powered RF signals via a broadband horn antenna using image processing methods. By examining the average lateral and vertical positions of the mosquitoes versus frequency and time, the data shows negligible consistency in the reactions of the mosquitoes toward the different frequencies ranging from 10 to 20,000.00 MHz, with a step of 10 MHz. This was done by examining 33 hours of spatiotemporal data, which was divided into three sessions. All three sessions showed totally different convolutions in the positions in arbitrary units based on the raster scan of the image processing output. Several frequencies apparently showed up to 0.2–70% shift in both lateral and vertical components along the spectrum, without repeatability for all three sessions. This study contributes to the following: A pilot study for establishing the collective effects of RF against mosquitoes, open-source use, and finally a low-cost and easily adaptable platform for the study of EM effects against any insects.

1. INTRODUCTION

A detailed understanding of the issues around mosquito control is crucial as current methods are still being put into question for further improvements [1]. Electronic mosquito repellents, as one of the supposed vector controls are mostly ultrasonic types, and have made way into the consumer electronics for quite some time although reviews have been made on the inefficacies of the products [2]. However, there is a huge gap in the conclusions made with several available publications, where only a few selected frequencies between 125–74,600 Hz was reported and the details on the electronic signals are ignored in most reports, i.e the type of waves, whether they were radio frequencies (RF), optical, or acoustic signals, the kind of antenna used if the signals were electromagnetic waves, the circuitry used, and so on. Therefore an attempt to systematically verify this assertion was made, and to establish a more detailed data-based approach for reporting purposes.

The biological effects of RF have been studied for quite some time. An unwritten law in the industry of RF technology was the assertion that all RF is safe provided they were kept below the limits, and on the whole, comprehensive standards have been established for EMF radiation in the non-optic region, from 0 to 300 GHz. The World Health Organization (WHO) and International Commission on Nonionizing Radiation Protection (ICNIRP), among others, are bodies which monitor the radiation safety policies in the industry, although several other bodies have differing policies such as the Eastern European (EE) standards [3]. ICNIRP has defined limits to the electric field levels (V/m) for varying frequencies, allowing higher power levels up to 100 kHz and lower toward 100 GHz, whereas IEEE delved further into the magnetic and electric field thresholds levels with emphasis on the 2.4 GHz region, which corresponds with conventional cell phone frequencies [3].

From 0 to 10^{22} Hz, potential effects of electromagnetic radiation have been generally divided into non-ionizing and ionizing types, with four general characteristics, namely the Extremely Low Frequencies (ELF) and Very Low Frequencies (VLF), which has non-thermal effects, and up to the microwave region, which has thermal effects, and infrared (IR) and ultraviolet (UV) which has photoelectric effects, and finally ionizing waves such as the X-Ray and gamma rays, which are hazardous [4].

Reviews have compiled effects of high-powered electromagnetic field effects in Extremely Low Frequencies (ELF), defined as ranging less than 100Hz, on several insect types including Drosophila and bees [5]. The effects range from cellular mitotic activity of cells to animal fatalities, with nuances to different insect types. Behavioral changes were also reported on several insect types, from drinking habits to interspecies killings. However, how these waves specifically affect the insects are not resolved. Adverse effects have also been reported on personnel working in high-voltage substations, said to affect their central nervous system and cardiovascular activity. However, none were reported for mosquitoes [5]. It is likely that the use of ELF for mosquitoes is unsafe even if any form of vector control can be observed.

As we go along the spectrum, dosimetric studies of microwave radiation, which range from very low frequencies up to 300 GHz, have demonstrated some concern, although not in an imminent manner [3]. Effects of cell phone RF radiation with frequency of 850 MHz on human semen have been observed in terms of oxidative effects [6]. Effects have been observed in the animals tracked using radio transmitters operating between 27–401 MHz, including oxidative damage, behavioral and genotoxic effects, fertility and navigational disruption [7]. Telecommunication antenna radiation operating between 800–2,600 MHz have been known to affect several insect species, included developmental delays and reproduction problems in fruit flies (*Drosophila melanogaster*), colony strength degradation, navigation interruption in honeybees and cockroaches which leads to straying, and disruption of the olfactory faculties in ants [8]. Although most reports are made in-lab, these recent findings prompted the call to investigate further the effects of commercial RF exposure on wildlife [9], which was answered in a recent study by comparing the abundance of pollinators in RF and non-RF radiated areas. For the first time, it has been shown that RF indeed affects the pollinators [8].

Although reputedly safe, microwave ovens commonly use 2.4 GHz albeit on a much higher power level, as industrial grades may rate up to 1800 W, which reportedly interfere with the conventional communication frequencies, which operates on the same frequency [10]. Terahertz (THz) frequencies were also reviewed, including the use against fruit flies (*Drosopholia melanogaster*), where effects included gene expression at 6.69 THz. Also, the use of IR was reported with somatic mutations observed, at 3.69 THz [11]. However, none of the frequencies ranging from Low Frequencies (LF) up to microwave have been used towards mosquitoes in literature.

Although a call to explain the biological effects of RF in individual research was made due to the numerous factors contributing to the peculiarities, there is a consensus on the thermal effects it induces [12]. Certain RF frequencies (and at a certain power level) is understood as a heating mechanism due to the oscillations of polarized molecules and charged ions while interacting with RF, which recent applications still find its way in postharvest control for disinfecting insect pests [13, 14]. To date, the biological effects of low-powered RF, for instance emitted by cell phones towards humans have been inconclusive [4, 6, 15]. However, a review has highlighted the possibility of cellular interactions with RF, including cell plasma membrane, genome, and the presence of water, which may the cause of the diversions from normal activities of various bacteria strains. This was articulated in the highlights of previous research, which observed altered ionic transport processes and enzymatic activity in the plasma membranes of bacteria when exposed to certain frequencies between 50–73 GHz [16]. The navigation of honeybees were said to be affected due to the magnetite structures in the bee, acting as a natural compass interfered by telecommunication antenna radiations [8].

Based on the previous studies, we would like to define several criterion on studying the effects of RF on mosquitoes, including:

1) Including electromagnetic frequencies excluding the ionizing regions which are unsafe,

2) RF Power levels kept within the thresholds of the governing bodies which conforms to local standards (WHO, ICNIRP, IEEE, and others), and

3) Performed by quantifying the mosquitoes' activities, either by tracking or swarm behavior within an enclosure

Based on these criterion mentioned above, the materials and methods to fulfill them will be discussed in the following section.

2. MATERIALS AND METHODS

In this section, there are three main components which we employ in this study. The first will be the chamber design used to house the mosquito swarms, where rationale, dimensions, components and materials used for the construction will be elaborated. The second will be the RF setup, which includes the function generator including methods of frequency management, the amplifiers, and the antenna chosen for the scheme. The third and final component will be the quantification method, which uses a custom-made software to map the two-dimensional mosquito positions in the chamber and to study the collective behaviour by averaging techniques.

2.1 Chamber setup

There are various designs for analysing insects in-lab with various justifications [17–19]. However, in this study, a more specific design is required as shown in Fig 1, which is a modified version from a previous closely related research [20].

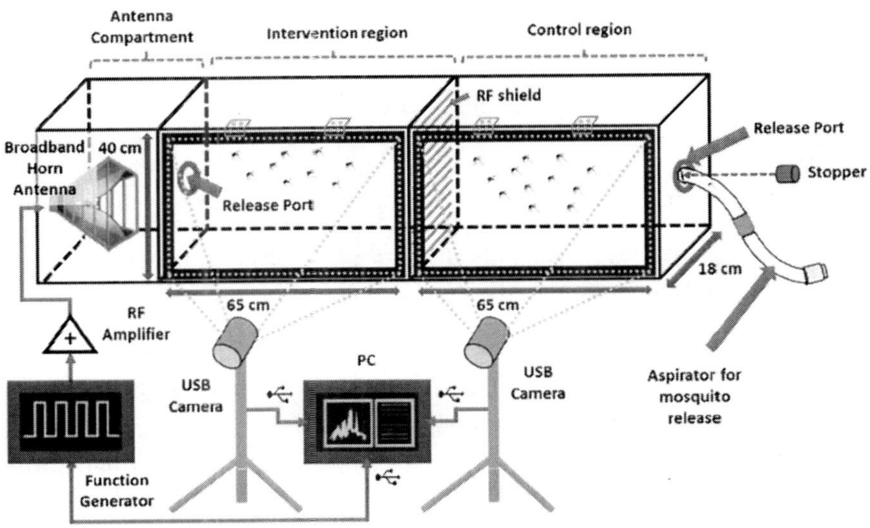

Figure 1. Evaluation chamber.

The chamber design is custom made with stacked plywood mounted using plastic brackets (dimensions 176.0 x 47.0 x 18.0 cm) enclosed with a transparent acrylic panel to access visuals on the subjects. The whole chamber is painted white, for contrasting of the mosquitoes and the background for digital quantification. The acrylic panel is mounted using hinges, which allows opening the chamber for cleaning inside the compartments of any residue from the experiments, since fungi may grow on the mosquitoes' faeces and affect the experiment. Three compartments are introduced, namely the antenna compartment for placing the antenna, an intervention compartment for testing of the radio frequencies against the mosquitoes, and a control compartment.

The control and intervention compartments are separated from each other by an RF shield to absorb or dispel any RF radiation towards the mosquitoes. Both compartments, including the gaps around the release ports are completely sealed with general purpose silicone sealant and reused elastomeric nitrile foamed rubber (Superlon™) for extra precaution. Airways for breathing are provided by allowing small gaps between the acrylic panel and the foamed rubber. The chamber area is provided with diffuse lighting for the imaging process to occur satisfactorily.

2.2 Antenna and detection setup

Different antennas are used for different applications with different properties, which primarily depend on the frequency range and output power. In previous studies for biological effects of RF, starting from ELF, long cable networks (part of the now ceased Project Sanguine [21]) as mentioned in a study on the biological effects of ELF were used [5, 8]. Omni-directional antennas were used for the study of effects of RF against wildlife tracking [7] and human semen [4, 6, 15], and base station antennas against pollinators [8]. In this study, a broadband horn antenna is chosen (LB7180, Ainfo™). The antenna has a frequency range of 0.7-20GHz, 12dB typical gain, linear polarization, and weighing around 1.5 kg. The antenna is used to deliver the generated RF wave to the subject of the study. The large bandwidth of the antenna provides enough flexibility for us to use a single antenna for variety of frequencies. The antenna was mounted into the designated compartment as shown in Fig 1 and oriented in the direction toward the intervention region directly.

Referring to Fig 1, a wideband function generator, capable of generating from 0.01–20 GHz (HMC-T2100, Hittite Microwave Corporation™) is used as the source of RF wave for this study. This signal generator is connected to a super-ultra-wideband amplifier (ZVA-213+, Mini Circuits™) using a both male SMA-to-SMA coaxial cable. A constant DC voltage of 12V supplies the amplifier 400 mA using a standard adjustable DC power supply. The amplifier is connected to the antenna using a both male SMA-to-N coaxial cable to the amplifier. There is a setup tradeoff which includes the use of frequencies lower than 0.7MHz for the broadband antenna, and the amplifier which operates from 0.8MHz. However, there is no evidence of a cutoff beyond the operating frequencies for both devices, and therefore with only 3% outlier beyond the operating frequencies, we choose to include the frequencies beyond the operating parameters for wider coverage.

Two units of off-the-shelf USB-connected web cameras are used for the mosquito motion recording (Sirius USB2.0 and Logitech S7500). The cameras are mounted on standard aluminum camera tripods and placed in front of the panels of the chamber. Pre-test alignment was performed by activating video mode and adjusting the best 3D orientation in order to allow the whole view of the respective compartments to come into view. In order to use these cameras for data collection, video applications are not required for installation, since the default driver installation upon USB connection is sufficient for the OpenCV™software to operate the captured feed.

2.3 Mosquito preparation & release

Mosquito preparation was carried out similar to our previous work [20]. For each testing sessions, approximately 150–250 female *Aedes aegypti* (Linnaeus) mosquitoes aging between 5–9 days were collected from Department of Parasitology, University of Malaya. The mosquitoes were colonized under controlled insectary settings with humidity levels of 75–90%, temperature 27±2°C, and equal light and dark cycles per a 24-hour day. The mosquitoes were nourished with a blend of powdered milk, cat fodder, bovine liver and mice chow with weight ratio of 1:1:1:2. Following the hatching, the adults were sustained with 10% sucrose solution. The mosquitoes are then mouth-exhaled into smaller containers, supplied with 10% sucrose solution wetted cotton and transported to the test chamber as shown in Fig 1. No blood meals were provided before the testing. The reason female *Aedes aegypti* mosquitoes were used was due to the fact that only females are known to take blood meals and thus turn into disease vectors, hence the importance of understanding their response against RF [22]. Also, blood meals were excluded to optimize the host-seeking behavior of the mosquitoes [23].

After cleaning the chamber from contaminants from mosquito residue and dust, the mosquitoes are then mouth-exhaled into the test chamber, and a stopper is immediately applied to both compartment openings. This is performed for both compartments, where each housed approximately 100–150 mosquitoes. The mosquitoes are then allowed to settle down for half an hour.

2.4 Software setup, data collection and treatment

Prior to data collection, there are three core software which requires initialization, namely a custom made software using OpenCV™ v3.1 coded within Visual Studio Community™ 2013 (VS2013™), and finally HMC-T2200 (Hittite Microwave Corporation™), which operates the function generator via a USB connection, and allows for automatic adjustments of a plethora of parameters including the frequency, delay times, signal power, and others. All software used in this study is open-source or free with plenty of tutorials for setup.

The mosquito behavior quantification is performed using OpenCV™ by utilizing Suzuki's method of topological structure extraction analysis by border following [24] algorithm, a function readily made available for use in the coding library, which was used in our previous research for quantifying aggregate mosquito positions in 2D [20]. Similar to the previous work, the image processing library allows for mapping the position of the mosquitoes behind the background, excluding the ones resting in the edges of the chamber, as also implemented in a previous study on mosquito flights against human odor [25]. The data output returns collective *x*- and *y*-coordinates as a function of response of the mosquitoes against the frequencies which will be radiated in the direction of the swarm. Similar to our previous study, the coordinates' origin start from the top left of the image matrix. We simplify the quantification collectively by averaging any 2D motion registered in the system since previous studies have established the relevancy of 2D mosquito flight quantification [26].

Upon the release of the mosquitoes, the quantification starts initialization from executing the OpenCV™ application through VS2013™ application (.exe) file, or via debugging the program in VS2013™ coding window tab. A boundary is defined within the raster shown in the main feed of the camera by adjusting the trackbars which correspond to the position and the size of the boundary in 2D. This will simplify the captured blobs in the video feed and registers a data collection with less noise (**SEE** S1 Video). The data output will appear according to the path input during the startup of the application in.txt files upon sliding item "Start" from 0 to 1. This code development is performed for one camera, therefore for the other camera another debugging file must be used with modifications to the code (**SEE** S1 Source code and files). Given the difficulty in maintaining precision on the detection of mosquitoes using machine vision, a basic noise level is expected from the data gathering, where lighting, camera resolution and placement plays a critical role in minimizing the errors. However, detected mosquitoes within the image feed will introduce the displacement in the spectrum of the detection, where the shifts will indicate a response. Initially before proceeding toward the testing, a complete run on the experimental setup with all the hardware in place without powering on the devices is performed by collecting data for approximately 40–80 minutes to observe the response, designated as the null test. The data output is collected for comparative purposes.

Next, the application HMC-T2200 (Hittite Microwave Corporation™) to control the frequency of the signal into the amplifier and the antenna is initialized. For the periodic sweep test, the starting frequency is fixed from 10 MHz, with an interval of 20 seconds, frequency step of 10 MHz and a fixed power output of 10 dBm in continuous sweep mode, whereas the input power of the amplifier is fixed to a maximum 4.8W (as per typical settings by fixing 12V and 400 mA in the ZVA 213+ amplifier datasheet). In this test, power levels from the source are kept constant due to limits of the power amplifier and also for variable limitation and to avoid data complexity. After the parameters have been set, the signal generator is turned on via checking the checkbox *"RF Output On"* on the HMC-T2200 GUI, and the time is recorded. Upon the completion of the frequency sweep, the recording is ceased. This procedure is performed three times in different sessions.

The output data comprises of three main parameters, including the timestamp in the YYYY-MM-DD.hh:mm:ss format, and the average x- and y-coordinates in arbitrary units (a.u.). The output data in.txt is transferred into an Excel format worksheet (.xlsx). Additional columns are added for conversion of the timestamps into seconds starting from 0, as the standard temporal reference, and the corresponding frequency based on the recorded time of turning on the RF output. The additional columns added and the headings are highlighted in the Excel worksheet as indication of the addition in the dataset. However, the original timestamp column is maintained for reference and data integrity. For the periodic sweep test, the system is left into autonomous data collection mode, where recording of the data may extend beyond the frequency sweep. The excess data output is truncated corresponding to the maximum frequency output at 20 GHz.

3. RESULTS & DISCUSSION

Four data collection sessions were conducted as per the methodology outlined in Section II, totaling four mosquito collection sessions and approximately 800 mosquitoes were used for all tests. The first is the null test while all hardware is disconnected from any power source. The following three tests comprises of mainly the periodic data collection. The following elaborates on the findings **(SEE** S1 Dataset).

3.1 Null test

In this test, the mosquitoes are released before turning on the apparatus for behavioral patterns to be observed. Upon the data collection, a graphical visualization using a statistical software (JMP Pro 12, SAS Institute Inc., 2015), of the average mosquito distance in arbitrary units (a.u.) as per defined by the video frame size, versus time in seconds. For this test, Fig 2 shows the average horizontal and vertical distance of the mosquitoes in arbitrary units (a.u.) from the rightmost side of the raster scan within the chamber captured from both intervention and control regions against time (seconds) for four cases:

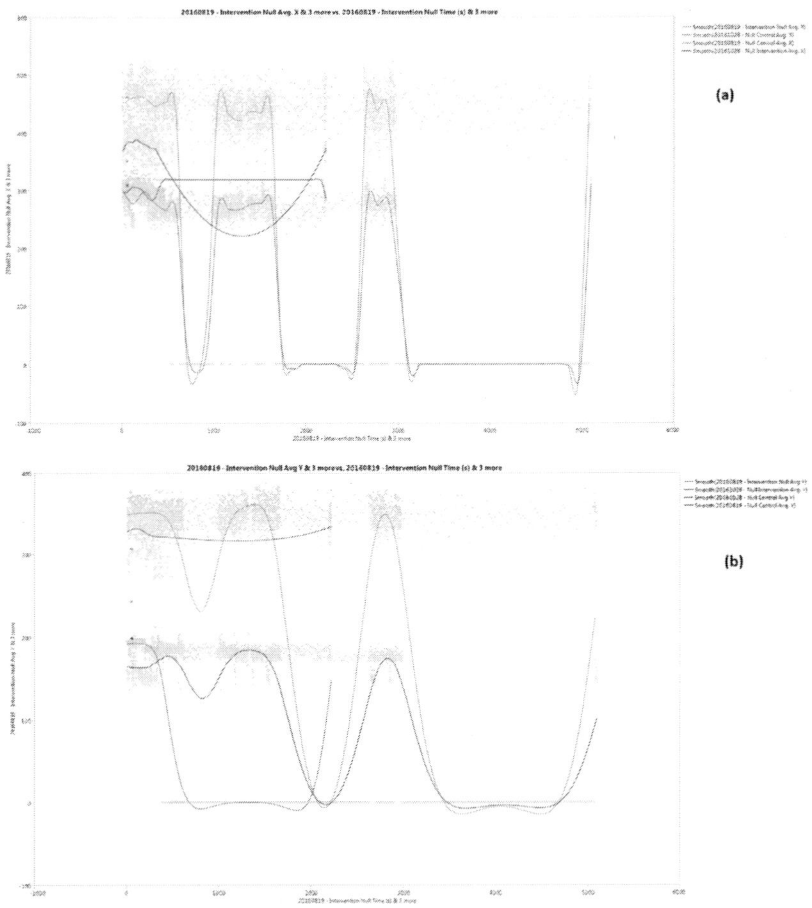

Figure 2. (a) Average Lateral Distance (a.u) and (b) Vertical Distance vs Time (s) for the null test.

From Fig 2(a) and 2(b), there are four series in the graph which shows the smoothed pattern of aggregated lateral and vertical distance from the left side of the chamber. Two of the series, where purple and red color in 2(a) and green and red in 2(b) show up to 36 minutes of behavior, and the remaining show up to 86 minutes. This comparison demonstrates the relative difference of the activities and predictably, the collectively random activity of the mosquitoes. Both control and intervention regions show non-uniform sporadic activities, most likely due to the fluctuations of the environment i.e. presence of human heat and temporal activity. As per visual observation, the mosquitoes have a high tendency to fly around during the initial period upon release into the chamber, before ultimately exhibiting dormancy.

During the periodic tests, three sessions of data logging were conducted, totaling approximately 33 hours of data collection period, where each session lasted up to 11 hours, covering mostly the scotophase period for the mosquitoes, as also performed in a previous study although the test was performed within a shorter period [25]. In this section, the aggregated spatiotemporal position is represented in smoothed plots, with the controls with each test for comparison. The output data was analyzed similar to the method in the previous section.

Periodic test 1: 01 September 2016. A smoothed plot comparing the aggregated lateral and vertical positions of the mosquitoes versus time on 1st September 2016 is shown in Fig 3(a) and 3(b):

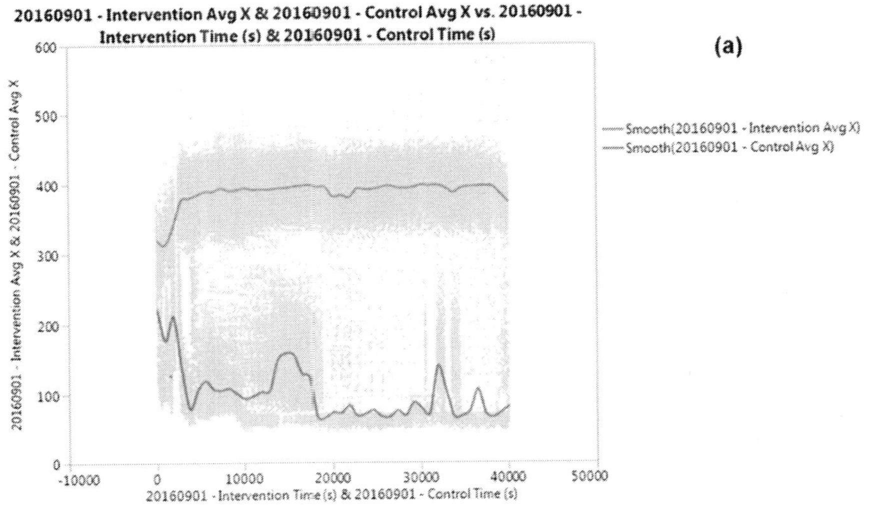

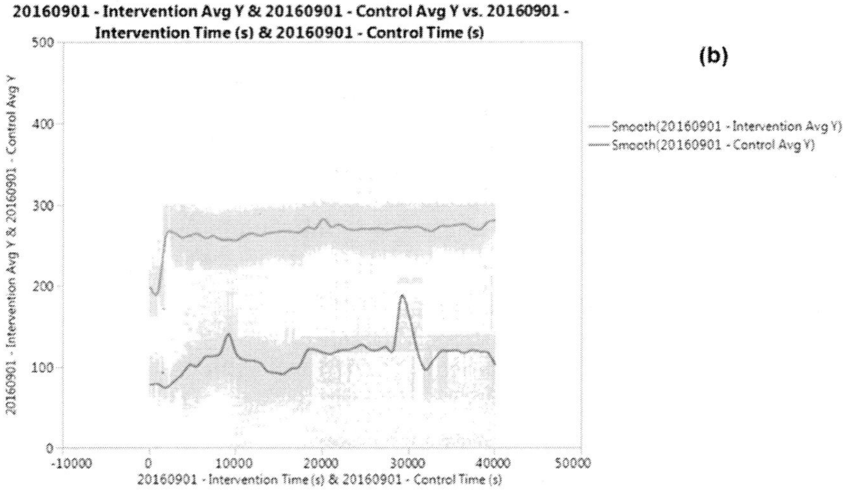

Figure 3. (a) Average Lateral Position (a.u) and (b) Vertical Distance vs Time (s) for the Periodic Test for intervention and control regions conducted on 01 September 2016.

Based on Fig 3(a), there is relatively more activity within the testing period in the Control region as shown from the blue smoother line, where during the first two hours indicate a more centred positioning of the mosquitoes. This excludes the ones resting outside the region of the camera view. After initial positions are assumed, from an initial position of 200 units, the readings indicated a left shift by approximately 100 units until the first hour of the test, followed by moderately constant fluctuations until the 4th hour. This was followed a shift toward the left again for an hour for another approximately 100 units until the 5th hour. This was followed by a relative monotony with less than 20% shift from the 5th hour. An approximately 80-unit shift was observed toward the right after approaching the 8th hour within the chamber until the end of the data collection. However, the orange-coloured series indicating activity in the intervention is significantly less observable, while only registering a slight change from the assumed positioning from the nearly monotonous spectrum during around the 5th and 9th hour toward the left of the chamber. This was probably due to very little mosquito activity, and also some might have escaped the detection algorithm due to occlusion, as also experienced in other studies where identification of mosquitoes may be lost after 10-20s [27]. Based on our observation, the mosquitoes tend to exhibit dormancy most of the time, resting on the edges of the chamber.

The vertical positions detected in the control region in Fig 3(b), however, shows increase during the 3rd hour up to about 50 units and fluctuates throughout the session until peaking sharply during the 8th hour up to around 180 units. The increase of activity during the end of the data collection may be attributed to the feeding behaviour as observed in female *Aedes Aegypti* in previous studies, which occurs in the early morning. As for the intervention region, however, shows very little vertical activity, except for a short period of movement towards the upper part of the chamber during the 5th hour. However, the activity of the mosquitoes within this region suggests a despondency towards the stream of RF waves through the region, as can be observed from both lateral and vertical components.

Periodic test 2: 05 October 2016. Smoothed plots comparing the aggregated lateral and vertical positions of the mosquitoes versus time on 5th October 2016 is shown in Fig 4(a) and 4(b):

Based on Fig 4(a), similarly convoluted patterns in the control region as shown by the blue line can be observed compared to the control region patterns in the previous Periodic Test 1. A rise in the lateral component occurred in relatively similar timing, which shows more leftward motion during the 4th and 7th hour and stagnating until the end of the test. However, the intervention region shows much less activity after the first hour, and stagnating during most of the period.

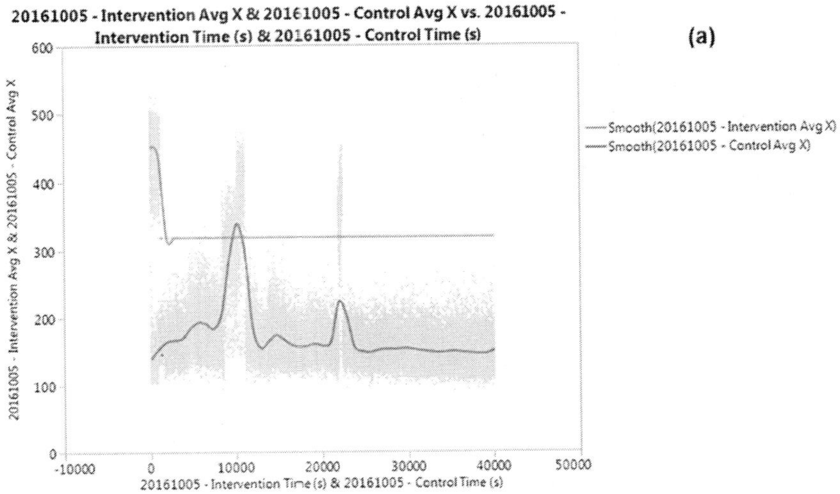

(a)

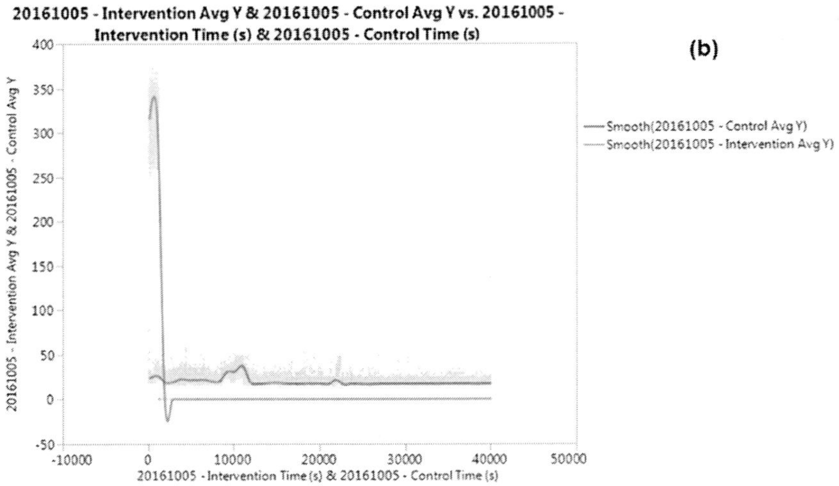

Figure 4. (a) Average Lateral Position (a.u) and (b) Vertical Distance vs Time (s) for the Periodic Test for intervention and control regions conducted on 05 October 2016.

The vertical motion for the control region as shown in Fig 5(b), however, shows much less significant change except a slight shift to the bottom of the chamber during the 3rd hour. The intervention region shows a similar sharp decrease in the vertical motion compared to the lateral component, most probably due to most of the mosquitoes resting on the edges of the chamber and continues to show virtually no observable activity afterward. Again, this may be due to occlusion and detection issues. However, our estimation also inclines toward the non-activity of the mosquitoes against the RF streams, which is prevalent in *Anopheles gambiae* mosquitoes where they tend to preserve energy when no stimuli is present [27].

Periodic test 3: 28th October 2016. Smoothed plots comparing the aggregated lateral and vertical positions of the mosquitoes versus time on 5th October 2016 is shown in Fig 5(a) and 5(b):

Fig 5(a) s howed the relatively stagnant lateral activity of the mosquitoes in the control region throughout the testing period, except a slight movement toward the right side of the chamber region during the 1st and 2nd hour of the period for approximately 20 units. However, the intervention region shows more convolution during the first half of the testing period. This can be seen from the rise of the lateral motion towards the right side of the chamber during the 2nd hour and slowly moving back left during the 3rd hour. Consequently, the movement spectrum shifted to the right side again until the 5th hour for around 50 units to the base of the line before fully stagnating until the end of the testing period. Again, this can be attributed to the mosquitoes moving out of the frame of observation to the edges of the chamber, as can be observed from Periodic Test 2.

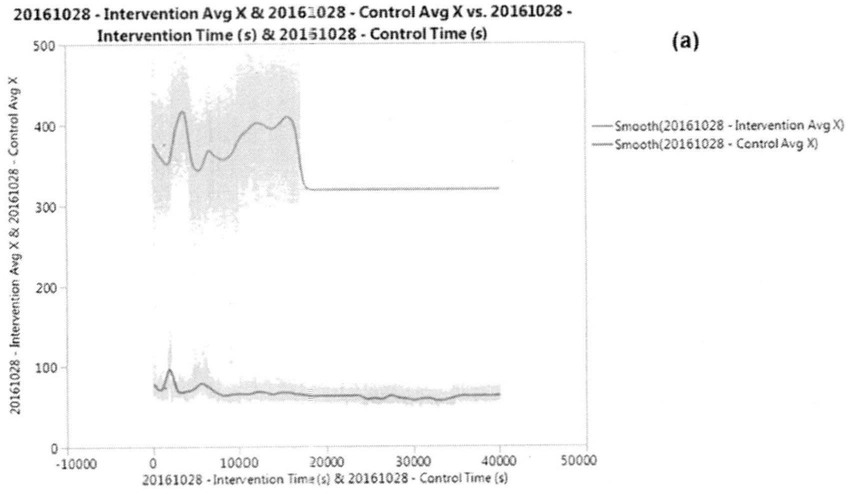

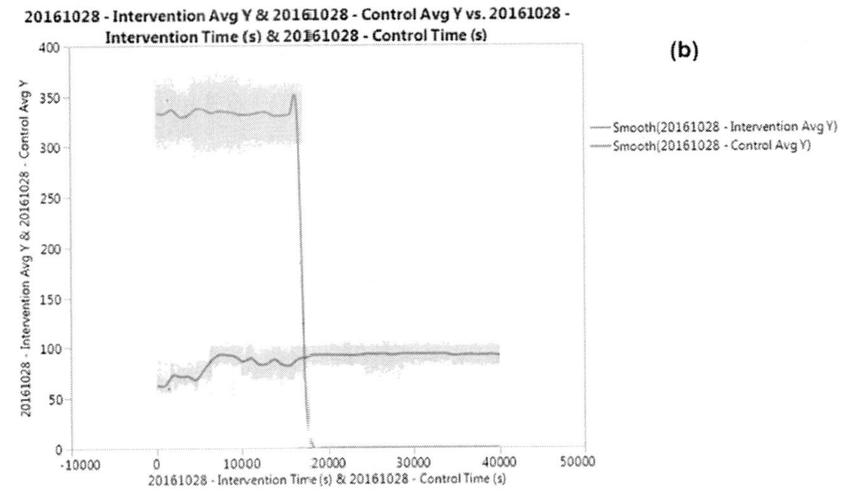

Figure 5. (a) Average Lateral Position (a.u) and (b) Vertical Distance vs Time (s) for the Periodic Test for intervention and control regions conducted on 28 October 2016.

The vertical components as shown in Fig 5(b) also showed similar patterns for the control region compared to its lateral counterpart with only about

30 units of maximum displacement, which suggests the mosquitoes were mostly in resting mode throughout the test. Part of the mosquitoes rested on the chamber panel within the observation raster, and the majority of others were resting on the edges of the chamber. The vertical component for the intervention region, however, shows a sharp drop from a nearly stable position at the lower part of the chamber edges of the observation from the start until around the 4th hour, with a drop of around 330 units, indicating a sharp leftward shift in the 5th hour. This indicates most of the mosquitoes chose to reside on the edges until the end of the experiment. Another explanation is due to the occlusion and lighting drift, which may also play a role in the absence of detection, which might be related the challenges in maintaining mosquito detection which was lost after some time in another study on *Anopheles gambiae* mosquitoes [27].

Collective positions versus frequency. The following figures show the collective positions of the mosquitoes versus frequency by compiling all three tests mentioned in all periodic tests into one. Since the frequency is a time-dependent function in this test, the pattern of the spectrum versus frequency is similar to the one versus time.

Based on Fig 6(a), by analyzing the pattern changes in the plots, it can be seen that there is little consistency in the flight patterns of the mosquitoes against all the frequencies for the three tests. This can be manually observed by examining the changes in the pattern, rather than the plot per se. Both lateral and vertical flights have very different reactions in the pattern changes. Fig 6(a) shows the plot on 28th October 2016 (red line), where the changes of the direction toward the right side of the chamber occurred between 880.00 until 1830.00 MHz (observation provided via JMP Pro 12 Graph Builder), which shows a furthering of the mosquitoes from the source of the RF stream. This was followed by a significant 65 unit distance shift towards the RF source until 2240 MHz. This is followed by another slight movement back against the antenna 4730.00 MHz. After another slight movement towards the antenna on the left, the spectrum shifted towards the right again until 8937.00 MHz. This is again followed by a left shift until 10101.00 MHz. However, this shift only occurred within <10 units, which is less than 2% of the observation raster scan. Another shift towards the right side of the chamber occurred until 11484.00 MHz for 14 units and sharply drops toward the left until the end of the test period.

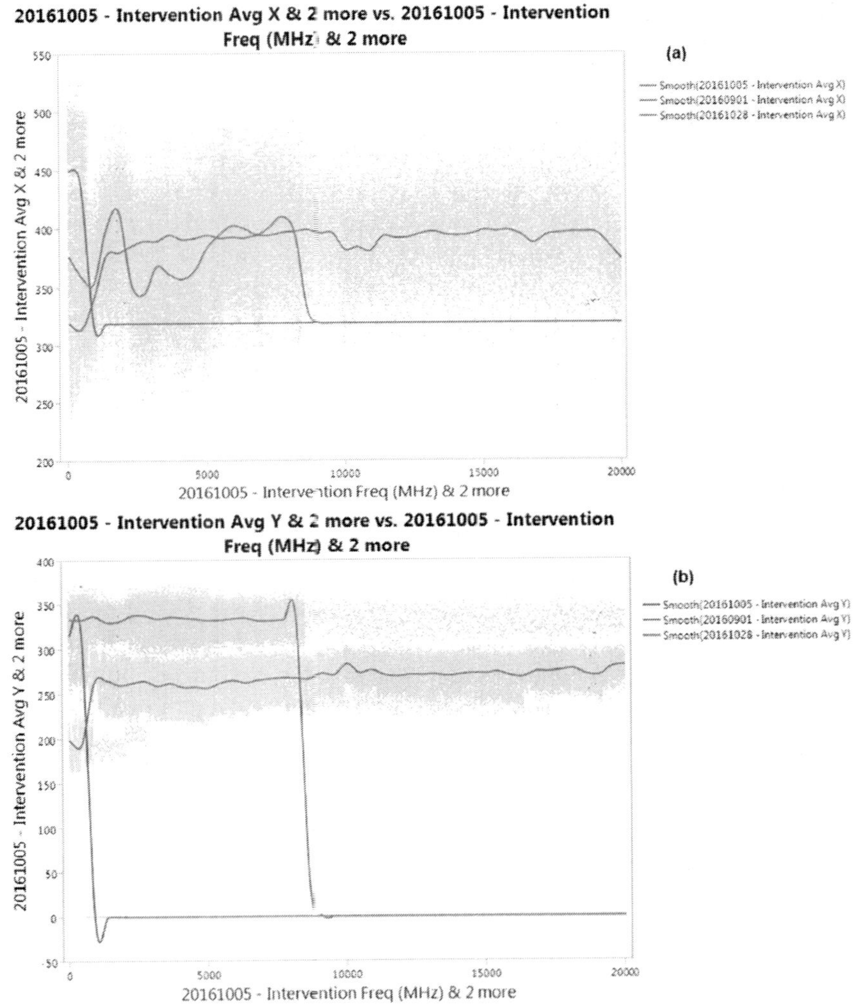

Figure 6. Collective (a) Average Lateral Distance (a.u) and (b) Vertical Distance vs Frequency (s) for the Periodic Test.

For the test occurring 1st of September 2016 (blue line), the results are much less convoluted. The captured positions shows a sharp rise between 440.00–1267.00 MHz of 105 units. However, following this sharp rise, there apparently is negligible change throughout the spectrum except a slight dip toward the left between 9610.00–11330.00 MHz for less than 2% of the region of observation. The pattern shown by the test conducted on 5th of October 2016 (purple line) shows a much less reaction from the test subjects, with only an initial movement towards the antenna during the first part between 150.00–1060.00 MHz for 162 units followed by the constant readings until the end of the test. These constant readings are, in our opinion, due to the inactivity of the mosquitoes which most frequently reside on the foam edges while occasionally moving.

Fig 6(b) shows the collective vertical flight patterns versus frequency, where both tests on 28 Oct 2016 and 01 Sept 2016 show similar patterns from 1170.00–7910.00 MHz, except the latter has an initial downward path in the observation from 0.00–1157.00 MHz for 78 units, whereas the test conducted on 5th Oct 2016 (purple line) showed a sharp drop from within the same spectrum range for 340 units. The flight patterns of all three series stayed constant without any significant changes throughout the frequency spectrum. Any constant readings are most likely due to the presence of noise in the matrix of the processed frame, however this also demonstrates the relevancy of analyzing the pattern using this method due to the necessity for a significant number of mosquitoes needed to overcome the noise spectrum. Therefore, the viability of this technique to assess the collective behavior of the mosquitoes in this paper is maintained.

According to previous studies on investigating mosquito behavior, it is likely that the resting behavior, observed in most of the period during the test, is non-genetic trait, as observed on *Anopheles arabiensis* mosquitoes when subjected to a genomic sequencing test for comparing human/cow attraction factor [28]. Another factor contributing to convoluted activity is the presence of ambient carbon dioxide (CO_2) as observed in *Anopheles gambiae* mosquitoes, which is probably the reason of the sporadic appearance in the spectrum as per the permeating CO_2 in the lab environment [29]. Another study on *Anopheles stephensi* mosquitoes showed that mosquito fitness decreased if blood meals were restricted or delayed, as per the mosquitoes were not provided with any blood meals in this study [30]. From this exercise, it can be safely concluded that the mosquitoes exhibit a non-uniform behavior in the chamber. A more prevalent form of behavior observed from the data is dormancy, perhaps due to the mosquitoes' adapting to the environment, as also reported in a previous study on the study of CO_2 plumes toward mosquitoes, where mosquitoes tend to conserve energy with the absence of human odor [29].

Comparing the findings in this study with the current literature, researchers suspected frequencies emitted by between 27.00–401 MHz may induce oxidative effects and navigational capabilities towards different animals, however none was reported for mosquitoes [9]. However, these frequencies as mentioned are not tested in-lab with more precision, therefore subject to further investigation. Frequencies emitted by telecommunication towers between 800 to 2600 MHz have been known to disrupt a few insect species, including fruit flies, honey bees and cockroaches [8], although also none reported for mosquitoes.

4. CONCLUSIONS

The findings from this study show that the frequencies between 0.01–20 GHz exhibit inconclusive effects, which was proven from the data-based approach of examining the mosquitoes' positions. To the authors' estimation, after examining the data and comparing the output with the control data, there is negligible consistency in the reaction of the mosquitoes towards the whole spectrum of RF waves used in this study. In future, other researchers may adopt this method by examining other frequencies beside the ones tested in this study, and delve into in-depth investigations in this area for confirmation. It is hoped that further use of the algorithms in the software development in this study undergo more improvement for future entomologists to use.

In conclusion, this study has managed to document the methods of in-depth investigation of effects of low-powered RF waves towards mosquitoes. The materials include a low-cost and technically simple chamber with three sections, namely the antenna compartment, an intervention and control region with openings for mosquito release and the use of two low-cost USB-connected cameras attached to tripods, one camera for each region of observation. The mosquitoes are handled using mouth aspirators for transfer-to-chamber purposes.

As for the quantification of the average position of the mosquitoes, an open-source software coding using OpenCV and Visual Studio 2013 (S1 Source code and files and S1 Video for software demonstration) is demonstrated using the same tools in a previous related research on studying effects of human odor on the mosquitoes, with a light data size by generating text files for the data output (**SE**E S1 Dataset). The RF waves are emitted at close proximity toward the intervention region for analysing the effects of different frequencies against the mosquitoes. The frequencies are adjusted using another free software HMC-T2200 to operate a function generator HMC-T2100, Hittite Microwave Corporation™. The data output is plotted and analyzed using the statistical software JMP Pro 12, SAS Institute Inc., 2015.

ACKNOWLEDGMENTS

The author would like to express utmost gratitude to the Department of Parasitology, University of Malaya for providing support, especially in form of mosquitoes, aspirators and handling techniques of the mosquitoes. Also to the staff of Development and Maintenance Unit of Engineering Faculty, University of Malaya for providing reused materials for the chamber reconstruction, Mr Christopher Fearday of Aerospace Lab, Faculty of Engineering, University Malaya, for equipment used for the chamber construction, Mr Nik Remeli for aiding the function generator access, and Mr Shahrilhafiz Zainol for supplying the web camera. Also, special thanks to Mr Jarrett Liner of Analog Devices for helping access the software for controlling the function generator, to Ms Yean Pin of JMP Technical Support for help with the statistical analytics and finally to the developers in the OpenCV forum who have been providing much info on the programming.

REFERENCES

1. Benelli G. Research in mosquito control: current challenges for a brighter future. Parasitology research. 2015;114(8):2801–5. pmid:26093499
2. Enayati AA, Hemingway J, Garner P. Electronic mosquito repellents for preventing mosquito bites and malaria infection. Cochrane Database Syst Rev. 2007;2.
3. Lopez-Iturri P, de Miguel-Bilbao S, Aguirre E, Azpilicueta L, Falcone F, Ramos V. Estimation of radiofrequency power leakage from microwave ovens for dosimetric assessment at nonionizing radiation exposure levels. BioMed research international. 2015;2015.
4. Zamanian A, Hardiman C. Electromagnetic radiation and human health: A review of sources and effects. High Frequency Electronics. 2005;4(3):16–26.
5. Marino AA, Becker RO. Biological effects of extremely low frequency electric and magnetic fields: a review. Physiol Chem Phys. 1977;9(2):131–47. pmid:414240
6. Agarwal A, Desai NR, Makker K, Varghese A, Mouradi R, Sabanegh E, et al. Effects of radiofrequency electromagnetic waves (RF-EMW) from cellular phones on human ejaculated semen: an in vitro pilot study. Fertility and sterility. 2009;92(4):1318–25. pmid:18804757
7. Balmori A. Radiotelemetry and wildlife: Highlighting a gap in the knowledge on radiofrequency radiation effects. Science of the Total Environment. 2016;543:662–9. pmid:26615484
8. Lázaro A, Chroni A, Tscheulin T, Devalez J, Matsoukas C, Petanidou T. Electromagnetic radiation of mobile telecommunication antennas affects the abundance and composition of wild pollinators. Journal of Insect Conservation. 2016;20(2):315–24.
9. Balmori A. Anthropogenic radiofrequency electromagnetic fields as an emerging threat to wildlife orientation. Science of the Total Environment. 2015;518:58–60. pmid:25747364
10. Kamerman A, Erkocevic N, editors. Microwave oven interference on wireless LANs operating in the 2.4 GHz ISM band. Personal, Indoor and Mobile Radio Communications, 1997 Waves of the Year 2000 PIMRC'97, The 8th IEEE International Symposium on; 1997: IEEE.
11. Wilmink GJ, Grundt JE. Invited review article: current state of research on biological effects of terahertz radiation. Journal of Infrared, Millimeter, and Terahertz Waves. 2011;32(10):1074–122.
12. Chou C. A need to provide explanations for observed biological effects of radiofrequency exposure. Electromagnetic biology and medicine. 2015;34(3):175–9. pmid:26444189
13. Hou L, Johnson JA, Wang S. Radio frequency heating for postharvest control of pests in agricultural products: A review. Postharvest Biology and Technology. 2016;113:106–18.
14. Alfaifi B, Tang J, Rasco B, Wang S, Sablani S. Computer simulation analyses to improve radio frequency (RF) heating uniformity in dried fruits for insect control. Innovative Food Science & Emerging Technologies. 2016;37:125–37.
15. Kim K-H, Kabir E, Jahan SA. The use of cell phone and insight into its potential human health impacts. Environmental monitoring and assessment. 2016;188(4):1–11.

16. Soghomonyan D, Trchounian K, Trchounian A. Millimeter waves or extremely high frequency electromagnetic fields in the environment: what are their effects on bacteria? Applied microbiology and biotechnology. 2016;100(11):4761–71. pmid:27087527
17. Dankert H, Wang L, Hoopfer ED, Anderson DJ, Perona P. Automated monitoring and analysis of social behavior in Drosophila. Nature methods. 2009;6(4):297–303. pmid:19270697
18. Cheng XE, Qian Z-M, Wang SH, Jiang N, Guo A, Chen YQ. A novel method for tracking individuals of fruit fly swarms flying in a laboratory flight arena. PloS one. 2015;10(6):e0129657. pmid:26083385
19. Chambers E, Bossin H, Ritchie S, Russell R, Dobson S. Landing response of Aedes (Stegomyia) polynesiensis mosquitoes to coloured targets. Medical and veterinary entomology. 2013;27(3):332–8. pmid:23336712
20. Poh AH, Moghavvemi M, Leong CS, Lau YL, Ghandari AS, Apau A, et al. Collective behavior quantification on human odor effects against female Aedes aegypti mosquitoes—Open source development. PloS one. 2017;12(2):e0171555. pmid:28152031
21. Wait J. Propagation of ELF electromagnetic waves and project Sanguine/Seafarer. IEEE Journal of Oceanic Engineering. 1977;2(2):161–72.
22. Ribeiro JM, Martin-Martin I, Arcà B, Calvo E. A Deep Insight into the Sialome of Male and Female Aedes aegypti Mosquitoes. PloS one. 2016;11(3):e0151400. pmid:26999592
23. Klowden MJ, Lea AO. Blood meal size as a factor affecting continued host-seeking by Aedes aegypti (L.). The American journal of tropical medicine and hygiene. 1978;27(4):827–31. pmid:686250
24. Suzuki S. Topological structural analysis of digitized binary images by border following. Computer Vision, Graphics, and Image Processing. 1985;30(1):32–46.
25. Spitzen J, Spoor CW, Grieco F, ter Braak C, Beeuwkes J, van Brugge SP, et al. A 3D analysis of flight behavior of Anopheles gambiae sensu stricto malaria mosquitoes in response to human odor and heat. PloS one. 2013;8(5):e62995. pmid:23658792
26. Wilkinson DA, Lebon C, Wood T, Rosser G, Gouagna LC. Straightforward multi-object video tracking for quantification of mosquito flight activity. Journal of insect physiology. 2014;71:114–21. pmid:25450566
27. Van Breugel F, Riffell J, Fairhall A, Dickinson MH. Mosquitoes use vision to associate odor plumes with thermal targets. Current Biology. 2015;25(16):2123–9. pmid:26190071
28. Main BJ, Lee Y, Ferguson HM, Kreppel KS, Kihonda A, Govella NJ, et al. The Genetic Basis of Host Preference and Resting Behavior in the Major African Malaria Vector, Anopheles arabiensis. PLoS Genet. 2016;12(9):e1006303. pmid:27631375
29. Webster B, Lacey ES, Cardé RT. Waiting with bated breath: opportunistic orientation to human odor in the malaria mosquito, Anopheles gambiae, is modulated by minute changes in carbon dioxide concentration. Journal of chemical ecology. 2015;41(1):59–66. pmid:25572756
30. Ohm JR, Teeple J, Nelson WA, Thomas MB, Read AF, Cator LJ. Fitness consequences of altered feeding behavior in immune-challenged mosquitoes. Parasites & vectors. 2016;9(1):1.

INDEX